International Law and the Resolution of Central and East European Transboundary Environmental Disputes

Also by Paul R. Williams

TREATMENT OF DETAINEES: Examination of Issues Relevant to Detention by the United Nations Human Rights Committee

International Law and the Resolution of Central and East European Transboundary Environmental Disputes

Paul R. Williams
Assistant Professor of Law and International Relations
American University
Washington, DC

First published in Great Britain 2000 by
MACMILLAN PRESS LTD
Houndmills, Basingstoke, Hampshire RG21 6XS and London
Companies and representatives throughout the world

A catalogue record for this book is available from the British Library.

ISBN 0-333-76495-1

 First published in the United States of America 2000 by
ST. MARTIN'S PRESS, LLC,
Scholarly and Reference Division,
175 Fifth Avenue, New York, N.Y. 10010

ISBN 0-312-22780-9

Library of Congress Cataloging-in-Publication Data
Williams, Paul R.
International law and the resolution of Central and East European transboundary environmental disputes / Paul R. Williams.
p. cm.
Includes bibliographical references and index.
ISBN 0-312-22780-9 (cloth)
1. Transboundary pollution—Law and legislation—Europe, Eastern. 2. Transboundary pollution—Europe, Eastern. 3. Environmental law, International. I. Title.
KJC6245 .W55   2000
341.7'62—dc21

00-035263

© Paul R. Williams 2000

All rights reserved. No reproduction, copy or transmission of this publication may be made without written permission.

No paragraph of this publication may be reproduced, copied or transmitted save with written permission or in accordance with the provisions of the Copyright, Designs and Patents Act 1988, or under the terms of any licence permitting limited copying issued by the Copyright Licensing Agency, 90 Tottenham Court Road, London W1P 0LP.

Any person who does any unauthorised act in relation to this publication may be liable to criminal prosecution and civil claims for damages.

The author has asserted his right to be identified as the author of this work in accordance with the Copyright, Designs and Patents Act 1988.

This book is printed on paper suitable for recycling and made from fully managed and sustained forest sources.

10 9 8 7 6 5 4 3 2 1
09 08 07 06 05 04 03 02 01 00

Printed and bound in Great Britain by
Antony Rowe Ltd, Chippenham, Wiltshire

For Kathy and Samantha Nicole

Contents

Acknowledgements   xi
List of Cases   xiv
List of Major Treaties and International Instruments   xvii
List of Abbreviations   xxi

Introduction   1

Part I Central and East European Transboundary Environmental Disputes from the Baltic Sea to the Black Sea   9

1. The Northern Tier: Polluting the Baltic Sea, Dark Clouds over the Black Triangle and Silesian Coal Basins, and East Meets West in Temelín   11
   - I. The Baltic Sea   12
   - II. The Black Triangle   16
   - III. The Silesian Coal Basin   22
   - IV. The Temelín nuclear power plant   25

2. The Southern Tier: Cleaning up after the Soviets, Dumping in the Danube, Dueling Nuclear Power Plants, and Suffocating the Black Sea   31
   - I. The environmental legacy of Soviet occupation   31
   - II. The Romanian–Bulgarian Danube River Gauntlet   35
   - III. The Romanian–Bulgarian Nuclear Corridor   38
   - IV. The Black Sea   42

Conclusion to Part I   45

Part II Using International Law to Resolve the Slovak–Hungarian Dispute concerning the Construction and Operation of the Gabčíkovo–Nagymaros Project   49

3. The Dispute-Formation Phase: Soviet-Inspired Designs to Harness the Power of the Danube and Post-transformation Second Thoughts   51

	I. Factors encouraging the negotiation and adoption of the 1977 Agreement	52
	II. The growing awareness of the environmental consequences of the Gabčíkovo–Nagymaros Project	55
	III. Hungary's initial proposals to modify the Gabčíkovo–Nagymaros Project	58
4	The Pre-Resolution Phase: Enter Sub-state Actors, Third Parties and International Law	60
	I. The involvement of sub-state actors	60
	II. The involvement of third parties	62
	III. Attempts to establish or identify appropriate dispute resolution mechanisms	64
	IV. The use of international law to frame negotiating positions	65
5	The Resolution Phase: Making the Case to the International Court of Justice	67
	I. Inquiry and mediation by the European Commission	68
	II. The submission of the dispute to the International Court of Justice	70
	III. The multifarious issues and options put before the International Court of Justice	73
	IV. The determination of certain factual and legal questions by the International Court of Justice	77
6	The Implementation Phase: Back to the Negotiating Table and Possibly Back to the Court	109
	I. The preliminary efforts of Slovakia and Hungary to negotiate a joint operating regime	109
	II. Situational circumstances affecting the negotiation and implementation of a joint operating regime	112
	III. Sub-state actors and interested third parties influencing the negotiation and implementation of a joint operating regime	113
	IV. Prospects for the enforcement and verification of a joint operating regime	116
	Conclusion to Part II	118

Part III Understanding the Role of International Law — 121

| 7 | The Regime of International Law: Its Nature and Function | 123 |

	I. The nature of international law as a constituent element of the international regime of transboundary environmental protection	123
	II. The functions served by international law during the four phases of the transboundary environmental dispute resolution process	127
8	Influencing the Utilization of International Law: Sub-State Actors, Interested Third Parties, Situational Circumstances and Factors of Functionality	135
	I. Sub-state actors	135
	II. Interested third parties	139
	III. Situational circumstances affecting the role of international law in the transboundary environmental dispute resolution process	145
	IV. Factors promoting the functionality of international law within the regime of transboundary environmental protection	154
Conclusion to Part III		162

Part IV Prospects for an Increasing Role for International Law in Promoting Central and East European Transboundary Environmental Dispute Resolution — 165

9	Gauging the Operability of International Law: the Evolving Circumstances	167
	I. Ecological circumstances	167
	II. Economic circumstances	170
	III. Domestic political circumstances	175
	IV. International political circumstances	179
	V. National minority circumstances	182
10	Predicting the Future: an Increasing Role for International Law?	188
	I. The Practice of Central and East European states applying international law to resolve international disputes	188
	II. The practice of Central and East European states applying municipal environmental law to resolve domestic resource disputes	191

III.	The correlation of Central and East European environmental law with municipal international environmental law	197
IV.	The accessibility of international judicial dispute resolution mechanisms to Central and East European states	199
V.	The incentive for Central and East European states to utilize international law to assist with the resolution of transboundary environmental disputes	204
VI.	The perceived capacity of international law to assist in the resolution of Central and East European transboundary environmental disputes	207

Conclusion to Part IV	215
Conclusion	218
Notes	228
Bibliography	280
Index	325

Acknowledgements

Although researching and writing a book is often perceived as a solitary adventure, the opportunity to exchange ideas and thoughts along the trail frequently provides the most rewarding experiences. I am thus greatly indebted to Dr Jiri Toman for the initial inspiration to pursue this publication and to Professor Malcolm Grant for his numerous insightful comments and gentle encouragement throughout the process. My work has benefited from the substantive and analytical comments of various individuals associated with the University of Cambridge, where I studied as a Fulbright Research Scholar from 1993 to 1996, including Dr Martin Dixon, Dr James Crawford and Dr Michael Byers. I am particularly indebted to Dr Cecilia Albin for her helpful assistance in understanding the nuances of negotiation theory, and to Dr Jack Shepherd for his service as an intellectual sounding-board and willingness to allow me to tap his numerous contacts in Central and Eastern Europe, as well as for providing me with a number of helpful hints for surviving the fieldwork experience. I am also indebted to Professor Dan Cole of the University of Illinois, Dr Alan Boyle, from the University of Edinburgh, and to Major Paul Wade of the United States Air Force.

For assistance with understanding the various transboundary environmental disputes facing Central and East European states and the role of central governments in resolving these disputes I am grateful for the assistance of Anna Violová, Ivan Závadský, Adela Ladzianska, Dr Július Maljkovič, Dr Viliam Klescht, Dr Božena Gašparíková, Dr Juraj Tözsér, Gabriela Fischerová, Dr Jozef Sloboda, Darina Dzurjaninová and Bohuslav Bezúch of the Slovak Ministry of the Environment, Peter Kollárik and Dr Cecília Kandráčová of the Slovak Ministry of Foreign Affairs, Dr Jacek Jaśkiewicz, Dr Stanisław Zadrożny, Dr Stanisław Wajda of the Polish Ministry of Environmental Protection, Dr Katarzyna Juda-Rezler of the Warsaw University of Technology, Ms Katherine Gorove, Consultant to the Hungarian Ministry of Foreign Affairs, Dr Jan Kára of the Czech Ministry of Foreign Affairs, Josef Vejvoda of the Czech Ministry of the Environment, Dr Bedřich Moldan, the former Minister of the Environment of the Czech Republic, René Pisinger, Adviser to the Minister of the Environment of the Czech Republic and Dr Ivan Rynda, former Member of the Presidium of the Czech Federal Assembly and the Chairman of the Environmental Committee of the Chamber of People.

xii Acknowledgements

For assistance with understanding the role of nongovernmental organizations, industrial actors and local government officials I thank János Zlinszky, Dr Ferenc Fehér and Stanislav Sitnicki of the Regional Environmental Center for Central and Eastern Europe, Dr Andrzej Kassenberg, President of the Institute for Sustainable Development, Dr Vilmos Kiszel, President of the Göncöl Alliance, Dr Juraj Králik, Chairman of the Slovak National Centre for Human Rights, Dr Lubomír Paroha, Director of the Foundation Project North in Ustí nad Labem, Karin Krchnak, Environmental Consultant to the Environmental Law Institute, Dr Krzysztof Görlich, Deputy Mayor for the City of Kraków, Milan Kunc, Mayor for the City of Děčín, Lubomír Novák, Chief Engineer for the City of Děčín, Stephen McCormick, Senior Associate at the National Civic League, Jaroslav Mráz, Regional Chairman of Euroregion NISA, Jaroslav Zámečník, Deputy Regional Chairman of Euroregion NISA, Dr Miroslav Liška, Chief Deputy of Public Relations, Vodohospodárska Výstavba, Miroslav Čomaj, Director of the Office of Hydraulic and Dams Construction for Hydroconsult, Dr Sláva Kubátová and Dr Roman Vyhnanek, Czech Environment Management Center, Dr Thomas Owen, Harvard Institute for International Development, Dr Maciej Nowicki, President of EcoFund and former Minister of Environmental Protection for Poland, Dr Václav Poštolka, Professor on the Faculty of Environment at the Jan Evangelista Purkyně University, Dr Jerzy Sommer, Professor of Law and President of the Polish Environmental Law Association, Steve Stec, Regional Coordinator, American Bar Association Central and East European Law Initiative, Dr Tomasz Stypka, Mechanical Engineer, Technical University of Kraków's Institute of Environmental Protection, and Dr Jindrich Tichy, Professor of Ecology at the University of Ústí nad Labem.

I am also thankful to the many United States government officials who provided assistance in arranging contacts in the field, including James Hooper, Ambassador Theodore Russell, Cameron Hunter, James Chamberlin, Steve Taylor, Jan Pisko, Loren Schulze, and Martin Brenosk, of the United States Department of State and US Agency for International Development, and Paul Almeida of the United States Environmental Protection Agency.

Special thanks also go to Sabi Ardalan who served as my Junior Associate at the Carnegie Endowment for International Peace, and to my research assistants at the American University, including Shazia Anwar, Vanessa Jimenez, Jennifer Harris, Jason Carmel, Ramzi Nemo, Elizabeth Finberg, Omar Vargas, Rose Parks, and Monica Talwar. I am also grateful for the administrative assistance provided by Peggy Bailey of the Carnegie Endowment and Leslie Scott of the American University, and to Jennifer Little and Kathleen Daly of the Carnegie Endowment Library for tracking down the most obscure journal articles.

Finally, I am eternally grateful for my wife Kathy's support and patience throughout the process and for Tricia McLaughlin's cheerful encouragement.

In addition to the support of these individuals, this book would not have been possible without the financial support of the United States/United Kingdom Fulbright Commission, the Institute for the Study of World Politics, the American Council of Learned Societies, and the Social Science Research Council.

List of Cases

Case Concerning the Gabčíkovo–Nagymaros Project (Hungary v. Slovakia), 1997 ICJ (Judgement, September 25).

Legality of the Threat or Use of Nuclear Weapons, 1996 ICJ 226 (Advisory Opinion, July 8).

Nuclear Tests (New Zealand v. France), 1995 ICJ 288 (Order, September 22).

Applicability of the Obligations to Arbitrate Under Section 21 of the United Nations Headquarters Agreement of 26 June 1947, 1988 ICJ 12 (Advisory Opinion, April 26).

Military and Paramilitary Activities in and against Nicaragua (Nicaragua v. United States of America), 1986 ICJ 14 (Merits, Judgement, June 27).

Continental Shelf (Tunisia v. Libyan Arab Jamahiriya), 1985 ICJ 192 (Application for Revision, December 10).

Military and Paramilitary Activities (Nicaragua v. United States), 1984 ICJ 392 (Jurisdiction and Admissibility, November 26).

Gulf of Maine (United States v. Canada), 1984 ICJ 246 (Judgement, October 12).

Continental Shelf (Libyan Arab Jamahiriya v. Malta), 1984 ICJ 3 (March 24).

Liamco v. Libya Arbitration, 20 ILM 1 (1981).

Interpretation of the Agreement of 25 March 1951 between the WHO and Egypt, 1980 ICJ 73 (Advisory Opinion, December 20).

United States Diplomatic and Consular Staff in Tehran (United States v. Iran), 1980 ICJ 3 (May 20).

Aegean Sea Continental Shelf (Greece v. Turkey), 1978 ICJ 3 (December 19).

Fisheries Jurisdiction (United Kingdom v. Iceland), 1974 ICJ 3 (Merits, July 25).

Legal Consequences for States of the Continued Presence of South Africa in Namibia (South-West Africa) Notwithstanding Security Council Resolution 276, 1971 ICJ 16 (Advisory Opinion, June 21).

North Sea Continental Shelf (Germany v. Netherlands & Denmark), 1969 ICJ 3 (February 20).

Lac Lanoux Arbitration (France v. Spain), 53 AJIL 156 (1959).

Admissibility of Hearings of Petitioners by the Committee on South West Africa, 1956 ICJ 46 (Advisory Opinion).

Ambatielos (Greece v. United Kingdom), 1953 ICJ 10 (Judgement on the Merits, May 19).

Haya de la Torre (Columbia v. Peru), 1951 ICJ 71 (June 13).

Interpretation of Peace Treaties with Bulgaria, Hungary and Romania (Second Phase), 1950 ICJ 221 (Advisory Opinion, July 18).

International Status of South West Africa, 1950 ICJ 128 (Advisory Opinion).

Interpretation of Peace Treaties with Bulgaria, Hungary and Romania (First Phase), 1950 ICJ 65 (Advisory Opinion, March 30).

Corfu Channel (United Kingdom v. Albania), 1949 ICJ 4 (April 9).

Air Service Agreement of March 27, 1946 between the United States of America and France, 18 RIAA 443 (December 9, 1978).

Trail Smelter (United States v. Canada), 33 AJIL 182 (1939), 35 AJIL 684 (1941).

Diversion of Water from the Meuse (Netherlands v. Belgium), 1937 PCIJ (ser. A/B), No. 70 (June 28).

Free Zones of Upper Savoy and the District of Gex (France v. Switzerland), 1930 PCIJ (ser. A), No. 24 (Order, December 6).

Case of Serbian Loans (France v. Serbia) 1929 PCIJ (ser. A.), No. 20 (July 12).

Case of Brazilian Loans (France v. Brazil) 1929 PCIJ (ser. A.), No. 21 (July 12).

International Commission of the River Oder, 1929 PCIJ (ser. A.), No. 23 (September 10).

Nauillaa Arbitration (Germany v. Portugal), 2 RIAA 1013 (1928).

Chorzów Factory (Germany v. Poland), 1928 PCIJ (ser. A.), No. 17 (September 13).

Chorzów Factory (Germany v. Poland), 1927 PCIJ (ser. A.), No. 13 (December 16).

Lotus Case (France v. Turkey), 1927 PCIJ (ser. A.), No. 10 (September 7).

Mavrommatis Palestine Concessions (Greece v. United Kingdom), 1924 PCIJ (ser. A.), No. 2 (August 30).

List of Major Treaties and International Instruments

Convention on the Law of Non-Navigational Uses of International Watercourses, 36 *ILM* 700 (1997) (adopted by the United Nations General Assembly on 21 May 1997). Not in force.

Convention on the International Commission for the Protection of the Oder River (1996). In force 11 April 1996.

Convention on Conciliation and Arbitration within the OSCE, 32 *ILM* 557 (1992). In force 5 December 1994.

Convention on Cooperation for the Protection and Sustainable Use of the Danube River (signed 29 June 1994), *Official Journal of the European Communities* No. L 342/18/1997.

Protocol to the 1979 Convention on Long Range Transboundary Air Pollution on the Further Reduction of Sulphur Dioxide Emissions, 33 *ILM* 1542 (1994). Not in force.

Convention on Nuclear Safety, 33 *ILM* 1518 (1994). In force 24 October 1996.

Draft Articles on State Responsibility, *ILC Yearbook Vol. II* (1977); *ILC Yearbook* (1994); 'Report of the International Law Commission on the work of its forty-eighth session, 6 May to 26 July 1996,' Official Records of the General Assembly, Fifty-first Session, Supplement No. 10 (A/51/10).

Ministerial Declaration on the Protection of the Black Sea (7 April 1993), 32 *ILM* 1101 (1993). Nonbinding.

Convention on Civil Liability for Damage Resulting from Activities Dangerous to the Environment, 32 *ILM* 1228 (1993). Not in force.

Brussels Agreement on the Gabčíkovo–Nagymaros Project between the Czech and Slovak Federation, the European Commission and Hungary (27 November 1992).

London Agreement on the Gabčíkovo–Nagymaros Project between the Czech and Slovak Federation, the European Commission and Hungary (28 October 1992), 32 *ILM* 1291 (1993).

Framework Convention on Climate Change, 31 *ILM* 849 (1992). In force 24 March 1994.

Convention on Biological Diversity, 31 *ILM* 822 (1992). In force 29 December 1993.

Convention on the Protection of the Black Sea against Pollution, 32 *ILM* 1101 (1992). In force 15 January 1994.

Convention on the Protection of the Marine Environment of the Baltic Sea Area (1992), 22 *Law of the Sea Bulletin* 54. Not in force.

Convention on the Transboundary Effects of Industrial Accidents, 31 *ILM* 1330 (1992). Not in force.

Convention on the Protection and Use of Transboundary Watercourses and International Lakes, 31 *ILM* 1312 (1992). Not in force.

Treaty on European Union, 31 *ILM* 247 (1992). In force 1 November 1993.

Declaration of the United Nations Conference on Environment and Development, 31 *ILM* 874 (1992). Nonbinding.

Statement of the Rules of International Law Applicable to Transfrontier Pollution, *ILA 60th Meeting*, 1 (1982).

Convention on Environmental Impact Assessment in a Transboundary Context, 30 *ILM* 802 (1991). Not in force.

Convention on the International Commission for the Protection of the Elbe, *Official Journal of the European Communities* No. L 321/25/1991. In force 18 November 1991.

Protocol to the 1979 Convention on Long Range Transboundary Air Pollution on the Control of Emissions of Volatile Organic Compounds and Their Transboundary Fluxes, 31 *ILM* 568 (1991). Not in force.

Convention on the Control of Transboundary Movements of Hazardous Wastes and Their Disposal, 28 *ILM* 657 (1989). In force 24 May 1992.

Protocol of 6 February 1989 amending the Agreement between the Government of the Hungarian People's Republic and the Government of the Czechoslovak Socialist Republic for Mutual Assistance in the Construction of the Gabčíkovo–Nagymaros System of Locks.

Protocol to the 1979 Convention on Long Range Transboundary Air Pollution Concerning the Control of Emissions of Nitrogen Oxides or

Their Transboundary Fluxes, 28 *ILM* 214 (1988). In force 14 February 1991.

Protocol to the 1979 Convention on Long Range Transboundary Air Pollution on the Reduction of Sulphur Emissions or Their Transboundary Fluxes by at least 30 per cent, 27 *ILM* 707 (1987). In force 2 September 1987.

Protocol on Substances that Deplete the Ozone Layer, 26 *ILM* 154 (1987).

Convention on the Early Notification of a Nuclear Accident, 25 *ILM* 1370 (1986). In force 27 October 1986.

Declaration on the Right to Development, *A Compilation of UN International Human Rights Instruments* 403 (1988). Adopted 4 December 1986, nonbinding.

Protocol to the Convention on Long Range Transboundary Air Pollution on Long Term Financing of the Cooperative Program for Monitoring and Evaluation of the Long Range Transmission of Air Pollutants in Europe, 24 *ILM* 484 (1985). In force 28 January 1988.

World Charter for Nature, 22 *ILM* 455 (1983). Nonbinding.

Vienna Convention for the Protection of the Ozone Layer, 26 *ILM* 1529 (1987). In force 22 September 1988.

Convention on the Law of the Sea, 21 *ILM* 1261 (1982). In force 16 November 1994.

Convention on Long Range Transboundary Air Pollution, 18 *ILM* 1442 (1979). In force 16 March 1983.

Treaty Between the Hungarian People's Republic and the Czechoslovak Socialist Republic Concerning the Construction and Operation of the Gabčíkovo–Nagymaros System of Locks, 16 September 1977, 1109 *UNTS* 236 (1978). In force on 30 June 1978.

Convention on the Protection of the Marine Environment of the Baltic Sea Area, 13 *ILM* 546 (1974). In force 3 May 1980.

Convention for the Prevention of Marine Pollution from Land-Based Sources, 13 *ILM* 352 (1974). In force 6 May 1978.

The International Covenant on Civil and Political Rights, *A Compilation of UN International Human Rights Instruments* 18 (1988). In force 23 March 1976.

The International Covenant on Economic, Social and Cultural Rights, *A Compilation of UN International Human Rights Instruments* 7 (1988). In force 3 January 1976.

Declaration of the United Nations Conference on the Human Environment, 11 *ILM* 1416 (1972). Nonbinding.

Declaration on Permanent Sovereignty Over Natural Resources, *A Compilation of UN International Human Rights Instruments* 49 (1988). Nonbinding.

Helsinki Rules on the Uses of the Waters of International Rivers, *52nd ILA Report* 477 (1966). Nonbinding.

Treaty Establishing the European Atomic Energy Community, 298 *UNTS* 171 (1957). In force 1 January 1958.

Convention Regarding the Regime of Navigation on the Danube, 1949 *UNTS* 197 (1948). In force 18 August 1948.

Peace Treaty of Trianon, *Treaties, Conventions, International Acts, Protocols, and Agreements between the United States of America and Other Powers, 1910–1923*, 3558 (1923).

List of Abbreviations

AFP	Agence France Presse
AJIL	*American Journal of International Law*
APSR	*American Political Science Review*
BBC World Broadcasts	British Broadcasting Corporation Summary of World Broadcasts
BCL	*Bulletin of Czechoslovak Law*
BNA	The Bureau of National Affairs
CEE	Central and Eastern Europe
CMEA	Council of Mutual Economic Assistance
CO_2	Carbon Dioxide
CTK News Wire	Czech News Agency National News Wire
CYIL	*Canadian Yearbook of International Law*
EBRD	European Bank for Reconstruction and Development
EIA	Environmental Impact Assessment
EIB	European Investment Bank
ELJ	*Environmental Law Journal*
ELP	Environmental Law and Policy
ELR	*Environmental Law Review*
EPA	United States Environmental Protection Agency
EU	European Union
EXIM Bank	United States Export–Import Bank
EYBCR	*Environmental Year-Book of the Czech Republic*
FBIS	Foreign Broadcast Information Service
FBIS-EEU	Foreign Broadcast Information Service – East Europe Daily Report
FBIS-WEU	Foreign Broadcast Information Service – West Europe Daily Report
G7	Group of Seven Industrialized Nations
GDP	Gross Domestic Product
GDR	German Democratic Republic
GEF	Global Environmental Facility
GNP	Gross National Product
IAEA	International Atomic Energy Agency
IELR	*International and Environmental Law Review*
ICJ	International Court of Justice
ICLQ	*International and Comparative Law Quarterly*

ICLR	*International and Comparative Law Review*
IER	International Environmental Reports
IFI	International Financial Institution
IIASA	International Institute for Applied Systems Analysis
ILA	International Law Association
ILC	International Law Commission
ILJ	*International Law Journal*
ILM	International Legal Materials
ILP	International Law and Policy
ILR	*International Law Review*
ILRPTS	International Law Reports
IMF	International Monetary Fund
JEL	*Journal of Environmental Law*
JICL	*Journal of International and Comparative Law*
JIELP	*Journal of International Environmental Law and Policy*
JIL	*Journal of International Law*
JILP	*Journal of International Law & Policy*
JILS	*Journal of International Legal Studies*
JPIL	*Journal of Public International Law*
JTL	*Journal of Transnational Law*
LJ	*Law Journal*
LQ	*Law Quarterly*
LR	*Law Review*
LRTRAP	Long Range Transboundary Air Pollution
MW	Mega-Watt
NGO	Nongovernmental Organization
NO_x	Nitrogen Oxide
NRJ	*Natural Resources Journal*
NYT	*New York Times*
ODD	OMRI Daily Digest
OECD	Organization for Economic Cooperation and Development
OMRI	Open Media Research Institute
OSCE	Organization for Security and Cooperation in Europe
PCIJ	Permanent Court of International Justice
PHARE	Poland–Hungary Assistance for the Reconstruction of the Economy
Proceed. ASIL	*Proceedings of the American Society of International Law*
RCADI	*Recueil des Cours d'Academie de Droit International*
RCEEL	*Review of Central and East European Law*

RECIEL	*Review of European Community & International Environmental Law*
RFE/RLN	Radio Free Europe/Radio Liberty Newsline
RIAA	Reports of International Arbitral Awards
RJUPR	*Revista Jurídica de la Universidad de Puerto Rico*
SECI	Southeast Europe Cooperation Initiative
SO_2	Sulphur Dioxide
TLR	*Transnational Law Review*
UNECE	United Nations Economic Commission for Europe
UNEP	United Nations Environmental Programme
UNTS	United Nations Treaty Service
USAID	United States Agency for International Development
WHO	World Health Organization
World Bank	International Bank for Reconstruction and Development
YIEL	*Yearbook of International Environmental Law*
YIL	*Yearbook of International Law*
YILC	*Yearbook of the International Law Commission*

Introduction

Soon after the initiation of the democratic transformation in Central and Eastern Europe (CEE), the European Union (EU) identified protection of the environment and the resolution of Central and East European transboundary environmental disputes as two of the most urgent political priorities in the intra-European dialogue for the foreseeable future.[1] As a result of the environmental degradation in CEE, the International Bank for Reconstruction and Development (World Bank) estimates over 30,000 people die each year.[2] Moreover, a 1997 American intelligence community report released in 1998 predicts CEE states will suffer the consequences of severe environmental degradation for the next five to seven years and will be unable to 'set the stage for sustained environmental progress', within the next ten to fifteen years.[3] The report concludes with the warning environmental degradation in border regions has given rise to a number of transboundary environmental disputes, which could be expected to cause significant and increasing tension among CEE states for the next several decades.[4]

The general purpose of this book is to explore the role of international law and the legal process within the international regime of transboundary environmental protection among CEE states, and more particularly that aspect of the regime relating to transboundary environmental dispute resolution.[5] With an increased understanding of how international law and the legal process operate to promote transboundary environmental dispute resolution, policymakers, international actors, nongovernmental organizations (NGOs) and public commentators may be able to improve the operation of international law and the legal process to better promote CEE transboundary environmental dispute resolution. Enhanced dispute resolution may then advance CEE transboundary environmental protection and reduce political tension among CEE states as they continue their economic and political transformations.

An inquiry into the role of international law and the legal process in promoting CEE transboundary environmental dispute resolution is particularly relevant given the political and economic changes occurring in CEE during the course of this decade. Notably, the dissolution of the former Soviet Union and the disintegration of Soviet hegemony in CEE, coupled with the transition of most CEE states from totalitarian regimes operating command economies to democratic regimes operating market economies has created an opportunity to resolve a number of transboundary environmental disputes. These changes, however, have also created a great deal of uncertainty concerning the relationships among CEE states and between CEE states and the rest of Europe, as well as an uncertainty about the role and purpose of international law.

The dissolution of the former Soviet Union and the disintegration of Soviet hegemony in CEE removed many obstacles to the initiation of efforts to resolve transboundary environmental disputes and increase transboundary environmental protection. Most important, prior to the revolutions of 1989, CEE states were seldom permitted to officially acknowledge the existence of environmental degradation as according to communist ideology, it was impossible for a socialist economic system to cause harm to the environment.[6] Moreover, in instances where the consequences of resource exploitation in border regions were apparent to the general public, attempts to raise the issue to the level of a transboundary dispute were stifled. The socialist governments believed it to be inherently impossible for the officially nonexistent environmental degradation to give rise to a transboundary dispute, and socialist doctrine held relations between socialist states represented a higher order of international relations and therefore 'in the nature of the socialist system [there could be] no objective reasons for conflicts between states'.[7] With their emergence from Soviet hegemony, CEE states are now free to acknowledge the consequences of the over exploitation of their natural resources or the environmental effects of hydroelectric, nuclear or military facilities and the emergence of transboundary environmental disputes associated with these consequences.

With their acknowledgment of the existence of transboundary environmental disputes, many CEE states are finding that particular disputes as well as their general relations with neighboring states are complicated by historical, political and economic circumstances. Thus at a time when CEE states are reacquainting themselves with their sovereign rights, they are also rediscovering the complicated nature of their mutual and interdependent relationships with other CEE states. CEE states are also finding as they reestablish their position in the general international political arena, the international political system is changing from a bi-polar system to an as yet undetermined system.[8]

As CEE states continue to reestablish their position in the international political system, pursue increased integration with Western political structures and institutions, and allow for more public participation in domestic decision making, they will find themselves increasingly called upon to employ international law and the legal process to resolve transboundary disputes. The ability of CEE states to use international law and the legal process effectively will be complicated by the CEE states' current conversion from a legal *Weltanschauung* characterized by a primary reliance on the utilization of socialist international law, which emphasized sovereignty and inter- and intra-state equality, to one based on the rule of general international law. Moreover, as CEE states acclimate to the norms and structures of general international law, they will find these norms and structures are themselves undergoing a transformation, which places less emphasis on traditional state sovereignty and state prerogative and the goal of resource exploitation, and increasing emphasis on the role of international organizations, the interests of sub-state actors, and the goals of resource conservation and even resource preservation.[9]

An examination and clarification of the role of international law in the process of CEE transboundary environmental dispute resolution may thus promote the more effective use of international law by articulating the functions served by international law during the dispute resolution process; by assessing the interests and capabilities of state actors, sub-state actors and interested third parties that influence how a state uses international law; by assessing the situational circumstances that structure the ability of these states, sub-state actors and third parties to influence the use of international law; and by evaluating certain factors that promote the functionality of international law within the regime of transboundary environmental protection.

To structure the inquiry into the role of law, this book employs the social science methodology of descriptive inference based on a qualitative assessment of secondary and primary case studies.[10] In order for the descriptive inference approach to be effective in explaining how international law operates to promote dispute resolution, the qualitative investigation has been designed in a manner which seeks to avoid biased inferences, inefficiency and inconsistency.[11]

Working under this approach, Part I (Chapters 1 and 2) examines eight secondary case studies of current transboundary environmental disputes. These eight disputes, geographically distributed across CEE from the Baltic Sea to the Black Sea, are situated at various stages of the dispute resolution process, and to varying degrees have been brought within the legal processes of inquiry, mediation, conciliation, arbitration or adjudication. These disputes also have been identified by sub-state actors and interested third parties consulted by the author as disputes likely to have a significant impact on bilateral and regional multilateral relations. Chapters 1 and 2 further the articulation and evaluation of how international law operates to promote transboundary environmental dispute resolution by utilizing the eight case studies to illustrate the wide variety of functions that international law and the legal process may serve. These functions include, for instance, acting as the catalyst for the exchange of information and the sharing of ideas, serving as the basis for the development of cooperative agreements, providing a set of rules and processes for resolving particular points of disagreement, and/or functioning as a means of enforcing and verifying steps taken consistent with an agreed course of action intended to resolve the dispute.

In instances where the role of international law appears to be minimal, the text indicates which circumstances or factors may be inhibiting the use of international law and identifies how an increased use of law might contribute to the resolution of the dispute. Furthermore, the close examination in each case study of situational circumstances, as well as the sub-state actors and interested third parties contributing to or involved in the dispute, provides an opportunity to assess the nature of the influence of these circumstances and actors on the operation of international law and the legal process.

To build upon the preliminary observations and inferences derived from the eight case studies, Part II (Chapters 3–6) conducts an empirical examination of the Slovak–Hungarian dispute relating to the construction and operation of the Gabčíkovo–Nagymaros Project (Project). The Gabčíkovo–Nagymaros Project dispute has been selected as the primary case study because it has significantly progressed through the phases of the dispute resolution process, the parties have utilized the various dispute resolution mechanisms of inquiry, mediation and adjudication, and the dispute and role of law within that dispute have been influenced by the interests and actions of a number of sub-state actors and interested third parties.

The exploration of the Gabčíkovo–Nagymaros Project dispute is designed in particular to further the understanding of the role of the legal processes of inquiry, mediation and adjudication in facilitating transboundary environmental dispute resolution, as well as an understanding of the various functions performed by international law and these processes during the dispute resolution phases. Notably, the Gabčíkovo–Nagymaros Project case study allows for the extensive investigation of how specific legal principles and various aspects of the legal process operate to address the multifarious elements of a transboundary environmental dispute in such a manner so as to allocate the respective rights and responsibilities of each party as necessary to produce a formal decision that can form the basis for the settlement of the dispute. The Gabčíkovo–Nagymaros Project case study, in addition, provides an opportunity to articulate precisely the influence of certain sub-state actors and third parties on the operation of international law and the legal process among the disputing parties, and the influence of situational circumstances on these sub-state actors and third parties.

Part III (Chapters 7 and 8) relies on the case studies investigated in Parts I and II to evaluate the role of international law within the international regime of transboundary environmental protection, which operates to modulate the otherwise power based relationships of CEE states. In particular, Chapter 7 explains the operation of international law as a constituent element of a regime and its relation to other constituent elements such as social norms, international organizations, state actors and state policies, and the identifiable interests of sub-state actors and sub-state actor transnational networks. Chapter 7 then combines the observations derived from the secondary and primary case studies with the select work of regional specialists, political and social scientists, and legal scholars to establish and articulate the functions served by international law during the different phases of the dispute resolution process. Chapter 8 similarly establishes and articulates the categories and capabilities of relevant sub-state actors and third parties influencing the use of international law; the categories and effect of situational circumstances that structure the ability of states, sub-state actors and third parties; and the factors promoting the functionality of international law within the regime of transboundary environmental protection.

Part IV (Chapters 9 and 10) then conducts a general review of the recent evolution and current status of the situational circumstances affecting the resolution of transboundary environmental disputes and the factors promoting the functionality of international law within the regime of transboundary environmental protection. Based on this examination, Chapter 9 assesses the extent to which these circumstances have promoted the interest in and ability of CEE states to resolve their transboundary environmental disputes, and examines the extent to which they have positively or negatively influenced the interests and abilities of states, sub-state actors and interested third parties. Chapter 10 then evaluates the prospects of an increasing role for international law in promoting the resolution of CEE transboundary environmental disputes. While predicting the likelihood of the increased operation of international law, Chapter 10 also identifies means by which CEE states, sub-state actors and interested third parties may foster the further positive evolution of the factors affecting the functionality of international law, and thereby promote the effective use of international law to resolve transboundary environmental disputes.

It should be noted here in the introduction that the term CEE is used throughout this book to refer to the states of Poland, the Czech Republic, Slovakia, Hungary, Romania and Bulgaria. Lithuania, Latvia, Estonia, Belarus and the former Soviet States of Ukraine and Russia are frequently parties to disputes with CEE states, and for the purposes of this book they are identified where relevant, but they are not included in substantive references to CEE states. Similarly, Turkey is involved in a limited number of transboundary environmental disputes with Romania and Bulgaria and other Black Sea Basin states, but is not considered a CEE state for the purposes of this book. In addition, although the geographic description of Central and Eastern Europe is not strictly delimited along national boundaries, it is possible to draw a number of ecological conclusions particular to CEE states, and the similarities between CEE states in terms of political, economic and social transformation, as well as the history of their industrial development and its environmental consequences are more definite than their individual differences.[12]

During the course of this book the term 'transboundary environmental dispute' is used to refer to a dispute or opportunity for cooperation between two or more neighboring states regarding the utilization, allocation and/or degradation of shared natural resources.[13] These disputes may range from a controversy over the construction of a particular project, such as a nuclear power plant or hydroelectric dam, to the pollution of a river crossing or forming an international boundary, to a conflict concerning the emission of airborne pollutants covering a range of hundreds of square miles.[14]

These disputes may also be currently situated within any of the four phases of a dispute, which are the dispute formation phase, pre-resolution phase, resolution phase and implementation phase. The environmental dispute formation phase occurs when one or more states begin to experience, become aware of or fear the effect of environmental degradation on their territory and/or population and bring this to the attention of the state believed to be responsible for this degradation. The dispute moves to the pre-resolution phase when the states engage in activity aimed at identifying the parameters of the dispute and possible options for resolving the dispute. In the dispute resolution phase, the parties undertake to agree upon a specific course of action designed to resolve the dispute. In the implementation phase the parties seek to implement and verify any agreement reached in the resolution phase. A more detailed discussion of the various phases of dispute resolution will be undertaken in Chapter 7.

As the book now turns to the discussion of eight case studies, it should be noted a secondary objective of this work is to demonstrate to CEE states, sub-state actors and interested third parties through academically rigorous and systematic analysis, international law and the legal process are capable of performing a wide variety of roles, and an increasing reliance on international law may positively benefit the resolution of transboundary environmental disputes. It is hoped sub-state actors and interested third parties capable of influencing the behavior of CEE states will take from the discussion contained within this book an understanding of how better to promote the use of law to achieve their common aims of reducing transboundary tensions arising from environmental disputes and advancing the spirit of regional cooperation.

Part I
Central and East European Transboundary Environmental Disputes from the Baltic Sea to the Black Sea

1
The Northern Tier: Polluting the Baltic Sea, Dark Clouds over the Black Triangle and Silesian Coal Basins, and East Meets West in Temelín

CEE states suffer from a number of bilateral and multilateral transboundary environmental disputes stretching from the Baltic Sea to the Black Sea and include disputes relating to water pollution, air pollution, the destruction of biodiversity, liability for hazardous waste and the siting and operation of nuclear facilities. These disputes exist at all four phases of the dispute resolution process, with some states just beginning to articulate the exact nature of their dispute and with other states striving to implement an agreed upon means for resolving the dispute. Similarly, some states have as yet failed to utilize a dispute resolution mechanism, while others are making active use of inquiry, mediation, conciliation, arbitration and in some instances, adjudication. The disputes involve not only states from CEE, but also states from the EU and former Soviet Union, and in at least one instance, Turkey. To structure the investigation of these case studies, which are arranged according to their geographic position on the North–South axis, each case study is divided into a discussion of the basic factual background necessary to understand the dispute, and a discussion of the current and potential role of international law and the legal process in promoting a resolution of the dispute.

To further an understanding of the role of international law and the legal process in promoting CEE transboundary environmental dispute resolution, this chapter examines four secondary case studies situated in the northern tier of CEE. These four disputes, like the four discussed below in Chapter 2, have been selected as secondary case studies

12 *Central and Eastern Europe*

because of their geographic distribution, substantive variety and because they are situated at various stages of the dispute resolution process, and to varying degrees have been brought within the legal processes of inquiry, mediation, conciliation, arbitration or adjudication. These disputes also have been identified by sub-state actors and interested third parties consulted by the author as disputes likely to have a significant impact on bilateral and regional multilateral relations.[15]

I. The Baltic Sea

A. Factual background

The semi-enclosed Baltic Sea is bordered and polluted by Sweden, Finland, Russia, Estonia, Latvia, Lithuania, Poland, Germany and Denmark and is further polluted by the nonriparian states of Slovakia, Norway, Ukraine and Belarus.[16] Because of the narrow exit of the Baltic Sea into the North Sea, there is little exogenous circulation of its waters and the pollution generated from the 80 million inhabitants of its drainage basin tends either continuously to recirculate or settle on the bottom of the sea. The Baltic Sea also absorbs sulphur pollution from land based sources, such as coal fired power plants, hazardous waste from direct dumping into the sea, and radioactive waste from former Soviet military sites in the region. The Sea is also under threat from a number of corroding Russian chemical weapons containers stored in the hulls of sunken ships on the seabed.[17] As a result of this pollution, over 100,000 square kilometers of the sea are unable to support any form of life. In areas of the sea where marine life still survives, the fish frequently contain large amounts of heavy metals, and organochloride residues three to ten times higher than fish taken from the North Sea. Such devastation poses a threat to the Baltic Sea's various ecological resources, including the marine ecosystem, wildlife and the tourism and fishing industries.[18]

In the past few years, the Baltic Sea Basin to a limited extent has benefited from the closure of many heavily polluting industrial enterprises in the former East Germany, which was made possible by the unification of Germany.[19] The Baltic Sea may also soon benefit from the establishment of a Baltic Sea Euroregion, which was agreed to on 22 February 1998 by local governments from all six states bordering the Baltic Sea. One of the primary purposes of the Baltic Sea Euroregion as expressed in the agreement is to promote cooperation in the field of environmental protection. In June 1998, the EU and 11 Baltic states also adopted

Baltic Agenda 21, which sets forth a cooperative plan for promoting sustainable development in the Baltic Sea Basin states.[20]

Moreover, as Poland and Estonia progress towards EU membership, and Latvia and Lithuania continue to seek an invitation to negotiate for EU membership, they may have an increasing incentive to adopt legal regulations necessary to reduce land based pollution affecting the sea. Estonia, for example, by 1997 had adopted over 80 per cent of the necessary laws for regulating the protection of the environment as required by EU Directives.[21] In addition, statements by Poland's Ministry of Environmental Protection indicate the Polish government is increasingly committed to observing and cooperatively implementing its international legal obligations relating to the Baltic Sea Basin. The Ministry also has embarked on separate bilateral efforts with Sweden, Denmark and Norway to deploy modern production and pollution control technologies.[22] Similarly, as Poland continues its transformation to a market economy, there will likely be an improvement in the efficiency of land based sources and the possibility Nordic states could purchase emission offsets for these sources under a regional permit scheme. The viability of emissions trading is enhanced by the fact that with the appropriate technological investment, it is predicted there could be a 20–30 per cent average reduction in SO_2 in CEE states.[23]

The Baltic Sea is, nonetheless, under new long term threats with the increased likelihood of oil drilling activity, as Russia seeks alternatives to Central Asian oil, the Baltic states in turn seek to reduce strategically their reliance on Russian oil, and Norway's Scandinavian neighbors seek to replicate in the Baltic Sea that nation's petroleum successes in the North Sea.[24] As a result, it is estimated by 2005 the Baltic Sea Basin states will construct ten new oil terminals in the region and expand nine existing terminals. This new construction would roughly quadruple the oil extracted from the Basin, considerably enhancing the risks of major and minor oil spills.[25]

B. The role of international law and the legal process

For the past 20 years, Baltic Sea Basin states have attempted to develop cooperative means for regulating and reducing pollution in the Baltic Sea.[26] A number of their efforts may be characterized as attempts to construct and operate within a regime of transboundary environmental protection, within which states could share obligations and coordinate efforts to improve the environmental condition of the Baltic Sea.[27] To date, the Baltic Sea Basin states have relied upon international law to clarify their obligations, establish and facilitate the implementation of requirements to abate pollution, and to assist with the movement of the dispute through the dispute formation and pre-resolution phases into the resolution and implementation phases.[28]

The first major use of international law to assist in the reduction of the emission of pollutants into the Baltic Sea was the adoption of the 1974 Convention on the Protection of the Marine Environment of the Baltic Sea Area.[29] This Convention, which represents the product of the resolution phase of the dispute, regulates pollution from land, air and maritime sources.[30] Specifically, the Convention prohibits the dumping of hazardous waste directly into the Baltic Sea, requires states to use the best practicable means to prevent the introduction of specific noxious pollutants into the sea, regulates pollution from maritime vessels and identifies specific actions to be taken to eliminate or minimize pollution of the sea by oil or other harmful substances.[31]

In order to institutionalize and perpetuate the authority of the regime, the Convention created the Marine Environmental Protection Commission (HELCOM) which was tasked with the preparation of additional proposals and recommendations for preventing and reducing pollution of the sea.[32] In 1988, at the Ministerial meeting of the HELCOM, the Environment Ministers of the Baltic Sea Basin states responded to the worsening situation in the Baltic Sea by agreeing to attempt to reduce pollution in the Baltic Sea by 50 per cent over the next ten years. The Ministers, however, failed to agree upon specific mechanisms for bringing about this reduction and left it to the member states to ascertain how best to meet their obligations.

In 1990, to enhance the lagging implementation of the 1974 Convention and strengthen the regime of transboundary environmental protection, the EU, operating both as an interested party and as a third party, established the Baltic Sea Joint Comprehensive Environmental Action Program.[33] This voluntary program sought to reduce pollution from agricultural runoff and atmospheric deposition by standardizing codes of practice and emission standards and integrating environmental policies.[34] Since its inception, the program has identified 132 environmental hot spots involving point and nonpoint sources and has concluded effective remediation will require $14.3 billion for environmental investment over the next 20 years. More recently, in April 1997, the Baltic Sea Basin states agreed to phase out the discharge of 'hazardous substances' by 2022.[35]

In April 1992, the Baltic Sea Basin states returned to the negotiating table and attempted to strengthen the regime further by agreeing to and signing the Convention on the Protection of the Marine Environment of the Baltic Sea, which revised and superseded the 1974 Convention. In order to strengthen the regime, the 1992 Convention adopted a number of international legal principles, including the polluter-pays principle;[36] the precautionary principle;[37] the requirement of Best Environmental Practice for all sources and Best Available Technology for point sources; an Environmental Impact Assessment (EIA) requirement,[38] which must include participation by the affected states; the duty to notify other state parties of accidents; the right of public access to permitting information, the result of scientific studies and stated water-quality objectives; and the requirement that states not comply with the agreement by merely transferring pollution from the Baltic Sea Basin to other water basins.[39] The 1992 Convention also calls upon the Basin states to prevent and eliminate all sources of pollution, includes within the jurisdiction of the Convention the internal waters of state parties and requires state parties to report to the HELCOM on regulatory measures they are taking to meet the Convention's requirements.[40]

Some commentators note with concern despite the inclusion of a number of contemporary principles of international environment law, the 1992 Convention may not be sufficient to improve the environmental quality of the Baltic Sea significantly because it fails to address the question of land based sources of pollution adequately, which contribute up to 80 per cent of Baltic Sea pollution.[41] These commentators suggest regulation of land based sources is neglected as it is costly to reduce their emissions, and as states are frequently reluctant and sometimes incapable of imposing such economic costs and mechanisms necessary for verifying compliance because they are perceived to impinge substantially upon the state parties' sovereignty.[42]

To address the weak implementation of the Conventions and the new threats posed to the environmental health of the Baltic Sea Basin, HELCOM and some academic commentators have noted Basin states may be able to strengthen further the regime of transboundary environmental protection governing the Basin by bringing within its purview existing international agreements, such as the Law of the Sea Convention, the United Nations Convention on the Protection and Use of Transboundary Water Courses and International Lakes, the Convention for the Prevention of Marine Pollution from Land-Based Sources, and the 1979 Convention on Long Range Transboundary Air Pollution.[43] These commentators, however, caution extra-regional conventions may lack the necessary relevance to the specific problems of the Baltic Sea, and although they may contribute to reducing the pollution of the

Baltic Sea, they are not sufficiently functional in and of themselves to constitute a legal framework for resolving the problem of transboundary pollution.[44]

Other commentators have considered the capacity of customary international law to promote protection of the Baltic Sea environment. The most relevant of those principles would be the prohibition against causing serious injury to neighboring states and the equitable utilization of shared natural resources.[45] These principles, however, appear to lack sufficient functionality as the causal links between specific types of pollution and specific harm are difficult to prove, and there is no generally acceptable criteria among the Basin states for determining an equitable allocation.

Thus, although Baltic Sea Basin states have made significant progress in relying upon international law to assist in creating an international regime to reduce the amount of pollutants entering the Sea, it appears the adequate protection of the Basin will require strict compliance with existing agreements, and further negotiations designed toward strengthening the regime to include regulation of additional sources of pollution and the development of customized rules and procedures for enhancing cooperation among the Basin states.[46]

II. The Black Triangle

A. Factual background

The Black Triangle is a region encompassing the Central-European Brown Coal Basin and includes areas in south-western Poland, northwestern Czech Republic and south-eastern Germany.[47] The degree of pollution in the Black Triangle region is so devastating that as a local Polish saying goes, 'the Black Triangle is a place where even the devil says good-bye'.

The Black Triangle and the neighboring Silesian Coal Basin contain over 50 per cent of Europe's lignite resources, generate over 30 per cent of Europe's SO_2 and are characterized by a contiguous industrial belt from Lower Silesia in Poland and Northern Bohemia in the Czech Republic into the Saxony region of Germany.[48] The primary cause of environmental degradation in the Black Triangle is the immense concentration of coal extraction, energy production and chemical industries operating obsolete emission control technologies and emitting substantial amounts of pollution that then permeates most of the European continent.[49] In 1995, for instance, between 1,500 and 2,000 tons of SO_2 were emitted into the atmosphere from the Black Triangle. This amount is eight times the amount emitted by France.[50]

The extensive operation of open pit coal mines to fuel the energy sector and heavy industrial enterprises[51] has created a region of environmental destruction unlike any other in Europe.[52] Specifically, this massive concentration of industrial enterprises emits significant levels of SO_2 and NO_x into the local and regional environment. In Poland and the Czech Republic, for example, coal fired power plants account for over 75 per cent of energy produced for industrial and household consumers, with Northern Bohemia accounting for between 50–60 per cent of the energy produced in the Czech Republic. As a result, Northern Bohemia itself accounts for 42.6 per cent of the Czech Republic's entire SO_2 emissions and 31.3 per cent of its NO_x emissions.[53]

Because of high levels of air pollution in the Black Triangle, its inhabitants suffer from poor living conditions, high infant mortality rates and extensive environmental damage.[54] Over 96 per cent of the population in Northern Bohemia lives in environmental conditions classified by the Czech government as 'unsuitable' for healthy living, and a full 100 per cent of the forests are appreciably damaged by air pollution and acid rain.[55] Moreover, median life expectancies in the area lag about five years behind the European average, and SO_2 pollution in the Black Triangle measures at 140 times the average level in Germany.[56] In Poland, acid rain generated by industrial enterprises in the Black Triangle is responsible for damage to thousands of hectares of forest and agricultural land, causing an estimated financial loss of over $150 million a year.[57]

The Black Triangle region also suffers from significant water pollution, in particular pollution of the Elbe, Nisa and Oder Rivers.[58] In Northern Bohemia, 64 per cent of drinking water does not meet Czech quality standards because of excessive nitrate concentrations in the rivers and frequent industrial accidents.[59] Upon entering Germany and Poland the level of pollution in the rivers is compounded by industrial enterprises and mining operations.[60] In the Elbe River, for instance, nitrate concentrations are 14–16 times higher than the international norm for drinking water.

Transboundary environmental disputes arising from the Black Triangle, which remain in the dispute formation and pre-resolution phases, relate to general transboundary environmental pollution as well as specific point sources. The general disputes are based upon the substantial importing and exporting of air and water pollution between the states. Poland, for instance, is responsible for approximately 58 per cent of SO_2 emissions deposited on its territory, while it imports 42 per cent from foreign states. The Czech Republic is responsible for approximately 54 per cent of SO_2 emissions deposited on its territory, while it imports 36 per cent from neighboring states (11 per cent from Poland, 15 per cent from Germany and 10 per cent from more distant states).[61]

Point source disputes relate to projects such as German plans to transform a coal fired power plant at Herschfeld into a hazardous waste incinerator and the continued operation of the Polish 2,000 MW Turów coal fired power plant. The Turów plant was constructed on a peninsula of land surrounded on three of four sides by territory of either Germany or the Czech Republic and contributes significant amounts of SO_2, NO_x and ash to the immediate environment, while the adjacent coal mine threatens a major Czech aquifer.[62]

The operation of the plant also has produced a mountain of toxic tailings that spill over into the Czech Republic during times of heavy rain and contribute to the salinity of local rivers. Polish press reports identify the plant as one of the two main sources of the region's grave SO_2 pollution. At the same time, however, they also cite economists discussing the risk of 25 per cent unemployment connected to wholesale changes in the Black Triangle's industrial and commercial infrastructures and claim the greatest support for closing Turów comes from competing German power plants which generate power at a higher cost and are more environmentally destructive.[63] Importantly, in 1995 the Polish government embarked on a plan of modernization at a cost of $1.2 billion in order to reduce sulphur and particulate matter emissions from the plant.[64]

In the short term, the ecosystem in the Black Triangle region is experiencing a respite with a reduction in emissions caused by the temporary decline in industrial production brought about by the collapse of the former economic system.[65] In addition, to some extent there is a degree of environmental investment in the Black Triangle: the Czech electric utility, for example, has invested over $250 million annually in Northern Bohemia.[66] This effort, however, appears to be unique as most large industrial enterprises cannot afford the investment required to meet environmental regulations, and the Czech privatization fund has a limited capacity to provide such assistance.

As a result of the unification of Germany and privatization of many East German power plants, the German central government intends for former East Germany to meet EU air emission standards. Plans to meet these standards have been substantially advanced by the 'clean collapse' of many noncompetitive former East German industrial enterprises. As a result, from 1990 to 1996, Germany reduced its production of energy in the Black Triangle from 12,000 MW to 9,000 MW and intends to reduce the number of power plants from seven to four. Polish central and local government officials generally believe the German share of pollution in the Black Triangle will be drastically reduced in the short term, and because Germany tends to suffer less from air pollution generated in the Black Triangle (as a result of the prevailing wind patterns), Germany will focus most of its attention on matters relating to pollution of the international watercourses entering Germany from the Black Triangle region.

B. The role of international law and the legal process

Despite the historic lack of cooperation on issues relating to the Black Triangle, the affected states are making efforts to develop regimes for the purpose of promoting cooperative means for reducing transboundary pollution and for resolving general and point source disputes.[67] The primary impetus for efforts to create these regimes comes from the involvement of the EU as a third party, which created and funded a program of inquiry.[68] With an initial allocation of $12.7 million for the Black Triangle Project, the EU established a tri-partite secretariat in Ustí nad Labem and charged it with developing a framework plan for environmental rehabilitation and establishing a joint air monitoring system.[69]

Transboundary cooperation is also enhanced by the EU sponsored creation of Euroregion NISA, which consists of representatives from Poland, the Czech Republic and Germany and engages in a variety of local programs to promote cooperative solutions to transboundary environmental problems. In October 1994, Polish and Czech environmental authorities agreed to begin combining their efforts, with special attention reserved for pollution monitoring in the Black Triangle.[70] Two months later, the Polish Government created a new cabinet level post to coordinate NISA activities, on the finding that without an effective Euroregion it was unlikely sufficient cross border cooperation would develop.[71] In November 1996, the Euroregion's member governments established an inter-parliamentary group to oversee transboundary cooperation, and in September 1997, the Euroregion allocated $3 million to promote reforestation in the Jizerskehory mountains in order to combat further erosion in the Nisa River Basin.[72]

Poland, the Czech Republic, Germany, and the European Union also have laid the foundation for the creation of a regime of transboundary environmental protection by adopting the Elbe River Convention and creating an international commission for the protection of the Elbe

River. The Convention requires the parties to endeavor to promote the use of the river for drinking water, to achieve as natural an ecosystem as possible and to reduce the emission of pollutants into the river.[73] The Commission is empowered to prepare surveys of point sources of pollution, propose limits on the discharge of effluents, propose water quality objectives, propose specific actions for the reduction in the discharge of harmful materials, and propose a uniform warning and alert system.[74] In 1996, Poland, the Czech Republic, Germany and the European Union also created a similar commission for the Oder River.[75]

Continued development of these regimes and institutions will depend upon the energy policies adopted by Black Triangle states in the next five to ten years. The most desirable policy would be for the parties to establish a regional energy program, with the creation of a multinational linkage between power plants.[76] The next most desirable policy would be to shift the source of energy to Russian oil and gas, with an emphasis on improving efficiency of energy production and consumption. The least desirable option, but most probable, is continuation of the status quo, with each state feigning cooperation with its neighbors and generating complex maps indicating the extent to which other states are responsible for regional pollution.[77]

The latter path is identified as the most probable because the parties have to date been unable to summon the political will to agree upon a long term strategy. The lack of political will results from the Czech Republic's concern it will invest in improving the quality of water exported from its territory, without Poland reciprocally investing in the reduction of air pollution; the Polish preference to allocate regional resources to the reduction of pollution from the Turów power plant, with a disinclination to concentrate on local sources as favored by the Czech Republic; and the general attitude that interests of individual states are advanced by a policy of economic competition and inhibited by a policy of ecological cooperation.

The element of law present in the rudimentary regime of transboundary environmental protection currently operating between Germany, Poland and the Czech Republic might further the generation of necessary political will and strengthen the regime through its substantive requirements that parties cooperate to resolve their disputes and share information and consult on various alternatives for reducing pollution. Notably, the Black Triangle states are currently in broad compliance with these international legal obligations. The Polish government has undertaken efforts to engage Germany in addressing Poland's environmental concerns, including more technical assistance and the reallocation of Polish debt to an EcoFund, and Poland and the Czech Republic have engaged in regular dialogue.[78]

The impetus for resolving the dispute and pursuing one of the two former paths may emanate from pressure brought to bear by the EU or through incentives offered by the United States and/or International Financial Institutions (IFIs), such as the International Monetary Fund (IMF), World Bank or European Investment Bank (EIB). An impetus for resolving the dispute may also arise if the parties expand the EU sponsored inquiry to investigate the technical and economic viability and environmental effect of fuel switching or the adoption of a regional energy program. Similarly, the parties may promote the resolution of specific point source disputes and possibly the broader dispute as well, if they seek the good offices of a third party or employ the mechanism of mediation.[79] The invocation of arbitration or adjudication would be unlikely to further the resolution of the dispute as the case requires less of an articulation of legal obligations or an assignation of responsibility/liability and more of a roadmap for the implementation of systems for transboundary cooperation.

If the parties do develop the political will to resolve the dispute or respond to third party pressure or incentives, international law and the legal process may play a powerful role in reducing pollution and resolving the Black Triangle dispute. Under the option of establishing a regional energy program, international law could provide the structure for an agreement to reduce pollution to specific levels, as in the United States/Canadian Acid Rain Agreement,[80] and for the creation of a regional tradable emission permit scheme.[81] Initial funding for regional schemes might be provided by the joint UN-World Bank Global Environmental Facility (GEF) fund or through joint implementation projects under the climate change convention.[82]

In order to construct the confidence necessary for a regional tradable permit scheme, Black Triangle states might first engage in more modest schemes, such as a local regional tradable permit scheme with a transparent verification process. Once this scheme is functional, the parties might develop increased confidence in the behavior of other parties, become familiar with the operation of the regime and assess the benefits of a broader scheme. Notably, the more frequently states engage in a certain type of rule-governed behavior yielding benefits, the more inclined they will be to expand the scope of that behavior.

Alternatively, if Black Triangle states pursue the shift to oil and gas sources, with an emphasis on improving efficiency of energy production and consumption, they can draw from the international law principles of sustainable development, polluter pays and precautionary action to develop legislation emphasizing environmental protection, devolution of environmental decision making authority to municipalities and pricing of energy resources at their market rate. International law might also provide a basis for an agreement on intra-regional substantive rules of liability for environmental harm and establishment of a transboundary regulatory commission.[83]

III. The Silesian Coal Basin

A. Factual background

The Silesian Coal Basin is one of the most polluted areas in all Europe, whose transboundary emissions significantly affect not only Poland and the Czech and Slovak Republics, but also Sweden, Finland and Germany.[84] The Silesian Coal Basin is comprised of the black coal and steel producing regions surrounding Ostrava in the Czech Republic and Katowice and Kraków in Poland, and together with the Black Triangle, forms the Sulphur Triangle stretching from Upper Silesia in Poland through Northern Bohemia in the Czech Republic to the Saxony Region of Germany.[85] The region suffers from low air quality, large scale mining devastation, accumulation of industrial waste and significant levels of water pollution caused by intense industrial development.[86] One local editorialist, lamenting the region's deposits of industrial waste, measured at 2 billion metric tons and growing, has suggested authorities should 'fence off the entire territory... set up an open air museum of large-scale industry in Europe, and demonstrate how industry should not have been practiced'.[87]

In the Ostrava region of the Czech Republic, there are over 370 sources of heavy pollution, most of which are coal burning thermal power stations and metallurgy factories. In 1995, industrial enterprises in the Silesian Coal Basin generated SO_2 amounts ten times higher than those generated in Germany.[88] As a consequence of this pollution, 97 per cent of Ostrava's population are currently considered by the Czech government to live in an 'extremely disturbed environment', and it is estimated by the year 2000, 96 per cent of the total acreage of forests in the region will be damaged.[89] Local press reports estimate remediation of the region would cost $200 billion and require 15–20 years.[90]

The Polish region of the Silesian Coal Basin suffers water and soil contamination from continued operation of antiquated industrial enterprises, air pollution from a disproportionate number of low stack industrial emitters and toxic waste pollution from the mining industry, which also annually discharges approximately four million tons of salt into the Vistula River.[91] As a consequence, near the historic town of Kraków, salt levels in the Vistula River exceed recommended levels by 2,400 per cent, and in the Katowice region, 65 per cent of the rivers are polluted beyond the range of classification.[92] Poland's Institute of Meteorology and Water Economy reports that during Silesian coal mines' periods of peak operation, they discharged 7,000 tons of salt into rivers daily, and as a result the Vistula exhibits a higher degree of salinity than the Baltic Sea.[93] Moreover, in areas such as Bukowno, Jaworzno and Olkusz the lead content in the soil is 99 times the allowable limit.[94]

In Katowice, Poland emissions from 3,829 industrial enterprises create ambient SO_2 levels exceeding safe limits by 200 per cent, CO_2 readings 21 times permissible health and safety levels and acid rain levels five times those of the German Rühr region. Life expectancy in the Silesian Coal Basin is four years less than the Polish average, with three-fifths of food produced in the Kraków region deemed by the Polish government to be unfit for human consumption.[95] Moreover, local press accounts report the combination of cross border environmental tensions, dislocations caused by economic inefficiencies, and the prevalence of workers from other regions residing locally and depending upon continued operation of polluting industries have combined to aggravate social tensions, and in the coming years the Silesian Coal Basin will be the primary hotbed of discontent that will radiate across Poland.[96]

B. The role of international law and the legal process

The development of an effective regional policy for reducing pollution and creating a regime for transboundary environmental protection in the Silesian Coal Basin is hampered by the lack of strong provincial governments in either Poland or the Czech Republic. In particular, the current structure of state government provides for a strong central government, with many powers delegated to municipal governments, but with few powers delegated to provincial governments. As such, no governmental organ outside of the central government possesses the competency to participate in using international law to craft and implement regional solutions to environmental degradation.

Since the development of a regional program of environmental protection is influenced by a myriad of other political considerations at the central government level, there is essentially no regional environmental policy for the Silesian Coal Basin. This lack of regional policy is complicated further by recent attempts to restructure the mining industry, which in 1996 led miners at 19 major Silesian coal mines to strike in protest of already low wages and the possibility the planned restructuring might result in layoffs and reduced social benefits. Despite these protests, in November 1998, the Polish Government passed legislation designed to restructure the coal mining sector, with the aim of reducing coal production from 137 million tons to 110 million tons and reducing the number of jobs from 245,000 to 138,000 by 2001.[97]

With the recognition that abating pollution in the Silesian Coal Basin would improve West European air quality to a greater extent than financial expenditure on site specific activities in Western Europe,[98] the EU is attempting to create the legal basis for a cooperative means of reducing pollution by sponsoring an inquiry program referred to as the Silesia Environmental Remediation Project. There is also some indication the EU may be willing to serve as a conciliator if the parties so desire. The project, initiated in June 1991, seeks to generate the necessary information and data to optimize cost benefit solutions to the most pressing environmental problems in the Silesian Coal Basin. The initial stages of the program have also identified actions needed to reduce health risks and improve environmental management capabilities and have evaluated various strategic options for reducing the emission of pollutants.[99]

The United States Environmental Protection Agency (EPA) operates a similar inquiry program designed to reduce pollution in the region by identifying industrial enterprises responsible for health risks and providing compensation for closure of plants or funding for installation of pollution abatement technology. The EPA program also sponsors an information center to promote the exchange and dissemination of information relating to the consequences of environmental degradation. The primary focus of the EPA program, however, relates to reducing pollution on a local basis.

Despite the good intentions of the EU and United States and their desire to move the Silesian Coal Basin dispute from the dispute formation into the pre-resolution and resolution phases, their efforts neglect to encourage the creation of a regime of transboundary environmental protection, and thus fail to promote any serious transboundary cooperation between Poland and the Czech Republic on this issue. Unless the EU and United States apply pressure to or create an incentive for Poland and the Czech Republic to engage in cooperative actions and unless the general public and environmental actors effectively generate the political will to resolve the dispute, there will likely be no significant operation of a regime, and international law will play only a minimal role in furthering a resolution of the dispute.

The Silesian Coal Basin dispute may benefit from more direct involvement of third parties as mediators or conciliators charged with the mandate of assisting Poland and the Czech Republic with developing a cooperative program for reducing transboundary emissions by engaging in fuel switching or a regional energy program. The mediators or conciliators might consider the inclusion of the Silesian Coal Basin into a regional emissions trading and joint investment program with the Black Triangle region. The development of intra-regional substantive rules of liability for environmental harm and the establishment of a transboundary regulatory commission might also serve to reduce pollution in the Silesian Coal Basin.

IV. The Temelín nuclear power plant

A. Factual background

The Czech Republic currently operates a 1,760 MW nuclear power plant of Soviet design at Dukovany near the Austrian border and is nearing completion of a second nuclear power plant with a 2,000 MW capacity at Temelín scheduled to be on line by May 2001.[100] Since the inception of the Temelín project, Austria has actively lobbied for a termination of the project on the basis it poses a grave threat to public safety because of its poor quality of construction and potentially unsafe operating procedures.[101]

Austrian objections are particularly robust as the project is located only 40 kilometers from the Austrian border and over 80 per cent of the Austrian population opposes its construction. On a number of occasions, Austrian and Czech nongovernmental organizations have staged mass protests at the construction site.[102] Supporters of the Czech plant charge while 'all of Austria's political parties and the majority of the population are opposed to nuclear energy ... the object of concentrated assaults is only Czech nuclear energy, despite the fact nuclear power plants exist in all countries bordering on Austria.'[103]

The Czech government itself has had doubts about completing the project, and in 1992, the Czech Minister of the Environment announced the Ministry was opposed to the use of atomic energy, and the project should be constructed only upon a decision of Parliament and a public referendum. The Ministry of the Environment favored discontinuing construction of the Temelín plant on the basis of the Czech

Republic's decreased consumption of energy. In March 1993, Prime Minister Vaclav Klaus decided to complete construction of the Temelín project on the basis it would improve the Czech environment by permitting an energy offset from the coal fired power plants in Northern Bohemia by 1999, reducing emissions by 23 per cent.[104] An alternative explanation is even in 1993 the Czech government considered the project, under construction since 1986, too expensive to abandon.[105] In either case, the Temelín project dispute became intertwined with the Black Triangle dispute.

In April 1998, the Czech Ministry of the Environment again proposed construction of the Temelín plant be abandoned and alternative sources be sought for providing the energy that would be produced by the plant.[106] Independent pollsters, however, have found over 60 per cent of Czech citizens support completion of the Temelín plant, despite a series of protests staged at the plant by environmental NGOs.[107] Moreover, in May 1998, the Czech Ministry of Industry and Trade emphasized the need to complete the project on the basis 'it would cost nearly as much to stop construction as to complete it'.[108]

In order to secure necessary support for the project from influential domestic political groups and Western governments, the Czech Republic redesigned the project in the early 1990s to include the latest in Western safety and operational technology and to operate as a completely closed system. These modifications brought the total cost of the project to an estimated $2.8 billion.[109] As a result, some environmental actors assert the project is now environmentally safe, yet it is economically untenable as the costs of additional safety and operational technology will vastly exceed any profits generated from energy production. It is alleged these excess costs will then likely be met either by increased taxes or by a reallocation of domestic revenue. In either event, there will be less public revenue available to invest in more necessary and reasonable environmental infrastructure.

Despite the Czech Republic's attempts to graft state of the art technology onto Soviet style reactors and commitments to permit frequent International Atomic Energy Agency (IAEA) inspections, Austria continues to object to the construction of the plant.[110] Austria, for example, sought to persuade the United States to deny $317 million in United States Export–Import Bank (EXIM Bank) loans to permit Westinghouse to build safety-related infrastructure at the plant.[111] The Austrian government also made alternative offers to compensate the Czech government approximately $50 million to halt construction of the project, assist in the conversion of the plant to a gas fired utility or to supply the

Czech Republic with energy in lieu of the amount to be generated by the Temelín plant.[112] These offers were largely viewed by the Czech government as insincere because Austria does not possess the surplus energy to meet the deficit and hypocritical because Austria purchases nuclear energy from France.

B. The role of international law and the legal process

Although Austria has not yet been able to move the Temelín dispute from the pre-resolution to the resolution phase, the Czech Republic and Austria generally operate within a strong regime of transboundary environmental protection with respect to nuclear facilities.[113] For example, Austria has signed and ratified, and the Czech Republic has signed and approved, the 1994 IAEA Convention on Nuclear Safety.[114] Among its terms, the convention requires enhanced state control of facilities, the formulation of emergency plans, strict radiation limits and a process of regular reporting to the IAEA.[115]

Employing the legal principle of an obligation to share information,[116] in 1995 Austrian Chancellor Franz Vranitzky requested the Czech government provide Austria with a complete assessment of the plant's safety mechanisms and asked for an official copy of the plant's IAEA safety certification.[117] In the future, Austria may also demand some type of international control of the project beyond the standard IAEA inspections and/or seek to participate as a partner in the operation and oversight of the project. Austria may either pursue these goals from within the IAEA forum, or it may seek recourse through the International Court of Justice (ICJ). Although Austria has not indicated which options it will pursue, the Czech Republic has indicated it might be willing to submit to the ICJ the limited issue of Austrian access to the project.[118]

Austria's negotiating position recently has been strengthened by its admission into the European Union and by the recent invitation of the EU to the Czech Republic to begin negotiations on accession. As a member of the EU, Austria could presumably condition the Czech Republic's membership on certain compromises relating to the operation of the Temelín plant. In fact, some Austrian government officials have publicly declared they believe Czech membership in the EU should be conditioned on the Czech Republic converting the Temelín plant into a gas fired power plant, while others have declared membership should be conditioned on the Temelín plant meeting EU safety standards for nuclear power plants.[119] In addition, the Upper Austrian Landtag (a Provincial Assembly) has created a special fund to finance unspecified public activities directed against the completion of the Temelín plant.[120]

Predictably, the Czech Republic has reacted negatively to the notion its membership in the EU be conditioned on its operation of the Temelín plant.[121] The EU also has indicated rather than conditioning Czech membership on the closure or reconfiguration of the plant, it will require the plant meet EU safety standards, and as of April 1998 had provided close to $110 million to assist the Czech Republic in meeting those standards.[122] The EU also has funded the testing of existing nuclear reactors in the Czech Republic by a private consortium of French and German nuclear safety analysts to ascertain whether they meet EU safety standards.[123]

As a result of these pressures and a recent change in the Czech government, the new Minister of Environment has called for a wholesale review of Czech energy policy, and as a basis for this review has cited a report of the Czech State Nuclear Safety Authority highlighting a number of safety concerns at the Temelín plant. In response, the Czech government decided in October 1998 to 'adopt a new energy concept' by March 1999 and to create a 12 member experts commission, including three representatives from the EU, to conduct a comprehensive safety and economic review of the Temelín plant.[124] The Czech government also has indicated it is willing to include an Austrian expert on the commission.[125] As a strategic complement to these measures, an association of environmentalists from Austria, Germany, France and the Czech Republic have initiated litigation in Czech Courts charging the construction of the plant violates the Czech Civil Code, which provides all private and public entities must behave in a manner that does not harm 'health, property, nature or the environment'.[126] The Ministry of Industry and Trade also has come under increasing pressure by the Czech Parliament to disclose surveys relating to the safety of the Temelín nuclear power plant and to its environmental impact.[127]

In addition to the Temelín Nuclear Power Plant dispute, a number of other transboundary disputes are related to nuclear power plants operated in the Czech Republic, Slovakia and Hungary.[128] The Dukovany nuclear power plant, which accounts for one quarter of the Czech Republic's electricity,[129] has been the subject of a dispute with Slovakia as the waste disposal site for the plant is situated on Slovak territory.[130] Since 1992, Slovakia has refused to accept shipments of spent nuclear fuel from the Czech Republic,[131] and in 1997 shipped a substantial quantity of spent fuel back to the Czech Republic.[132] As the Soviet Union also has declined to accept spent nuclear fuel from other states, the Czech Republic has constructed an intermediate storage facility at the plant, which has been the subject of much criticism by the local population and by the Austrian Government.[133] To move the dispute from the pre-resolution to the resolution phase, the Czech Republic made a formal claim against Slovakia for its share of the funding for construction of the disposal site, but offered to withdraw the claim if Slovakia resumes its acceptance of spent nuclear fuel.

In the Slovak Republic, the government continues operation of the Soviet model Bohunice nuclear power plant near Austria,[134] and is pursuing the construction of the Mochovce nuclear power plant near the border with Hungary and Austria, in which Electricité de France holds a 51 per cent ownership share.[135] Austria, Poland and Hungary consistently object to both the continued operation of the Bohunice plant and the construction of the Mochovce plant, but with little effect other than the reaching of an agreement with the former Czechoslovakia, which continues with Slovakia, relating to the creation of an early warning system.[136] Despite these concerns, on 28 September 1995, Slovakia announced it would complete construction of the Mochovce nuclear power plant, with the final aspects of construction being carried out by the Czech firm Skoda Praha, the German engineering firm Siemens, and the French firm Framatome. Although the European Bank for Reconstruction and Development (EBRD) had initially offered to provide funding for completion of the plant and the installation of safety related equipment, Slovakia rejected EBRD's offer when EBRD conditioned its assistance on an early phase out of the Bohunice nuclear power plant and the increase in energy prices by 30 per cent. Slovakia was then able to arrange funding through Czech and Russian banks.[137]

In May 1998, the Austrian head of an international team of experts who had inspected the Mochovce plant declared the plant should not be permitted to become operational as initiating the first chain reaction could contaminate the entire reactor core. To enhance its bargaining position, the Austrian Foreign Ministry forwarded the report to the European Commission and to the European Presidency, at that time held by the United Kingdom, and requested the EU consider 'appropriate action'. The European Parliament also became seized of the issue and adopted a resolution noting reservations about Mochovce's safety had not been refuted and called for the postponement of Mochovce's activation until Western nuclear safety experts had certified it as safe. Despite these objections, on 8 June 1998, with the support of the United States and Russia, Slovakia started up the first reactor.[138]

Notably, the Gabčíkovo–Nagymaros Project dispute, which is discussed in detail in Part II, and the Slovak–Austrian nuclear power plant disputes are interrelated, as the President of Slovakia has indicated once the Mochovce nuclear power plant and Gabčíkovo power station become operational, Slovakia would close down the Bohunice nuclear power plant.[139]

Finally, Hungary operates the Paks Soviet model nuclear power plant, which produces over one third of Hungary's power and is located near the geographic center of Hungary but could potentially affect Austria, Serbia, Croatia and Romania. Electricité de France has made Hungary a standing offer to build two additional 960 MW blocks at Paks at a cost of $3 billion.[140] Although the Paks plant is of general concern to Hungary's neighbors and to the EU, it has not yet become the basis of a fully formed dispute, and there is speculation Hungary is making plans to dismantle the plant by the year 2015 at a cost of over $1 billion.[141]

2
The Southern Tier: Cleaning up after the Soviets, Dumping in the Danube, Dueling Nuclear Power Plants and Suffocating the Black Sea

Like the northern tier CEE states, the southern tier states suffer from a number of transboundary environmental disputes that threaten public health, degrade the environment and inhibit the normal conduct of international relations. Also, like the northern tier states, the southern tier disputes involve the pollution of transboundary watercourses, extensive air pollution, the destruction of biodiversity, the transport and storage of hazardous and radioactive waste, the siting and operation of nuclear facilities, and the land based and sea based pollution of regional seas. It should be noted the first dispute discussed, that of the environmental legacy of Soviet occupation, affects northern tier as well as southern tier states.

I. The environmental legacy of Soviet occupation

A. Factual background

Prior to 1989, the former Soviet Union stationed over 600,000 troops throughout CEE at approximately 950 sites covering 2,600 square kilometers.[142] Without fail, each military site was subject to severe environmental strain, and the estimated cost to abate or clean up these environmental hazards exceeds $4–5 billion.[143] The American intelligence community estimates in Poland alone it will cost $2 billion to remediate the environmental pollution on the 59 former Soviet bases within its territory, and it will cost Hungary over $1 billion to clean the 171 former Soviet bases within its territory.[144]

In 1989, there were widespread protests against the continued stationing of Soviet troops in the former Warsaw Pact states and against the environmental degradation caused by the activities of those forces. In recent years, the parties have entered the dispute formation phase with specific disputes erupting near Turon, Poland at the site of an ex-Soviet military garrison and ammunition depot, in the Olomouc Region of the Czech Republic at sites of ex-Soviet garrisons, and at the Milovice site of the former Soviet command base where pollution of groundwater threatens the drinking water resources for Prague.[145]

Public disputes have also occurred in Swinoujscie, Poland where hydrocarbon pollution from a naval base polluted municipal water supplies, in Choana, Poland, where sewage contaminated local irrigation canals, in Hradcany, Czech Republic where Soviet MiG Squadrons left behind a waste dump of 6,500 tons of jet fuel, and in Petkus, Germany, where residents protested against environmental destruction caused by training operations.[146] Additional disputes will continue to form over the next ten years as local agencies uncover environmental hazards simply buried by the departing Soviet forces. In many instances, Soviet forces buried antiquated ammunition or fuel rather than transporting it back to the former Soviet Union. In one notable case, Bulgaria discovered a sunken Soviet nonnuclear submarine located near its coast, where it poses a danger to navigation and a potential environmental hazard.

In their attempts to clean up environmental degradation left by Soviet forces, many states have sought Western assistance, including military assistance, and have tried to involve private industry.[147] Hungary, for example, has proposed to convert former Soviet airfields into civilian airports, jet maintenance facilities and convention centers.[148] The Hungarian government would sell the property for a minimal fee in return for a commitment by the purchasers to remedy environmental degradation. Private investment in ex-military bases is discouraged, however, as the Soviet forces frequently destroyed or transported to Russia any assets of value, the degree of contamination is difficult to determine and the obsolete laws governing future liability are currently in a state of flux.[149] In addition, municipal investment is limited by the allocation of municipal resources to clean up operations at other industrial sites, which may be of more productive commercial use.

B. The role of international law and the legal process

In 1993 and 1994, CEE states affected by military pollution initiated the pre-resolution phase of the dispute resolution process by making claims for substantial compensation from Russia.[150] It will, however, be difficult for these disputes to move to the resolution stage as the legal arrangements contained in the Status of Forces Agreements between the former Soviet Union and the Warsaw Pact states include provisions relating neither to compensation for environmental damage nor to dispute resolution. The agreements, in fact, contain no provisions for the withdrawal of the Soviet forces. Assessing claims for environmental compensation is complicated by additional claims relating to illegal seizure of property and damage to pre-existing facilities, and the fact in many instances, that the military facilities were previously used by the host state's national forces or were jointly shared by those forces and Soviet forces.

To date, Russia has responded to claims for compensation by submitting counterclaims for the improvement of facilities and alternatively suggesting Russia and the host states enter into joint ventures for operation of the bases as recreational or industrial zones. The Russians also pleaded simple poverty. In addition, it is suspected some host-states entered into secret treaties with the former Soviet Union pledging to forgo claims of compensation in return for the early withdrawal of Soviet troops from their territory prior to the 1994 deadline. For example, it is believed in April 1992, Czechoslovakia and Russia entered into a 'Zero-Variant' agreement, whereby Czechoslovakia acquired title to the existing facilities in exchange for an agreement to release Russia from an obligation to pay compensation for any environmental damage.

In fact, local communities frequently supported the adoption of zero-variant agreements as they simply wanted Soviet troops withdrawn from their district as soon as possible. Now, however, after the extent of environmental degradation is coming to light, these local communities are asserting claims for compensation. As a result, in late 1992, Czechoslovakia rejected a proposal by Russia to exchange certain military assets in return for the construction of prefabricated houses in Russia for returning soldiers. The primary reason for rejecting this proposal was the desire of Czechoslovakia to sell the remaining equipment and use the proceeds for environmental regeneration.[151]

The unresolved issues of liability for environmental harm caused by the stationing of Soviet military troops have not faded with their withdrawal. In fact, as municipalities and industrial enterprises attempt to utilize the facilities, the degree of contamination will become more widely known, and looming questions of liability will need to be resolved. Claims for liability may include industrial interests seeking compensation from local municipalities, municipalities seeking compensation from the central government, and the central government in turn revisiting its claims for compensation from Russia. Industrial interests or municipalities may try to use the judicial system to place liens on unrelated Russian public property in the host state as a means of securing compensation for environmental damage. If NATO extends its operations into CEE, either through NATO expansion or through extensive Partnership for Peace contacts, and utilizes former Soviet facilities, there will be substantial pressure from the West to resolve the outstanding claims in order to avoid a situation where NATO forces are targeted for liability as deep financial pockets.

Given the history of the relationship between Russia and CEE states and Russia's lagging emergence from its doctrinal and practical lacuna of nonsocialist international law, it is unlikely Russia will participate within a regime of transboundary environmental protection and respond to attempts by the CEE states to use international law or the legal process to resolve its potential outstanding liability unless CEE states create an incentive for Russia to participate within such a regime. The most likely means for creating such an incentive would lie with CEE states, in particular Poland, Slovakia and the Czech Republic, threatening to restrict oil and gas transshipment from Russia to Western Europe or for them to attach a transit tax for the purpose of creating an environmental remediation fund. If CEE states threaten to take this action, it is possible Russia might then opt to create a formal regime for resolving the dispute, and agree that principles of international law should play a prominent role in this regime.

The CEE states may also enhance their ability to move the dispute into the resolution phase, and to encourage Russia to participate in the formation of a regime, by either enticing Russia into agreeing to a joint inquiry into the extent of contamination left behind, or by undertaking a unilateral inquiry with the assistance of a third party such as the EU, US or NATO. The results of a unilateral inquiry may then be used to augment international awareness of the gravity of the threat posed to environmental health in CEE and possibly increase international pressure on Russia to enter negotiations with the CEE states either to remediate this damage or to cover a certain share of the remediation cost.

With respect to NATO or Western private enterprise involvement in the use or renovation of former Soviet military facilities, it is very likely the parties involved will rely upon international environmental law and domestic environmental law to resolve their disputes, as Western states and enterprises have a significant past practice of using international law and the legal process, and CEE states are likely to be interested in using international environmental law and their own domestic law to protect themselves against claims from Western states or enterprises. CEE states will also be interested in providing a structure for resolving potential disputes so they may proceed with privatizing and developing the former military sites.[152]

II. The Romanian–Bulgarian Danube River Gauntlet

A. Factual background

The Danube River forms the boundary between much of Romania and Bulgaria. Throughout most of its length, the river is heavily polluted by both states either by direct discharge or by tributaries flowing through the two states, which collect industrial effluent and agricultural runoff.[153] As a result of the numerous industrial plants located along the river, many areas of the river basin also suffer from substantial air pollution.[154]

Particularly serious air and water pollution occur at the Giurgiu–Ruse Gauntlet where over 110 industrial units directly or indirectly pollute the Danube River and its environs.[155] These tenants emit substantial amounts of chlorine, alkalis, hydrochloric acid, chlorinated organic solvents, epichlorohydrin, benzidine and municipal sewage into the river and local environment. In addition, atmospheric emissions from a combination of Bulgarian and Romanian industries create conditions of toxic fog during much of the winter.[156] Other point-source disputes include the pollution of Silistra, Bulgaria by industrial enterprises in Câlăradi, Romania,[157] the pollution of Nikopol, Bulgaria by a chemical plant constructed on the river banks in Turnu Magurele, Romania and the pollution exchange between Vidin, Bulgaria and Calafat, Romania.[158]

The Romanian–Bulgarian Danube River Gauntlet dispute is complicated by a variety of political concerns. The Bulgarian political opposition to communist rule, for example, emerged during the late 1980s in Ruse as a result of the substantial environmental devastation brought on by the economic policies of the communist regime.[159] And, until the November 1997 change in the Romanian government, many Bulgarians were of the opinion the government in Romania was controlled by hold-over communists, and it would be fruitless to attempt to persuade Romanian officials to reduce the amount of environmental pollution emitted along the Danube River.[160]

As an indication of the sensitivity of this issue and its link to relations between Romania and Bulgaria, the Bulgarian ecological party regularly organizes sit-ins on the international bridge connecting Giurgiu and Ruse in order to pressure the Bulgarian government to take action against Romanian polluters.[161] And on at least one occasion, the President of Bulgaria has canceled a state visit to Romania and threatened to recall its Ambassador as a result of increased levels of pollution from the industrial zones in Giurgiu. On numerous other occasions, local groups have referred to the pollution emanating from Romania as 'environmental aggression', an attempt at 'genocide' and an 'undeclared war'.[162]

B. The role of international law and the legal process

In an attempt to build the foundation of a regime of transboundary environmental protection designed to bring about a reduction of pollution in the area and reduce tension between Romania and Bulgaria, the EU, World Health Organization (WHO) and United Nations Environmental Programme (UNEP) have undertaken efforts to mediate the dispute.[163] The EU also has conditioned continued Poland–Hungary Assistance for the Reconstruction of the Economy (PHARE) assistance upon a resolution of the dispute.[164] In the early 1990s, these efforts produced the outlines of a regime based on the legal principles of a duty to cooperate and a duty to share information in the form of a bilateral cooperation agreement of December 1991. This agreement provided for the exchange of information on matters relating to Giurgiu–Ruse point-source disputes, operation of Bulgarian and Romanian nuclear power plants and pollution of the Danube River.[165]

Romania and Bulgaria are also parties to a bilateral Environmental Cooperation Agreement designed specifically to address issues arising from air and water pollution in the Giurgiu–Ruse region.[166] Efforts related to the entire Danube River Basin that might contribute to promoting greater cooperation on the Romanian–Bulgarian stretch of the river include attempts since 1985 to create an international water quality monitoring network,[167] the creation in 1991, of an Environmental Program for the Danube River Basin sponsored by EU, GEF, EBRD, US Agency for International Development (USAID), Austria, the Netherlands and the World Bank, and the negotiation of the Convention on Cooperation for the Protection and Sustainable Use of the Danube River.[168] The Danube River Convention aims to ensure waste products dumped into the Danube River do not endanger human health. The operational focus of the Convention is on promoting sustainable development and ascertaining means by which the protection of one environmental element will not lead to stress on other elements.[169]

As for the operation of this nascent regime, the Giurgiu–Ruse Environmental Cooperation Commission met on a number of occasions in 1992, wherein the parties exchanged information on the environmental conditions in their respective areas and discussed preliminary plans for promoting environmental protection. The parties also invited UNEP representatives to become involved in the work of the Commission. In an attempt to measure and trace the pollutants in order to establish their sources, the parties engaged in a joint monitoring program in the Silistra–Câlâradi region. As a result of these activities, Romania agreed to consider closing one of the major chemical plants in Ruse and moving another plant, and the Bulgarians considered reconfiguring one of their chemical plants to produce soap. By 1993, however, cooperation between Romania and Bulgaria deteriorated as a result of demands by Bulgaria for action on the part of Romania in the Giurgiu region and public agitation with the slow pace of progress.[170] Cooperation within the Convention on Environmental Protection similarly stalled.[171]

In 1995, Romania and Bulgaria attempted to reinvigorate the dispute resolution process by moving the dispute from the pre-resolution back into the resolution phase, when both Foreign Ministers recognized and agreed that development of their bilateral relations rested with the improvement of each state's physical infrastructure relating to environmental protection.[172] Bulgaria and Romania also ratified the Danube River Protection Convention, which may serve as a framework of principles to assist them in resolving the dispute.[173]

The Japanese Overseas Economic Cooperation Fund also has attempted to assist in the resolution of the dispute by extending to Bulgaria a loan package issued in 1995, for the purpose of adding pollution control technology to industrial plants that had previously discharged raw effluents into the Danube.[174] The Parliamentary Assembly of the Council of Europe also has added to the impetus to resolve the dispute by taking more general action in March 1997 to call for a European Danube Charter which would provide a comprehensive statute and framing mechanisms to resolve disputes in the Danube River Basin.[175] If such a Charter is adopted, it will naturally form the basis for a more general regime within which Romania and Bulgaria may seek to resolve their specific Danube River Basin disputes.

In addition to such a Charter, to create a stable regime capable of promoting the long term reduction of environmental pollution in the Danube River Basin between Romania and Bulgaria, and in particular in the Giurgiu–Ruse area, it will be necessary to revitalize the current cooperation agreements. The United States and EU may play an important role in creating an incentive for the reestablishment of this cooperation and might propose they serve as mediators or conciliators in an effort to keep the dispute resolution process from again slipping back into the pre-resolution phase. The aim of these cooperative endeavors should be to create an early warning system as well as mechanisms for the timely sharing of information relating to the environmental consequences of industrial pollution and an agreed schedule for reducing specific emissions.

Romania and Bulgaria must additionally continue the process of transformation from a communist plan of industrialization to a market economy, taking the opportunity of the transition to restructure their industrial base in accordance with the prevailing principles of sustainable development.[176] In particular, Romania and Bulgaria may benefit from legal provisions providing for public participation in future industrial zoning and requiring the employment of the precautionary principle when approving new industrial development.[177] With respect to existing sources of pollution, the application of domestic and transboundary rules of liability and the use of the principle of equitable utilization of shared natural resources might promote the protection of pollution.[178] Here, the mechanisms of the Danube River Convention may prove useful in supporting the application of some of these principles.

Although the prospects for the creation of an effective regime are currently limited given the slow rate of privatization in Romania and Bulgaria and the comparative lack of Western investment, the strong public interest in these issues and the position of environmental actors within the Bulgarian government, coupled with a recent change in both governments, may provide an initial opportunity to attempt to create an effective transboundary regime for reducing pollution along the Danube River Gauntlet.

III. The Romanian–Bulgarian Nuclear Corridor

A. Factual background

Both Romania and Bulgaria have located nuclear power plants on or near the Danube River where it forms their mutual international border.[179] The construction and continued operation of these plants give rise to substantial tension between the two states and cause concern among EU member states.[180] There is also the growing problem of nuclear waste disposal as Russia is insisting on hard currency payments to continue to accept nuclear waste, and local governments are prohibiting the creation of new waste facilities in their regions. There is thus the additional risk temporary storage sites at nuclear facilities will overflow and contaminate the Danube River.[181]

Romania has recently begun operation of the first unit of the 3,425 MW Cernavoda nuclear power plant, which is expected to meet 30 per cent of its electricity needs. The $1.6 billion plant is of Canadian design, with some of the construction supervised by the Canadian Atomic Energy Agency and the US firm, General Electric. The construction of the plant was funded in part by the Canadian Export Development Corporation and Mediocredito Centrale of Italy.[182] In Bulgaria, the Silistra town council strongly objects to the plant on the grounds it is located only 25 miles from the border and will utilize Romanian processed fuel, which, according to Bulgaria, does not meet international standards.[183] In an effort to reduce these tensions, Romania has granted limited permission for Bulgarian nuclear experts to visit the Cernavoda site.[184] Yet despite the continued objections of Bulgaria, in August 1997, the Romanian government authorized plant administrators to begin operating the plant on 30 May 1998.[185] In February 1998, Moldova rejected a suggestion by Romania it participate in the financing of a second reactor at the Cernavoda plant in exchange for an energy entitlement.[186]

Bulgaria currently operates a 3,760 MW nuclear power plant at Kozloduy, which came on-line in 1974 and provides 40 per cent of Bulgaria's electricity.[187] The Kozloduy reactors are of Soviet design and their continued operation is subject to substantial criticism by the IAEA and Romania.[188] In a US intelligence report, two of the reactors at Kozloduy were listed among the ten most dangerous Soviet designed reactors still operating.[189] In August 1994, the director of the Kozloduy plant requested $150 million from the Bulgarian government for modernization of the plant's safety and operational equipment; this request was denied due to a lack of domestic financial resources.[190] Yet, during the autumn of 1995, Bulgarian Prime Minister Zhan Videnov announced the Kozloduy plant's Russian builders believed they could provide Bulgaria with maintenance services and upgrades to extend the reactors' service life by ten years.[191]

Bulgaria is also in the process of constructing the Belene nuclear power plant, which would produce 4,000 MW from Soviet designed reactors. Construction on this project, which began in 1985, was temporarily halted in 1990, in part because 56 per cent of the Bulgarian population openly opposed the project. Government proposals in 1996 to restart construction of the plant were met with public demonstrations, and a price tag of $2 billion. Nuclear power, however, will likely continue to play an important role in Bulgarian energy production as the Bulgarian government estimates Bulgarian coal reserves will last only until 2010, and it is quite reluctant to rely on imported sources of energy.[192]

B. The role of international law and the legal process

The dispute relating to the operation of these nuclear power plants began to move from the dispute formation phase to the pre-resolution phase, when in February 1995, Bulgaria shut down Reactor No.1 at the Kozloduy nuclear power plant for refueling, repairs and domestic and IAEA inspections. On 21 September 1995, the ambassadors of the Group of Seven Industrialized Nations (G7) informed Bulgaria that, due to safety reasons, Reactor No. 1 at the Kozloduy nuclear power plant should be permanently shut down. Germany, in particular, protested against reactivation of the reactor, asserting 'everything must be done to prevent this reactor starting up without sufficient safety precautions'.[193]

Bulgaria rejected the G7 claims as based on outdated analysis, which did not consider recent improvements to the plant and informed the G7 it would reopen the reactor at the Kozloduy nuclear power plant if the Bulgarian Atomic Energy Commission determined the reactor met technical inspection standards. Bulgaria pointedly declared during this announcement the continued operation of the plant would not be dictated by foreign interests or safety concerns. Bulgaria further asserted not only was its nuclear power industry safe, but in addition to implementing the Convention on Nuclear Safety, Bulgaria was one of the few states to implement a legal framework regulating the use of atomic energy.[194]

Declining to move the dispute further from the pre-resolution to the resolution phase, Bulgaria restarted the reactor on 6 October 1995, with the approval of the Bulgarian Atomic Energy Committee. The EU formally protested this action and attempted to initiate the creation of a regime for resolving the dispute by urging Bulgaria to shut down the reactor for the winter of 1996 in return for energy supplies provided by the EU. After meeting with the EU Environmental Ministers, Bulgaria rejected the offer on the basis it was not in written form and it would be technically difficult to ensure sufficient energy supplies from the EU for the course of the entire winter.[195] Again, in April 1998, the EU called upon Bulgaria to close the Kozloduy plant. Bulgaria again rejected this call and declared it intended to operate the first two reactors until 2005, and the third and fourth reactors until 2010. In reply, an EU spokeswomen declared the uncooperative nature of Bulgaria with respect to nuclear power 'could hinder Bulgaria's chances of joining the EU'.[196] Bulgaria responded in May 1998, by claiming Germany now supported operation of the plant, and by requesting that the EU conduct a new inspection of the facility, and in particular the recent upgrades relating to safety.[197]

In addition to the EU, a number of other interested third parties have played a role in the Kozloduy dispute. The IAEA has conducted a number of inspections of the plant and in 1997 certified its general compliance with international legal standards relating to nuclear safety.[198] In April 1995, Bulgarian authorities petitioned the EBRD for $200 million in credits to update the plant's two 1,000 MW reactors, to ensure they safely complete their projected service term to 2005–10. Although this request was denied, the EBRD did provide Bulgaria $26 million in financial assistance to improve the safety of the reactors, and by 1998 Bulgaria secured over $150 million from other sources.[199] Notably, in 1995 Japan announced the expansion of bilateral cooperation in the civil-nuclear field with Bulgaria. In addition to continuing the training of Bulgarian technicians in nuclear safety, the Japanese program includes the transfer of safety technologies. Local press reports speculated the Japanese transfer program indicated their interest in future contracts for the Belene facility.[200] More recently in 1998, Westinghouse Electric Company secured a contract to provide instrumentation and control upgrades to the Kozloduy reactors.[201]

Romania has tried to discourage the continued operation of the Kozloduy plant by intermittently seizing nuclear fuel shipments for the plant on the basis they do not possess the appropriate permits to transit Romanian territory.[202] Bulgaria has protested against these actions on the basis they contravene the Danube River Convention and the transport of such nuclear fuel, as opposed to nuclear waste, does not require permits as it is not radioactive.[203] This element of the dispute, however, was assuaged in November 1997 with the adoption of a multilateral agreement among Bulgaria, Romania, Russia and Moldova governing the transportation by land of nuclear material for the Kozloduy nuclear reactors, and a similar bilateral agreement between Bulgaria and Romania. Despite significant public protest, on 14 August 1998, Moldova agreed to permit the transit of radioactive materials from Bulgaria to Russia – on the condition two Moldovan experts be permitted to observe the loading of the radioactive materials at the plant.[204]

As part of a broader cooperation agreement, Romania and Bulgaria have agreed to share information relating to the operation of the Kozloduy and Cernavoda nuclear power plants and have established a process of joint inquiry. In July 1998, Bulgaria also agreed to allocate $500 million for safety improvements at the plant.[205] There, however, is little substantial progress being made on the dispute and no serious consideration of submitting the issue to conciliation or international arbitration or adjudication. Yet, as the EU becomes more involved and more aggressive with respect to this dispute and as the G7 continues to pursue its opposition to the Kozloduy plant, pressure will likely build to create a regime for resolving the dispute.[206] This regime may entail an initial submission of the dispute to the ICJ to ascertain the rights and obligations of Romania and Bulgaria and may further entail the creation of a joint Romanian–Bulgarian management council, with possible EU participation.

IV. The Black Sea

A. Factual background

The Black Sea, which once supported five times the number of fish as the Mediterranean Sea, is now classified as the most degraded sea on the planet. The Black Sea is an almost completely closed system, with one outlet to the Eastern Mediterranean, which itself is a virtually closed sea.[207] As a result, much of the pollution discharged into the Black Sea, either from the Danube River or from other rivers carrying pollution generated from the sixteen states and 162 million inhabitants of its drainage basin, continuously circulates in the sea. These pollutants include nitrates and phosphates from agricultural runoff, pathogenic bacteria from sewage effluent and toxic chemicals from industrial runoff and industrial accidents. On 29 February 1996, for instance, 250 tons of gasoline were spilled into the Black Sea while being off loaded from an oil tanker in the Romanian harbor of Constanta.[208] The effect of this pollution is augmented by the reduction of the flows of major rivers into the Black Sea, which upsets the Sea's delicate hydrological balance and reduces its ability to self purify.

The Black Sea is also under threat by the announcement in April 1998, of an agreement between Russia and Turkey to put into operation by July 2000, a 396 km gas pipeline from Djubga, Russia to Samsum, Turkey. According to objections raised by the Bulgarian Ministry of the Environment and Georgian environmentalists who have called for a joint international EIA for the project, a rupture of this pipeline could significantly threaten the remaining fisheries resources in the eastern section of the sea.[209] In addition, Romania and France agreed on 22 April 1998, to spend $15 million in joint exploration for oil in the Black Sea, which some environmentalists fear could lead to pollution problems similar to those caused by oil production in the Baltic Sea.[210]

The consequences of pollution emitted into the Black Sea include extreme eutrophication and a decline in fish populations and biodiversity, which has caused a collapse of the regional fishing industry (resulting in a loss of over 150,000 jobs) and brought the tourism industry under severe threat (resulting in a loss of an estimated $300 million in annual revenue).[211] Attempts to move the dispute from the resolution phase into the implementation phase through the creation of a common management plan for the Black Sea are complicated by the dissolution of the former Soviet Union and the tension between Bulgaria and Turkey arising from Bulgaria's complex relationship with its Turkish national minority.[212]

B. The role of international law and the legal process

To reduce the amount of pollution flowing into the Black Sea, Romania, Bulgaria, Georgia, Ukraine, Russia and Turkey have made a number of minor attempts to improve the environmental management of the sea by laying the foundation for the operation of an international regime for environmental protection. Thus, for instance, on 22 April 1992, all the riparian states negotiated and signed the Convention on the Protection of the Black Sea Against Pollution, which entered into force on 15 January 1994.[213] The Convention bans the disposal of radioactive waste into the Black Sea,[214] regulates pollution from both land-based and maritime sources, encourages efforts to combat the illegal traffic of hazardous waste,[215] and requires states to adopt rules and regulations relating to liability for harm caused to the marine environment and to provide judicial recourse to those seeking to establish liability. There is the further obligation to harmonize rules and regulations providing for the attachment of liability, but it is unclear whether recourse is limited to a state's own nationals or whether it includes all state parties' nationals. Concerning disputes arising from the interpretation of the convention, the document provides the parties shall pursue 'negotiations or any other peaceful means of [the parties'] choice'.[216]

On 7 April 1993, prior to the entry into force of the Convention, the Black Sea riparian states signed the nonbinding Black Sea Declaration, which established a Black Sea Commission; declared the parties' intent to apply the precautionary approach, to pursue economic and environmental integration and to participate in joint EIAs; and called for the development of additional protocols to the Convention relating to the transboundary movement of hazardous waste, pollution from ships, conservation of marine resources and the development of an emergency response plan.[217]

Similar to the case of the Baltic Sea, the applicability of customary international law principles such as the prohibition against causing serious injury to neighboring states and the equitable utilization of shared natural resources is limited since the causal links between specific types of pollution and specific harm are difficult to prove and since there is no generally acceptable criteria among the basin states for determining an equitable allocation. Similarly, the possibility for using existing international agreements to facilitate a reduction of pollution is limited as these extra-regional agreements do not appear to be sufficient in and of themselves to constitute a stable legal framework for resolving the pollution problems of the Black Sea. Efforts at mediation, and conciliation, and even adjudication, might also prove limited given the great number of parties to the dispute. Interestingly, however, a number of scientists and environmentalists from Black Sea states have begun to consider whether it might be possible to pursue litigation against Germany and Austria in their domestic courts, or in the European Court of Justice for the failure of these states to meet their obligations under certain EU directives.[218]

The ultimate construction and operation of any lasting regime for transboundary environmental protection in the Black Sea Basin will likely be inhibited in the near term by the Black Sea states' asymmetric degrees of transition to market economies based on the rule of law, lack of external pressure to resolve the dispute, and lack of a past practice of using international law to structure their relations. Although the EU may have some role to play in promoting the creation of a lasting regime, it has little direct interest in the Black Sea, and its ability to influence the behavior of Russia, Ukraine and Georgia is limited, with its relations with Turkey primarily focused on resolving human rights issues and moderating the geopolitical confrontation between Greece and Turkey. The Black Sea Basin states also lack the necessary resources to fund a major cooperative effort to reduce environmental pollution.

# Conclusion to Part I

CEE states suffer from a wide diversity of transboundary environmental disputes stretching from the Baltic Sea to the Black Sea and involving water and air pollution, the destruction of biodiversity, the disposal of hazardous waste and the threat of radioactive contamination, which affect the quality of the environment, public health and economic resources throughout CEE. Although many of these disputes are bilateral between CEE states, a number are multilateral and may involve non-CEE states in Western Europe and the former Soviet Union.

To varying degrees CEE states have looked to international law and the legal process as a means of promoting the resolution of these disputes. In cases such as the Baltic Sea dispute, the basin states have relied extensively on international law to construct a treaty regime to regulate their respective rights and obligations with respect to the sea and the other Basin states. The Baltic Sea Basin states have also relied upon the principles and mechanisms of international law to construct the HELCOM regime for monitoring the execution of the Basin states' rights and obligations. In the Black Sea dispute, the affected states are attempting to adopt a similar treaty regime for the purpose of regulating the emission of pollutants into the Black Sea. Although the Black Sea Basin states have to date been less successful than the Baltic Sea Basin states, the efforts of the Black Sea Basin states to construct a regime of transboundary environmental protection are still embryonic, and they have demonstrated a commitment to redefining and adapting the elements of the regime as necessary.

In other instances, such as the Danube River Gauntlet, the affected states have attempted to use international law and the legal process to create a set of formal mechanisms to facilitate the exchange of information and the creation of early warning systems. Although these efforts have met with very limited success to date, with the November 1997 change in governments, there is a strong likelihood they will begin to have a positive effect on the dispute. In still other instances, CEE states have recognized the existence of the dispute and have indicated a willingness to engage in a dialogue with other affected states, as in the Black Triangle and Romanian–Bulgarian Nuclear Corridor disputes, and have used international law to frame their perceptions of the conflict. In these cases, the affected states have also relied upon international law to establish some preliminary arrangements for cooperation and inquiry. Finally, in certain situations, such as the Silesian Coal Basin and Soviet Environmental Legacy disputes, CEE states have recognized the existence of a dispute, but have not taken meaningful action to employ international law to assist in resolving the disputes.

A number of sub-state actors and interested third parties have also influenced the various disputes among CEE states. In the Black Triangle dispute, for instance, sub-state actors, with the assistance of interested third parties, initiated the creation of the Euroregion NISA and have begun to develop transnational NGO networks. In the Temelín Nuclear Power Plant and the Danube River Gauntlet disputes, energy concerns and industrial enterprises have sought to encourage their respective governments to comply with existing principles of international law, in order to prevent attempts to limit their activities. These sub-state actors, however, have also dissuaded their governments from creating formal review mechanisms, which might provide an opportunity for critics of the activities to restrict their operation. As such, while refusing to create a multinational oversight regime beyond the current IAEA regime, the Czech Republic relies upon international law to the extent it can declare the operation of the Temelín nuclear power plant is consistent with its legal rights and does not contravene its legal obligations.

In most CEE transboundary environmental disputes, third parties have played a prominent role, either by providing funding for the projects giving rise to the dispute, as in the many nuclear power disputes, or by facilitating inquiry or offering their good offices, as in the Black Triangle and Danube River Gauntlet disputes. Notably, the role and influence of the United States will decrease as CEE states continue the process of integration with European institutions, in particular the EU. The role of international lenders may become more prominent as states seek financial resources to implement agreed-on legal regimes, as in the case of the Black Sea dispute.

Finally, a number of situational circumstances appear to influence the willingness of CEE states to use international law. The ecological circumstances giving rise to the disputes and dictating which substantive areas of international law may be relied upon appear to be the most important. In cases involving transboundary air and water pollution, the states may invoke large bodies of international law, which may, however, be somewhat indeterminate. In other cases, such as the Soviet Environmental Legacy dispute, there is a more limited body of law which the parties may invoke. The economic circumstances, such as the shift away from heavy industry, have also influenced the interest of CEE states in employing international law. As a result of these shifting production configurations, some states are reducing their emissions at the expense of domestic employment, but are not experiencing improved environmental conditions because they suffer the consequences of pollution or other environmental degradation originating in a neighboring state. Furthermore, the existence or lack of economic resources influences a state's willingness to consent to new international legal obligations, which may require the commitment of significant economic resources, which might otherwise be allocated to social welfare or the process of economic transformation.

Political circumstances, both domestic and international, also appear to impact the willingness and ability of CEE states to use international law to promote a resolution of their transboundary environmental disputes. The democratic transformation, in particular, coupled with increased public attention to environmental issues and opportunities for public advocacy, has motivated CEE states to move certain disputes into the pre-resolution and resolution phases. Most notably, public activism and democratic responsiveness have recently led to an increasing willingness of Romania and Bulgaria to resolve the Danube River Gauntlet dispute and to manage their Nuclear Corridor dispute. International circumstances, such as increasing integration into EU institutions and EU insistence on the rule of law, have led many CEE states to rely upon legal principles to accommodate their population's and their neighbor's interests, and in some instances, to consider the invocation of the legal process, most notably with the submission of the Gabčíkovo–Nagymaros Project dispute to the ICJ.

Building on these initial observations, Part II investigates the role of law in the Slovak–Hungarian dispute relating to the construction and operation of the Gabčíkovo–Nagymaros Project on the Danube River to provide additional insight into the functions of international law and the legal process, the influence of interested third parties and sub-state actors, and the circumstances affecting the behavior of these actors and their respective states. The Gabčíkovo–Nagymaros Project dispute was selected as a case study warranting detailed examination because the construction of the Project has given rise to one of the most controversial bilateral environmental disputes in Europe, involving issues relating to environmental degradation, economic development, political transformation, inviolability of international borders, development of sustainable energy resources and the continuing influence of the former Soviet economic and political culture in CEE. Moreover, as noted by the Agent of Slovakia in his oral pleading before the ICJ, the Gabčíkovo–Nagymaros Project dispute is the first dispute between two CEE states to be submitted to the Court.[219]

Part II

Using International Law to Resolve the Slovak–Hungarian Dispute concerning the Construction and Operation of the Gabčíkovo–Nagymaros Project

3
The Dispute-Formation Phase: Soviet-Inspired Designs to Harness the Power of the Danube and Post-transformation Second Thoughts

To initiate the inquiry into the role of international law and the legal process in the Gabčíkovo–Nagymaros Project dispute, this chapter begins with the dispute formation phase. The Gabčíkovo–Nagymaros Project dispute serves as a particularly useful case study since the parties have attempted to resolve the dispute through negotiation, inquiry, mediation and adjudication.[220] During each of these processes, the parties have invoked legal rationales ranging from well established treaty and transboundary resource law to more recent principles of international environmental law, and throughout the dispute, the behavior of the parties has been significantly influenced by a wide range of social and political issues.

During the dispute formation stage of the Gabčíkovo–Nagymaros Project dispute, Czechoslovakia and Hungary, under the influence of the Soviet Union, adopted a formal plan for modification of the Danube River based on a mutual needs assessment, including the need for increased production of energy, improved navigability and flood control, and increased socialist integration.[221] Soon after the initial implementation of this plan, Czechoslovakia and Hungary experienced certain political and economic changes, which redefined both their individual needs and the ability of the Project, as designed, to meet those needs. Moreover, Czechoslovakia and Hungary became aware of certain environmental consequences that would result from the construction and operation of the Gabčíkovo–Nagymaros Project, and which might possibly outweigh the Project's benefits.

Reacting to these changes, Hungary sought to delay construction of the Project to conduct additional needs and damages assessments. Czechoslovakia, fearing the cost of delaying the Project and determining the changed circumstances did not significantly affect its desire to complete the Project, consented to some delay, but declined to consider the possibility of terminating the Project. At this point a dispute formed between Czechoslovakia and Hungary.

As will be discussed in this chapter, the dispute formation phase was characterized by disagreement over the implementation of obligations arising from a binding international legal instrument; the perceived failure of Czechoslovakia to share information as required by that instrument and by general principles of international law; and a perceived refusal of Hungary to comply with the binding obligations and timetables contained in the instrument. The dispute also formed around allegations by Hungary that Czechoslovakia failed to abide by its international legal obligation to provide due notice of its plans to implement a Provisional Solution and failed either to assess or provide adequate information about the environmental and other impacts of this unilateral act.

I. Factors encouraging the negotiation and adoption of the 1977 Agreement

In the early 1950s the Soviet Union and Czechoslovakia developed a plan for improving navigability of the Danube River and for increasing hydroelectric production by diverting the Danube from its natural watercourse into a navigable artificial waterway located within the borders of Czechoslovakia. By 1963, Hungary had been enticed to join the plan, and the governments of Hungary and Czechoslovakia began to develop a joint investment program and construction schemes for the Project.[222] On 6 May 1976, Hungary and Czechoslovakia agreed to adopt a Joint Contractual Plan setting forth the technical specifications for the project and the division of responsibility between the parties.[223]

In 1977, under the supervision of the Soviet Union, Czechoslovakia and Hungary entered into a bilateral agreement to construct along the stretch of Danube River forming the boundary between the two states a series of dams, hydro-electric plants and navigational locks, to divert the Danube out of its natural watercourse and into an artificial power canal for 25 kilometers.[224] The 1977 Agreement further provided Czechoslovakia and Hungary would equitably share the construction costs and value of the hydroelectric energy,[225] as well as share in the benefits of improved navigation and flood control.[226]

A. The perceived need for the increased generation of electricity for industrial consumption

The Gabčíkovo–Nagymaros Project was intended to generate substantial amounts of electricity at a relatively low environmental cost.[227] At the time of the agreement, both the Czechoslovakian and Hungarian economies depended on high energy consumption heavy industries, whose energy needs were met by coal fired power plants that emitted a significant amount of pollution.[228] The Gabčíkovo–Nagymaros Project was thus intended to provide an alternative, renewable, domestic source of electric energy, which would improve the regional environment by reducing industrial dependence on coal-generated electricity.[229] This increased capacity for the generation of hydroelectric energy was particularly important for Czechoslovakia as, unlike most of the other CEE states, Slovakia possesses very little lignite and black coal resources, and is thus required to import significant amounts of coal to support its industrial output. Hungary too perceived significant benefit in the project as all of its main hydroelectric resources are located on boundary-watercourses.[230]

During the early and mid-1980s Hungary, and to a lesser extent Czechoslovakia, initiated a change in economic structure from a command economy to a market economy, which entailed both a shift from heavy industry to light and medium industry and the closure of some large scale inefficient industrial enterprises. As a result, Hungary determined the additional electricity generated from the Gabčíkovo–Nagymaros Project might be surplus to its needs.[231] In addition, Hungary calculated, based upon an energy entitlement owed to Austria in exchange for a loan made to complete the Nagymaros works, that the completion of the project might not result in an increased supply of energy to Hungary's national grid. In fact, although at the time of the agreement it was projected the electricity payment to Austria would amount to two-thirds of Hungary's entitlement, in the late 1980s Hungary feared that in order to meet its energy entitlement obligations, it would have to build additional coal fired power plants.[232] Hungary also began to perceive the economic costs of maintaining and operating the Gabčíkovo–Nagymaros Project, coupled with the costs of measures to moderate environmental damage and the value of lost agricultural production – which could no longer be treated as an unaccountable externality[233] – would dwarf the value of energy produced over the life of the Project.[234]

54 *The Slovak–Hungarian Dispute*

Although Czechoslovakia experienced a similar reduction in need for energy, and domestic and international pressure for the development of noncapital intensive alternative energy sources, Czechoslovakia determined it wished to reduce its dependence on imported coal and projected future economic growth required the increased capacity. Czechoslovakia also held the view as it had invested substantial sums of capital into the Project, it was prudent to operate the project in order to generate at least a minimal return on its investment.

B.   The perceived need for improved navigation

The Gabčíkovo–Nagymaros Project was further intended to enhance the navigability of the Danube River substantially by circumventing a series of sandbars, narrows, and fords downstream of Bratislava, which all but prohibit navigation three months of the year, and at other times restrict the carrying capacity of transit barges. With the construction of the Project, fully loaded barges would be able to traverse this portion of the river at any time of the year. This improved level of navigation was intended both to reduce the shipping costs of goods from the Soviet Union and to enhance the ability of the Soviet Black Sea fleet to navigate the Danube.[235] The resulting increase in barge traffic was also intended to improve the possibility Bratislava could establish a major industrial port.[236]

As Hungary's trade with the Soviet Union was not affected by the decreased navigability upstream of Budapest, it had less interest in improving navigability than Czechoslovakia. Furthermore, with the dissolution of the Warsaw Pact, Hungary was less interested in pursuing the military benefits of the project. Although it might be expected Hungary would seek to facilitate improved navigability north of Budapest as its trade relationships developed with Western Europe, Hungary in fact held a counter economic incentive in keeping this region of the Danube intermittently non-navigable as it received substantial revenue from trucks carrying goods through Hungary, which might otherwise travel on Danube barges.

C.   The perceived need for improved flood control

The Gabčíkovo–Nagymaros Project was also originally intended to increase the flood control level of the region, which lies below the high-water level of the Danube, in order to prevent a repetition of floods similar to the 1954 flood, which inundated two hundred square kilometers of the Danube plain, and the 1965 deluge, which flooded approximately

1000 square kilometers of the plain, forcing the evacuation of some 55,000 people.[237] Czechoslovakia believed the lack of economic development in the area was a direct result of the frequent flooding of the Danube Plain near Gabčíkovo.

Although agreeing with the need to improve flood control, Hungary soon began to consider whether it might be possible to achieve flood control in the region by cheaper and safer means,[238] and whether the Project itself might destroy the existing natural flood control regime and even heighten the chances of flooding since the geological and seismological characteristics of the region were little understood when the Project was conceived.[239] Despite Hungary's growing concerns, Czechoslovakia maintained the Gabčíkovo–Nagymaros Project was the most economically efficient means of controlling floods in the adjacent Danube plain, citing a number of comprehensive studies attesting to the geological stability of the region and the structural integrity of the Project facilities.[240]

D. The perceived need for increased socialist integration

In light of the need to maintain Soviet hegemony in CEE during the Cold War, the preamble of the 1977 Agreement declares the Gabčíkovo–Nagymaros Project 'will further strengthen the fraternal relations of the two States and significantly contribute to bringing about the socialist integration of the States members of the Council for Mutual Economic Cooperation'.[241] After the initial stages of the Project, however, the authority of the Soviet Union and institutions of socialist integration began to wane, and eventually in 1991, the Soviet Union dissolved, and the Warsaw Pact and Council of Mutual Economic Cooperation disbanded, thus removing the need for socialist integration. Moreover, according to Hungary, the Project in fact 'brought about a virtually permanent dispute between the two countries'.[242]

II. The growing awareness of the environmental consequences of the Gabčíkovo–Nagymaros Project

As the construction of the Gabčíkovo–Nagymaros Project progressed, Hungary began to fear the Project would harm the environment of the Danube River region by adversely affecting the hydro-dynamic characteristics and pollution levels of the river.[243] Czechoslovakia, however, believed Hungary grossly overestimated the environmental dangers of the Gabčíkovo–Nagymaros Project, it was possible to mitigate much of the Project's feared environmental damage, and in fact, the Project would have significant environmental benefits.[244]

A. Concerns relating to water quality and quantity

Hungary's first concern was based on the fact the Danube River is capable of supplying approximately 60 per cent of Hungary's drinking water,[245] and this water is of particularly high quality because of a bank-filtering process in which water seeps from the Danube through a naturally filtering bio-active pebble bed into an area aquifer hundreds of meters deep.[246] Hungary feared the reservoir would substantially reduce the rate of water flow, causing sediment to accumulate both along the floor of the reservoir and the river,[247] and toxins such as iron, manganese and other industrial pollutants which would normally be transported downstream would settle along the floor of the reservoir.[248] The sediment could also provide a breeding ground for harmful bacteria, which would combine with the toxins, infiltrate the aquifer and enter drinking water supplies, polluting the ground water reserves within ten years.[249]

Hungary developed an additional concern that as the sediment accumulated, it would have to be dredged, which would reintroduce the harmful pollutants into the surface water, while also destroying the bio-active pebble bed responsible for the natural filtration of the river water.[250] Destruction of the pebble bed would then permit additional organic micropollutants and microbes to enter and contaminate the aquifer.[251] Moreover, Hungary was concerned the leaching of hydrocarbon compounds from the asphalt layer lining the power canal might further pollute the river water,[252] and the diversion of the river channel would disconnect the main river from a complex system of tributaries, which play an essential role in the self-purification of the Danube River.[253] In general, Hungary's concern over water quality was particularly heightened as over 95 per cent of its surface water resources enter Hungarian territory from neighboring states, with most of these rivers entering Hungary in a seriously polluted condition.[254]

Czechoslovakia asserted the Dunakiliti reservoir would actually improve water quality by decreasing erosion and slowing the rate of the Danube, thereby permitting larger quantities of water to pass through the bio-active pebble layer and dilute the currently polluted aquifer.[255] According to Czechoslovakia, the Hungarian projections of reduced water quality due to the decrease in oxygen content of the river, the increasing siltation of the reservoir and the infiltration of toxins were based on scant scientific research, and were contrary to reports from international experts.[256]

B. Concerns relating to agricultural production and ecological integrity

Hungary also became concerned that, as a result of the Project, the water table upstream of the Dunakiliti Dam would rise, causing soil saturation and reducing the productivity of farm land.[257] Below the dam, the water table would recede, reducing crop production and increasing drought susceptibility.[258] The corresponding fluctuation in the water table would then not only affect the forest habitat of the region and potentially result in the destruction of 50,000 acres of flood plain, but would also jeopardize the freshwater supply of local inhabitants.[259] Czechoslovakia did not share Hungary's concern over the fluctuating water table. Rather, Czechoslovakia believed as the pre-Project water table in the Danube Region fluctuated by as much as seven meters, creating less than ideal conditions for agricultural production, the Gabčíkovo–Nagymaros Project would stabilize the aquifer level by raising the water table where crops had been susceptible to drought and lowering the water table where crops had been damaged by inundation.[260]

Given the complex ecological nature of the river basin in the area of the Gabčíkovo–Nagymaros Project,[261] Hungary also became concerned the extensive ecological disruption caused by diverting the river from its natural channel for 25 kilometers would likely destroy an estimated 80–90 per cent of the 5,000 animal species inhabiting the area, with the general fish population falling by 17 per cent and the commercial fish yields falling by 90 per cent.[262]

Czechoslovakia seemed to acknowledge a change of the hydraulic regime would most likely alter the wildlife habitat in the meadow-forests and fish habitat, but believed the impact on the meadow-forest could be mitigated by timing the discharge of water from the Dunakiliti reservoir and the construction of 'low stony barrages' in the old river bed.[263] Czechoslovakia also believed the Project would bring water to regions of the Danube meadow-forests in Hungary that were previously drought-ridden throughout much of the year.[264]

C. Concerns relating to the impact on the Hungarian national minority in Slovakia

Hungary also expressed concern the Project would disproportionately affect the Hungarian national minority in Slovakia by physically isolating the three predominantly Hungarian villages of Dobrohost, Vojca and Bodiky from the rest of Slovakia. Hungary believed this isolation, coupled with the environmental devastation and reduced agricultural production would give rise to an exodus of Hungarians from these towns. Czechoslovakia responded by asserting as the Project would bring numerous environmental as well as economic benefits to the region, it would in fact improve the situation of the Hungarian national minority resident in that area, and encourage their continued cultural prosperity.[265]

III. Hungary's initial proposals to modify the Gabčíkovo–Nagymaros Project

As early as 1981, just three years after construction of the Gabčíkovo–Nagymaros Project began, Hungarian economists, hydrologists and biologists, for the reasons noted above, began to question the wisdom of the project, fearing the parties were risking 'human and economic disaster' with financial losses running 'ten times the cost of the whole investment'.[266] In response to these concerns, the Central Committee of the governing Hungarian Socialist Workers' Party requested the Hungarian Academy of Sciences create an ad hoc committee to investigate the environmental and economic consequences of the Project.[267]

The report of the ad hoc committee found the Gabčíkovo–Nagymaros Project, as designed in the 1977 Agreement, did not comprehensively account for the ecological consequences of diverting the Danube River and did not adequately consider the technical, ecological and economic risks posed by the Project. The ad hoc committee therefore recommended the project be postponed for a significant period of time or abandoned altogether.[268] The concerns raised by the ad hoc committee, coupled with the effects of the economic recession of the period, led Hungary to suspend construction of its share of the project and initiate a further study of the Project's ecological and economic consequences. Based on these studies, Hungary proposed Czechoslovakia unilaterally construct the project, with Hungary compensating Czechoslovakia for its half of the construction cost through a returned electricity payment.[269]

Czechoslovakia refused to assume full responsibility for the completion of the Project, but was willing to assume some of the construction responsibilities and provide Hungary with a five-year delay with which to secure financing for construction of its remaining share of the Project.[270] The parties formalized this agreement in a Protocol to the 1977 Agreement on 10 October 1983, and anticipated the project would be operational by 1995.[271]

With the start of construction of the Nagymaros works in 1988, the Hungarian Parliament developed a more focused interest in the Project and adopted a resolution on 7 October 1988, declaring ecological interests must have priority over economic interests during the construction and operation of the Project.[272] The Parliament also authorized a complete review of the potential consequences of constructing the project as designed. This review confirmed for the Parliament there were a number of deficiencies in the Project's design and recommended Hungary suspend its participation in the joint endeavor.[273]

Concerned with the growing domestic opposition to the Gabčíkovo–Nagymaros Project, the Hungarian government sought to present any subsequent Hungarian government with a *fait accompli* by signing a Protocol to the 1977 Agreement on 6 February 1989 to accelerate the construction of the Nagymaros and Gabčíkovo works by 15 months.[274]

On 13 May 1989, contrary to the 1989 Protocol and in the midst of growing economic and environmental concerns, the Hungarian government unilaterally announced it would suspend construction work on the Nagymaros works to undertake even further economic, environmental and legal reviews of the entire project.[275] In a meeting of the Czechoslovakian and Hungarian Prime Ministers on 24 May 1989, Hungary proposed the parties conduct a joint analysis of the ecological risks of the Project. By some estimates, the former Czechoslovakia had completed two-thirds of the Project by this time.[276] In mid June 1989, Czechoslovakian and Hungarian environmental and economic experts held a series of meetings wherein they exchanged information and acknowledged the need to protect the environment, but could not agree on the most reasonable means of doing so.

As a result of this further review of the project, on 20 July 1989, Hungary extended its construction moratorium until 3 October 1989 and expanded the moratorium to include its contribution to construction at the Dunakiliti and Gabčíkovo sites.[277] Hungary also proposed Czechoslovakia and Hungary continue to meet at the expert level and they jointly agree to suspend construction of the project while they evaluated feasible alternatives. In August 1989, Czechoslovakia formally rejected this proposal and warned a continued suspension of construction activity by Hungary could lead Czechoslovakia to implement a Provisional Solution unilaterally.[278] Hungary objected to Czechoslovakia's warning and noted the implementation of a Provisional Solution would violate the general norms of international law.[279]

4
The Pre-Resolution Phase: Enter Sub-state Actors, Third Parties and International Law

During the ensuing pre-resolution phase of the Gabčíkovo–Nagymaros Project dispute, Hungary and Czechoslovakia further developed their positions with respect to the Project through considerable contact with various sub-state actors, which as a result of the 1989 political revolutions were provided greater opportunity to influence the dispute resolution behavior of their respective state entities.[280] Czechoslovakia and Hungary also initiated contact with European states, the European Commission and the United States to assess their interest in facilitating a resolution of the dispute. Hungary, in addition, pursued contact with Austria to attempt to resolve Austria's interest in the dispute in order to facilitate dispute resolution. Czechoslovakia and Hungary then invoked norms and principles of international law to frame their initial negotiating positions, to indicate they were willing to settle the dispute by legal means and to identify the dispute resolution mechanisms available to assist with resolving the dispute.

I. The involvement of sub-state actors

In the autumn of 1989, both Czechoslovakia and Hungary underwent 'velvet revolutions' whereby they began the process of transforming their state structures from totalitarian communist regimes to democratic regimes. Because of the high profile role of environmental organizations in the May 1989 suspension of Hungary's participation in the Project and their influence on Hungary's transformation to a democratic regime, many environmentalists were elected to Parliament and were appointed to positions in the new government.[281] Many of these Hungarian environmental organizations also established stable bases of public support and developed links with international environmental NGOs.²⁸²

As noted in Chapter 3, in the late 1980s, the Hungarian Parliament had taken an active interest in the controversy surrounding the Project, and on 31 October 1989, the Hungarian Parliament further aligned its interests with environmental actors and elements of the general public opposing the project, halting indefinitely construction of the Nagymaros works.²⁸³ The Hungarian Parliament additionally authorized the Hungarian Council of Ministers to enter negotiations with Czechoslovakia for the modification of the 1977 Agreement.²⁸⁴ Then, in November 1989, Hungary proposed Czechoslovakia be permitted to continue construction of the Gabčíkovo plant as a run-of-the-river operation with appropriate ecological guarantees, with Hungary being permitted to abandon construction of the Nagymaros works.²⁸⁵

The Hungarian government anticipated, given the strong opposition to the Gabčíkovo–Nagymaros Project expressed by the leaders of the 'velvet revolution', the new Czechoslovakian government would be receptive to a modification of the Agreement.²⁸⁶ The Hungarian government also anticipated as Czechoslovakia evolved democratically, Slovak environmental actors would be able to pressurize the federal government into canceling the project.

In Czechoslovakia, the initial decision to build the Gabčíkovo–Nagymaros Project was significantly influenced by the economic and political interests of the highly profitable and powerful water and construction industries. Between 1988 and 1989, however, the Project became a focal point and symbol for dissenting political discourse and a catalyst for a number of mass demonstrations, as the Project was perceived as a symptom of a system that permitted the undertaking of such massive and detrimental projects.²⁸⁷ After the 1989 revolution, however, the mass movement faded, but the powerful water and construction industries remained. And, although the new Czechoslovakian government virtually excluded members of the old government and was inclined to terminate the project,²⁸⁸ the Slovak Republic government retained many pre-revolution political actors and remained resistant to abandoning the project.²⁸⁹ Moreover, the federal government was unwilling to upset the delicate situation of Czech–Slovak relations by terminating the project.²⁹⁰ Thus although Czechoslovakia temporarily suspended its unilateral construction of the Project in December 1989, it soon restarted construction, declined Hungary's offer to modify the Project and again warned Hungary of its intention to implement the Provisional Solution and seek damages.

For a brief period of time from 1990 to 1992, however, environmental actors appeared to have a second opportunity to shape the Slovak government's policy with respect to the Project as the Green Party held six of the Parliamentary deputy positions, and the Minister of the Environment held the position of Vice-Prime Minister, despite the fact a Ministry of the Environment had yet to be created.[291] After 1992, however, the prospects for meaningful influence began to wane as the Green Party lost all its seats in the Slovak Parliament and as most of the other parties contesting the election supported the construction of the Project – with the leader of the popular HZDS party proclaiming the Project to be a 'fundamental Slovak achievement'.[292] And, although the Slovak government eventually created a Ministry of the Environment, the Ministry of Agriculture continued to keep tight control over the administration of water resources. In part because of this lack of an institutional power base, none of the 19 recommendations made by the Slovak Commission for the Environment with respect to reducing the environmental consequences of the Gabčíkovo–Nagymaros Project were implemented by the Slovak Republic government.[293]

II. The involvement of third parties

Both Austria and Germany supported the Gabčíkovo–Nagymaros Project, as they believed the energy produced by the Project would offset the need for Czechoslovakia to operate the Soviet-era Bohunice nuclear power plant and assist in reducing harmful SO_2 emissions by replacing a certain amount of capacity from coal fired power plants.[294] In 1986, Austria became directly involved in the Gabčíkovo–Nagymaros Project by agreeing to fund construction of the Hungarian dam at Nagymaros in return for an energy entitlement running from 1996 to 2016, and an agreement 70 per cent of the construction work would be awarded to the Austrian firm Donaukraftwerke – which was barred from constructing the Hainburg hydroelectric plant in Austria because of environmental protests.[295]

As noted above, although at the time of the agreement, it was projected the electricity payment to Austria would amount to two-thirds of Hungary's entitlement, in the early 1990s, Hungary concluded in order to meet its energy entitlement obligations, Hungary would have to build additional coal fired power plants. When Hungary temporarily suspended construction of the Nagymaros works in May 1989, Austria sought to persuade Hungary to restart construction as Austria had already calculated the energy entitlement into its energy plan and it was keenly interested in providing for the continued economic viability of the Donaukraftwerke construction company. When it became clear to Austria that Hungary was seriously considering the complete abandonment of the project, Austria sought to increase political and economic pressure on Hungary by making a claim on behalf of Donaukraftwerke for $360 million.[296]

Hungary was unwilling to restart construction on the Nagymaros works and terminated the contract in 1989, counter-proposing to allow the Austrian company to transfer some of the investment to other joint Austrian–Hungarian projects.[297] When the Austrian firm rejected this proposal, Hungary sought to downsize the Austrian claim for damages and reserve a right to pay damages with returned energy.[298] At this point, the Austrian Green Party intervened in an attempt to persuade the Austrian government to waive the compensation claim, while many other political parties began to support the company vocally. Concerned about the international attention attracted by the Gabčíkovo–Nagymaros Project and about Austria's role in promoting ecologically insensitive development, the Austrian government persuaded the construction company to settle for a $255 million payment.[299] Certain Austrian banks also sought to provide financing for the Czechoslovakian share of the Project, but the Austrian government declined to guarantee such loans.[300]

In addition to the Hungarian contacts with Austria, both Czechoslovakia and Hungary began to approach European heads of state, the European Commission and the United States and considered the possibility of approaching the Soviet Union in an attempt to persuade these parties of the merits of their respective opinions and to ascertain whether any of these third parties might be willing to serve as mediators or whether they might be able to identify applicable dispute resolution mechanisms. The United States declined to serve as a mediator, but suggested the parties consider submitting their case to the ICJ. The European Commission encouraged the parties to try and settle their dispute on a bilateral basis, but indicated if this was not possible, the Commission might be able to provide the parties with technical assistance to facilitate an identification of the potential consequences of the Project and to assess possible modifications.

At a later time during the dispute, Hungary again contacted the European heads of state and the European Commission attempting to convince them of the environmental and legal merits of Hungary's abandonment of the Gabčíkovo–Nagymaros Project and requesting they intervene in the conflict to prevent Czechoslovakia from implementing the

Provisional Solution.[301] Although none of the European states were willing to intervene in the conflict, this wide canvassing of international opinion led the European Commission to agree to participate in a trilateral commission as discussed below.

III. Attempts to establish or identify appropriate dispute resolution mechanisms

Acting on the advice of the European Commission and the United States, in October 1989, Hungary proposed the dispute should be resolved by Hungary and Czechoslovakia on the basis of the results of scientific investigations to be made by independent experts and international institutions.[302] Shortly thereafter, in May 1990, the post-revolution Hungarian government published its political program declaring it 'considers the construction of the Danube Barrage System a mistake and will initiate, as soon as possible, negotiations on the rehabilitation and the sharing of the damages with the Czecho–Slovak government to be elected'.[303] Responding to growing public awareness of the economic and environmental consequences of the project and increasing domestic pressure to abandon the project, in April 1991, the Hungarian Parliament authorized the negotiated termination of the 1977 Agreement and the formation of a new agreement providing for 'the consequences of the abandonment of the project and the rehabilitation of the Danube area'.[304]

From April 1991 to July 1991, Czechoslovakia and Hungary attempted to create a number of working committees to resolve the dispute, but were unsuccessful as Hungary proposed the creation of committees for the purpose of deconstructing the Project and rehabilitating the environment, while Czechoslovakia proposed the creation of committees charged with determining which structural adjustments to the Project were necessary in order to ensure adequate environmental protection.[305] During this time the parties exchanged information relating to their respective positions as to the extent of environmental harm caused by the project and in the case of Czechoslovakia, the technical measures which it believed could alleviate this harm.

On 15 July 1991, Czechoslovakia proposed a trilateral expert committee be created with the participation of the European Commission to suggest to the governments a technical solution for the operation of the Gabčíkovo power plant. Czechoslovakia further noted, in the event the proposal for the trilateral expert committee was rejected, Czechoslovakia would be forced to implement the Provisional Solution. Hungary counter-proposed the parties establish a bilateral committee to assess the ecological risks of the Gabčíkovo–Nagymaros Project and requested a suspension of construction on Czechoslovakian territory. In partial response to Hungary's rejection of the offer to establish a trilateral expert committee, on 25 July 1991, the Czechoslovak government voted to implement the Provisional Solution.[306]

In a subsequent meeting on 2 December 1991, Hungary accepted Czechoslovakia's proposal to establish a trilateral committee, but insisted this commission review the operation of the entire project and Czechoslovakia suspend its construction activities.[307] Czechoslovakia welcomed Hungary's consent to establish a trilateral committee, but rejected the request to reverse its 15 July decision or to suspend construction of the Project.

Shortly thereafter, on 12 December 1991, Czechoslovakia approved the technical plans for the Provisional Solution and initiated its construction with the intent to bring it into operation by autumn 1992. Hungary objected to the implementation of the Provisional Solution on the basis it would likely prejudice any decision of a trilateral commission and it would not be possible for the parties to agree to be bound by the decision of the commission if one of the parties had already taken action which might be contrary to that decision.[308] Hungary further warned if Czechoslovakia did not heed its calls to suspend construction, it would be forced to consider terminating the 1977 Agreement and taking counter-measures.

In April 1992, the Vice President of the European Commission notified both parties the Commission was willing to participate in a trilateral expert committee on the condition the parties would refrain from taking actions, including the implementation of the Provisional Solution, which could influence or anticipate the recommendations of the committee.[309] Hungary accepted this condition, while Czechoslovakia rejected it. In the face of Czechoslovakia's rejection of the Commission's offer, Hungary began to signal its interest in referring the dispute to the ICJ.[310]

IV. The use of international law to frame negotiating positions

On 19 May 1992, the Hungarian government, recognizing the trilateral commission would not be formed in the immediate future and fearing Czechoslovakia's imminent implementation of the Provisional Solution, unilaterally declared a termination of the 1977 Agreement. As a basis for this termination, Hungary asserted the completion of the project would be economically unfeasible and would cause substantial and irreparable environmental damage. Hungary also cited a number of specific legal norms and principles to support its termination of the Agreement, to support its objection to the implementation of the Provisional Solution and to signal it wished to resolve the dispute within the parameters of international law.

Hungary based its termination of the Agreement on the legal rationales of an ecological state of necessity, the impossibility of performance, a fundamental change of circumstances, reciprocal breach and the subsequent intervention of international environmental norms. Hungary objected to the implementation of the Provisional Solution on the grounds it constituted a breach of the sovereignty and territorial integrity of Hungary, it transgressed the inviolability of Hungary's territorial frontiers, it contravened the 1976 Boundary Waters Convention and the 1948 Danube River Convention, which required the consent of both parties before one party could alter the natural flow of the Danube River, and it violated customary international law principles regulating the reasonable and equitable utilization of international environmental resources and prohibiting transboundary harm.[311] Hungary then signaled its willingness to settle the dispute within the parameters of international law by proposing Hungary and Czechoslovakia enter into negotiations to conclude a new treaty that would develop mechanisms to provide for the rehabilitation and maintenance of the ecological and natural resources in the region, and improve flood control and navigation.

Czechoslovakia refused to accept the Hungarian declaration, arguing it did not state sufficient legal grounds for terminating the 1977 Agreement and Hungarian concerns were primarily political and not environmental. Czechoslovakia declared it thus considered the 1977 Agreement to continue in force and therefore Czechoslovakia was entitled to implement the Provisional Solution, which it officially announced to the Danube River Commission on 5 August 1992.[312] Czechoslovakia also sought to counter the Hungarian claims of negative environmental consequences by asserting the potential harm to flora and fauna could be substantially mitigated by timing the discharge of water from the Dunakiliti reservoir and construction of weirs in the old river bed. Czechoslovakia further argued the project would, in fact, improve water quality, stabilize the fluctuating water table, increase crop productivity, control floods and slow the erosion of the Danube River bed.[313]

5
The Resolution Phase: Making the Case to the International Court of Justice

Moving to the dispute resolution phase, this chapter assesses the perceived functionality of international law as an element of the regime of transboundary environmental protection, noting in particular the ability of both the substantive principles of international law and the processes of inquiry, mediation and then adjudication before the International Court of Justice to assist the parties in accomplishing their objectives. This chapter also affords an opportunity for an extensive exploration of how specific legal principles and various aspects of the legal process operate to address the multifarious elements of a transboundary environmental dispute in such a manner so as to allocate the respective rights and responsibilities of each party as necessary to produce a formal decision capable of forming the basis for the settlement of the dispute.

Subsequent to the failure of Czechoslovakia and Hungary to establish a bilateral or trilateral expert commission to resolve the dispute and Hungary's unilateral termination of the 1977 Agreement, the parties sought to enlist the assistance of a more formal dispute resolution mechanism, which could aid the parties in more clearly identifying their conflicting as well as common interests and thus facilitate their evaluation of settlement options. The parties also sought a mechanism that would assist them in resolving certain factual questions and clarify their international rights and obligations, while also providing an opportunity to garner support for their respective positions from interested third parties.[314] Czechoslovakia and Hungary thus invited the European Commission to mediate the dispute. Reluctant to involve itself too deeply in the dispute, the European Commission limited its mediation efforts to finding a technical solution and agreeing upon an equitable division of the river. Compelled to seek a determination of a

68 *The Slovak–Hungarian Dispute*

number of legal issues, the parties, with the encouragement of the European Commission, agreed to submit the dispute to the ICJ.

I. Inquiry and mediation by the European Commission

In talks between the Prime Minister of Hungary and the Czechoslovakian Federal and Slovak Republic Prime Ministers in August and September 1992, the parties agreed they should ask the European Commission to undertake an inquiry to establish a factual basis for resolving certain aspects of the dispute and to serve as a mediator with the authority to propose technical options which might assist in the settlement of the dispute. At this time the Prime Ministers also discussed the possibility of submitting the dispute to the ICJ.

The European Commission accepted the parties' invitation to conduct an inquiry with the participation of the parties and to mediate the dispute. In undertaking such involvement, the Commission was motivated in large part by a desire to demonstrate itself capable of playing an active role in promoting peaceful relations in CEE and by a desire to resolve any major transboundary disputes that might affect potential EU membership for certain CEE states or that might be imported into the EU if one of these states was to become a member. The Commission also feared the dispute might negatively affect the ability of European shipping interests to navigate the Danube River if the Provisional Solution were unilaterally implemented and if Hungary engaged in some form of physical countermeasures.[315] The Commission believed it might be able to influence positively the behavior of the parties by bringing to bear its considerable technical expertise, by offering or withholding future financial assistance and by modulating the speed of EU association and potential membership.[316]

To initiate the inquiry, the European Commission appointed a committee of experts, including representatives from Czechoslovakia and Hungary, to examine the ecological claims of Hungary and identify technical options available to the parties.[317] This committee, however, limited its recommendations to narrow technical suggestions, which were not in and of themselves sufficient to assist in promoting a resolution of the dispute. At a 22 October 1989 meeting in Brussels, Czechoslovakia declined to accept the proposal of the European Commission that all construction activities be halted while the committee continued its investigation. Concerned the European Commission would not compel Czechoslovakia to abandon the pursuit of the Provisional Solution, Hungary submitted an application to the ICJ seeking to have the Court issue an order suspending construction of the Provisional Solution.

During this time, Czechoslovakia determined Hungary intended to stall the mediation efforts until after January when it would become impossible to fill the Dunakiliti reservoir again until the autumn of 1993. Czechoslovakia thus began diverting the Danube River into the power canal on 24 October 1992,[318] asserting the failure of Hungary to construct the Nagymaros project was a politically motivated violation of international law and demanding Hungary complete the dam or compensate Czechoslovakia $500 million.[319] Some experts believe Czechoslovakia diverted the Danube in order to present Hungary, the European Commission and any future arbitrator or adjudicator with a *fait accompli*.[320]

Fearing the dispute was detracting from the European Union's efforts to promote a peaceful and relatively prompt transition in CEE, the European Commission invited Czechoslovakia and Hungary to the London meeting of European Community Prime Ministers on 28 October 1992. At the meeting, under significant pressure from the European leaders, Hungary and Czechoslovakia agreed to maintain not less than 95 per cent of the water quantity in the old river bed and refrain from operating the Provisional Solution. The parties further agreed the European Commission would establish two committees, one to conduct a fact-finding mission to consider concerns relating to navigation, ecological damage and flood control, and the second to investigate possible solutions to the dispute.[321] The Agreement also provided the parties would submit their dispute, including legal, financial and ecological elements, either to binding arbitration or the ICJ.[322] As a result of an apparent divergence of interests between Czechoslovakia's bureaucratic and national government actors, Czechoslovakia failed to fulfill its obligation to maintain the traditional flow of the river bed.

Throughout 1992, in addition to being involved in the Gabčíkovo–Nagymaros Project dispute with Hungary, the Czech Republic and the Slovak Republic were also involved in negotiations between themselves concerning the possible dissolution of Czechoslovakia. Then, on 25 November 1992, the Federal Assembly of the Czech and Slovak Federal Republics adopted a law providing as of midnight, 1 January 1993, the state of Czechoslovakia would cease to exist and would be succeeded by the independent states of the Czech Republic and the Republic of Slovakia.[323]

With the further assistance of the European Commission, on 27 November 1992, the parties agreed to work with the technical committee set up under the London Agreement to establish a temporary regime of water management.[324] Despite numerous meetings and reports of the technical committee, no temporary regime was established, and Czechoslovakia continued to divert almost the entire flow of the Danube into the power canal.[325] On 19 January 1993, the European Commission recommended 800 m^3/s should be allocated to the old river bed along with other technical measures. Hungary accepted this proposal, but Slovakia rejected it. Finally, on 15 April 1995, the parties entered into an agreement providing an average annual flow of 400 cubic meters per second into the old river bed and 43 cubic meters per second into the Mosoni Danube. The agreement also provided Hungary would construct an underwater weir near Dunakiliti to improve the water supply to the side-arms of the Danube River. In order not to prejudice any decision of the ICJ, this agreement was specified to terminate 14 days after the first Judgment of the Court.[326]

In addition to the work of the European Commission, the European Parliament tangentially involved itself in the dispute by passing two resolutions. The first called on Slovakia to be more flexible in its negotiations with Hungary, and the second criticized Slovakia for failing to agree to the European Commission's January 1993 recommendation for a temporary water management regime.[327] It is unclear whether the resolutions of the European Parliament had any appreciable impact on the behavior of Slovakia and Hungary.

II. The submission of the dispute to the International Court of Justice

With the assistance of the European Commission, on 7 April 1993, Hungary and Slovakia concluded a special agreement concerning submission of the legal aspects of the Gabčíkovo–Nagymaros Project dispute to the ICJ.[328] In addition to the pressure brought to bear by the European Commission, the decision of Czechoslovakia and Hungary to submit the dispute to the Court was influenced by a variety of circumstances and factors.

As this section will discuss, the decision to submit the dispute to the Court initially was influenced by Czechoslovakia's and Hungary's basic familiarity with using international law to manage resource conflicts, a growing awareness of the ability of environmental law principles to assist in resolving resource disputes, the perceived relevancy and accessibility of the Court as a dispute resolution mechanism, increasing pressure from the European Union, and the dissolution of Czechoslovakia.

Although the use of international law by CEE states prior to the 1989 transformation to democratic regimes was fairly limited, they did develop a basic familiarity with the use of international law to manage resource conflicts. Most notably, both Czechoslovakia and Hungary participated in the Council of Mutual Economic Assistance (CMEA) Program for Cooperation Concerning Environmental Protection, which was in part designed to explore legal and organizational means for reducing transboundary air pollution.[329] Czechoslovakia and Hungary also had an established practice of concluding bilateral and multilateral agreements to regulate the use of transboundary water resources.[330]

Despite the fact that prior to the 1989 transformation, domestic environmental law primarily served a symbolic purpose,[331] from early 1990 both Czechoslovakia and Hungary began to explore how environmental law might be used to resolve domestic resource conflicts, and from early 1991, both began the process of substantially restructuring their environmental legislation.[332] Although the early legislative attempts were subject to criticism for failing to include a number of well established principles of environmental protection,[333] both Czechoslovakia and Hungary continued to revise their legislation in an attempt to improve its effectiveness and to harmonize it with European Union standards as required by their EU Association Agreements.

As the European Commission's mediation efforts proved unable to promote a technical solution to the dispute and the parties acknowledged they had reached a political stalemate, the Commission suggested to the parties they consider pursuing arbitration or adjudication before the ICJ in order to clarify the applicable legal rules and add impetus to the dispute resolution process. Hungary was particularly receptive to the Commission's recommendation since Hungary had independently considered the merits of submitting the dispute to the Court. Prior to the recommendation, Hungary had also begun to rely increasingly upon international law to frame its objections to Czechoslovakia's threats to implement a Provisional Solution.

Notably, the parties appeared to prefer submission of the dispute to the Court over arbitration as the parties considered the Court to possess sufficient international prestige to persuade the other party to comply with a decision of the Court, or in the event the decision was adverse, the losing party could use the Court's prestige as cover for implementing its decision and extricating itself from the dispute.

Slovakia and Hungary also perceived the use of international law as having endemic value in that if it could resolve certain primary points of contention, such as whether Slovakia was entitled to or prohibited from constructing and operating the Provisional Solution, then they could fairly readily resolve the most contentious aspects of the dispute and begin to negotiate over damages. Although the issue of damages would be strongly contested, it would not cause as significant a disruption in the Slovak–Hungarian bilateral relations as the continued unregulated diversion of the Danube River.

Slovakia and Hungary also perceived the use of international law as having external value since the European Commission, various European states and the United States all indicated they expected the parties to resolve their dispute peacefully and within the confines of international law. The opinion of the European Commission and certain European states was particularly important as both Slovakia and Hungary wished to become members of the European Union, and it was subtly made clear to both these states their future relationship with the Union would in some part be dictated by their approach to resolving the Gabčíkovo–Nagymaros dispute. Slovakia and Hungary were further concerned that a disregard of the principles of international law might influence some states or international organizations to restrict their assistance programs,[334] although it is unclear whether any states or organizations actually informed Slovakia or Hungary they would be risking their assistance if they failed to resolve the dispute within the parameters of international law.

The general willingness of Slovakia to use international law, and more specifically, to submit the dispute to the ICJ, was influenced by the dissolution of Czechoslovakia on 1 January 1993, into the Czech and Slovak Republics. Had Czechoslovakia not dissolved, it is likely the central government, which was disproportionately influenced by the Czech portion of the state, would have agreed to abandon the Gabčíkovo–Nagymaros Project and would have been able to compensate financially those responsible for its construction. Slovakia, however, as an independent state was no longer subject to the political eclipse of the Czech Republic and perceived itself as being less financially able to absorb the cost of abandoning the Project. As a new state, Slovakia was also exposed to international pressure, but at the same time was more likely to resist any attempt to infringe upon its national pride and newly acquired sovereignty. Slovakia thus turned to international law in part because of its susceptibility to pressure from certain European states and the United States, and in part to prevent Hungary from abandoning the Project and blocking the implementation of the Provisional Solution.

Notably, while the case was before the ICJ, the European Commission and member states of the European Union continued their efforts to assist with the resolution of the dispute. In particular, the European Commission continued to investigate the environmental consequences of the Project for the purpose of proposing a temporary water management regime.[335] Based on these reports, the European Commission proposed a temporary water management regime in January 1993, which was rejected by Slovakia, and then proposed another temporary water management regime in April 1995, which was acceptable to both parties. The European Commission also made available its good offices to host negotiations between the parties while the case progressed before the Court.[336]

III. The multifarious issues and options put before the International Court of Justice

To assist in promoting a resolution of the Gabčíkovo–Nagymaros Project dispute, the Court was called upon to render a decision resolving certain factual and legal questions, the effect of which would be to narrow the points of factual disagreement within the dispute and to articulate legal conclusions that would lead to a limited number of acceptable outcomes. Slovakia and Hungary then intended to use the decision of the Court as a framework from within which they could negotiate a final resolution of the dispute. As detailed above, the factual and legal points of disagreement inhibiting the resolution of the dispute included the actual environmental and economic consequences and viability of the original Project and the Provisional Solution, whether Hungary was entitled to suspend and then terminate the 1977 Agreement, whether Slovakia was entitled to respond legitimately with the construction and implementation of the Provisional Solution, and for what damages, if any, was each party responsible.

To phrase these questions in a manner conducive to adjudication and to identify the international laws governing the adjudication, Slovakia and Hungary entered into a Special Agreement, which they then submitted to the Court. The Special Agreement asked the Court to determine on the basis of the 1977 Agreement, the rules and principles of general international law and other treaties as deemed applicable by the Court:

> 1) whether the Republic of Hungary was lawfully entitled to suspend and subsequently abandon, in 1989, the works on the Nagymaros

Project and on the part of the Gabčíkovo Project for which the Treaty attributed responsibility to the Republic of Hungary;

2) whether the Czech and Slovak Federal Republic was entitled to proceed, in November 1992, to the 'Provisional Solution' and to put into operation from October 1992 this system;

3) what [were] the legal effects of the notification, on 19 May 1992, of the termination of the Treaty by the Republic of Hungary;

4) [what where] the legal consequences, including the rights and obligations for the Parties, arising from [the Court's] Judgment on the questions [above].[337]

The Special Agreement then provided Slovakia and Hungary would accept the Judgment of the Court and immediately enter into negotiations on the modalities for its execution. If the parties, however, were 'unable to reach an agreement on implementation within six months, either Party [could] request the Court to render an additional Judgment to determine the modalities for executing its Judgment'.[338]

In making their representations before the Court, each party sought to persuade the Court to adopt its view pertaining to the factual and legal questions in dispute, and thus render a decision that would direct the parties toward a final arrangement proposed by that party. Thus, Hungary sought to persuade the Court of the significance of the ecological harm and economic ruin caused by the original project and by the Provisional Solution, and of Hungary's legitimate right to terminate the treaty, thus denying Slovakia a legitimate basis for constructing the Provisional Solution, and relieving Hungary of any obligation to compensate Slovakia for damages beyond those incurred before 1989. Slovakia, however, sought to persuade the Court the original project and the Provisional Solution in fact promoted environmental protection and were economically rational, and Hungary had no legal basis for terminating the 1977 Agreement. As such, Slovakia should be permitted to construct the Provisional Solution, with Hungary compensating it for damages suffered as a result of its illegitimate breach of the Agreement.

Before the Court, Hungary thus asserted by terminating the 1977 Agreement it had relieved itself of the obligation to construct the Nagymaros works, and it had rescinded its consent for Slovakia to divert the Danube River from its natural watercourse. In Hungary's view Slovakia could not therefore implement the Provisional Solution or any other physical project that diverted or significantly altered the flow of the Danube River.

Hungary based its justification for the termination of the 1977 Agreement on the legal grounds of ecological necessity, impossibility of performance, fundamental change of circumstances, material breach and supervening custom. Hungary also asserted the treaty lapsed with the dissolution of the former Czechoslovakia as there was no international rule of automatic continuation of bilateral treaties in the case of the dissolution of a state. In the view of Hungary, Slovakia thus had no legitimate basis for diverting the Danube River via the implementation of the Provisional Solution.

Hungary further contended the construction and operation of the Provisional Solution violated Slovakia's legal obligations to cooperate, provide prior notification to and engage in consultation with Hungary, to protect the environment, not to cause damage to the environment beyond one's borders, to engage in the equitable use of shared natural resources, to respect another state's permanent sovereignty over its natural resources, and to abide by the principle of nondiscrimination. Hungary therefore sought to have the Court order Slovakia to cease the operation of the Provisional Solution, restore the entire flow of the Danube River to its natural watercourse, undertake measures to restore the integrity of the environment in the region affected by the Provisional Solution and make reparations for any damages that could not be remediated.

Slovakia argued before the Court Hungary had no valid legal grounds for terminating the 1977 Agreement as there was no state of ecological necessity, fundamental change of circumstances, intervening custom, or material breach by Slovakia and it was possible to perform the obligations required by the Agreement. Moreover, Slovakia asserted the failure of Hungary to comply with its obligations under the Agreement created a situation requiring Slovakia to design and implement the Provisional Solution in order to ensure the protection of the environment. Slovakia also contended the 1977 Agreement did not lapse with the dissolution of the former Czechoslovakia, but rather under the norms of international law automatically continued in force between Slovakia and Hungary. As Hungary had no legitimate basis to terminate the Agreement, and as it did not lapse with the dissolution of Czechoslovakia, Slovakia believed it was entitled to implement the Provisional Solution as a means of achieving the objectives of the Agreement under the doctrine of approximate application. Slovakia also contended it was obligated to implement the Provisional Solution as a means of mitigating the losses suffered as a result of Hungary's breach of the 1977 Agreement.

In the view of Slovakia, even absent the continuing consent of Hungary as preserved by the approximate application doctrine, the Provisional

Solution was consistent with its various international legal obligations, including those contained in the 1977 Agreement, the 1948 Danube River Convention and the 1976 Boundary Waters Convention, and did not conflict with its obligations under international law to respect Hungary's frontiers and permanent sovereignty over its natural resources, cooperate with Hungary on matters relating to the Danube River, protect the environment and equitably utilize shared natural resources. Slovakia thus sought to have the Court order Hungary to complete construction of the Nagymaros works and permit the bringing into operation of the Dunakiliti dam in order to create the situation envisioned by the 1977 Agreement. Slovakia also requested the Court order Hungary to make reparations, including interest and loss of profits, in an amount sufficient to compensate Slovakia for loss and damage caused by Hungary's failure to implement its responsibilities under the Agreement and for the cost of constructing and operating the Provisional Solution.

In light of these arguments, the parties and the Court were faced with eleven possible options for resolving the dispute. These options included the following findings:

1) Hungary's termination of the 1977 Agreement was unlawful, and Slovakia is entitled to specific performance requiring Hungary to construct the Nagymaros works.
2) Hungary's termination of the 1977 Agreement was unlawful, and Slovakia may continue its operation of the Provisional Solution, but Hungary may not be compelled to construct the Nagymaros works, rather Hungary must compensate Slovakia for expectancy damages caused by the reduced output of the Gabčíkovo hydroelectric plant.
3) Hungary's termination of the 1977 Agreement was unlawful, and Slovakia may continue the Provisional Solution, but Hungary may not be compelled to construct the Nagymaros works, rather Hungary must compensate Slovakia for expectancy damages caused by the reduced output of the Gabčíkovo hydroelectric plant, minus the cost of environmental measures undertaken by Hungary.
4) Hungary's termination of the 1977 Agreement was unlawful, and Slovakia may continue to operate the Provisional Solution, but Hungary is not required to construct the Nagymaros works or pay compensation.
5) Hungary's termination of the 1977 Agreement was lawful, but Slovakia is entitled to operate the Provisional Solution, and Hungary

is not required to construct the Nagymaros works or pay compensation.
6) Hungary's termination of the 1977 Agreement was lawful, but Slovakia is entitled to operate the Provisional Solution so long as it undertakes sufficient measures to protect the environment and provides Hungary a share of the generated electricity.
7) Hungary's termination of the 1977 Agreement was unlawful, but Slovakia may not operate the Provisional Solution, rather Hungary must compensate Slovakia for lost energy output.
8) Hungary's termination of the 1977 Agreement was lawful, and Slovakia is not entitled to operate the Provisional Solution, but Hungary must compensate Slovakia for this lost income for lost energy output.
9) Hungary's termination of the 1977 Agreement was unlawful, but Slovakia may not operate the Provisional Solution, rather Hungary must compensate Slovakia for lost energy output, minus the cost of environmental measures that will no longer need to be undertaken.
10) Hungary's termination of the 1977 Agreement was lawful, and Slovakia may not operate the Provisional solution.
11) Hungary's termination of the 1977 Agreement was lawful, Slovakia may not operate the Provisional Solution, and Slovakia must pay damages to Hungary for the harm caused by the temporary operation of the Provisional Solution.

Faced with these options resulting from the position of the parties, the task of the Court was to render a judicial determination of certain factual and legal findings which would narrow these options in such a manner as to permit the parties to resolve their dispute.

IV. The determination of certain factual and legal questions by the International Court of Justice

To explore how the Court employed specific legal principles to allocate the respective rights and responsibilities of each party in order to narrow the legitimate options available to the parties, the discussion of the case before the Court will be organized around four issues: (1) whether Hungary was lawfully entitled to suspend and subsequently to abandon, in 1989, the Nagymaros works and its share of the Gabčíkovo works; (2) whether Hungary was lawfully entitled to terminate the 1977 Agreement; (3) whether in response to Hungary's suspension and abandonment of works on the Gabčíkovo–Nagymaros Project and subsequent termination of the 1977 Agreement, Czechoslovakia was lawfully entitled to construct and implement the Provisional Solution in an attempt to secure the objectives of the Agreement;[339] and (4) what are the legal consequences arising from the determination of these legal issues before the Court.[340]

A. Whether Hungary was lawfully entitled to suspend and subsequently to abandon, in 1989, the Nagymaros works and its share of the Gabčíkovo works

To justify its suspension and subsequent abandonment of the Nagymaros works and its share of the Gabčíkovo works, Hungary argued the ecological dangers associated with the construction of these works constituted a state of necessity requiring the works be suspended until the environmental effects could be adequately assessed and a means of avoiding such harm could be designed and implemented.[341] In crafting its arguments of necessity, Hungary asserted its suspension of work did not constitute a suspension of the 1977 Agreement,[342] and its acts were governed by the international law of state responsibility as well as relevant treaty law.[343]

Hungary further asserted its suspension and abandonment of the works was necessitated by the fact Czechoslovakia breached various provisions of the 1977 Agreement prior to 1989 by failing to consider the environmental dangers associated with the Gabčíkovo–Nagymaros Project and by insisting on the construction of the Nagymaros works. Specifically, Hungary alleged under Article 3, para. 2 of the 6 May 1976 Joint Contractual Plan, Czechoslovakia was required to assess the environmental effects of the Project adequately and no such assessments were carried out.[344]

In order to preserve the integrity of the doctrine of *pacta sunt servanda* and not unnecessarily unsettle treaty relationships among states, international law permits the invocation of the claim of necessity in only very limited circumstances. As codified by the International Law Commission (ILC) in draft Article 33 of the Draft Articles on State Responsibility, a state may not invoke the claim of a state of necessity unless the illegal act was the 'only means of safeguarding an essential interest of the state against a grave and imminent peril', and the illegal act 'did not seriously impair an essential interest of the state towards which the obligation existed'. Draft Article 33 further requires even if these two conditions are met, a state of necessity may not be invoked if the state seeking to invoke the claim of necessity contributed to the occurrence of the state of necessity.[345]

Invoking this draft article, Hungary contended it was necessary to suspend and subsequently to abandon construction of the Nagymaros works and its share of the Gabčíkovo works because the Project posed a 'major economic and environmental threat to [its] population', in the form of 'severe damage to Hungarian agriculture and [forestry]', and significant pollution of surface and subsurface waters, the source of potable water for millions of Hungarians.[346] The Project also posed a significant threat to Hungary's natural environment, which was essential for aesthetic purposes, its high tourism value and a source of natural resources, and these threats infringed on the Hungarian population's 'right to environment'.[347] Moreover, these threats were imminent, as on 23 April 1992, the Czechoslovak Prime Minister had informed Hungary it would begin implementation of the Provisional Solution on 31 October 1992,[348] and it was apparent to Hungary from the moment of diversion of the Danube River, Hungary would begin to suffer damage to its agriculture, forestry, water supply and general environment.[349] Hungary further asserted the suspension and subsequent abandonment of work was unavoidable as although it had warned Czechoslovakia of the necessity of modifying or terminating the Agreement, Czechoslovakia refused to participate in meaningful negotiations and had unilaterally prepared plans for the diversion of the Danube.[350]

With respect to the requirement the invocation of necessity not seriously impair an essential interest of another state, Hungary argued its suspension and subsequent abandonment of works in fact protected the essential interest of Czechoslovakia as it too would have, and had already suffered significant environmental harm as a result of the diversion of the Danube River. Hungary also added it did not in any way contribute to the occurrence of the state of necessity.[351]

Slovakia responded by first contesting the validity of the doctrine of necessity as a separate basis for justifying Hungary's suspension and subsequent abandonment of the Nagymaros works and its participation in the Gabčíkovo works, on the basis the doctrine of necessity derived from the law of state responsibility, which was inapplicable as the performance of treaty obligations could only be judged on the basis of treaty law.[352]

Slovakia then argued, in the event the Court determined Hungary could properly seek to justify its suspension and abandonment of works on the grounds of necessity, Hungary had not in fact met the requirements of that doctrine.[353] Specifically, Slovakia argued a state of necessity did not exist since there was no 'existence of an imminent ecological disaster',[354] as the project did not pose a significant threat to Hungarian agriculture, forestry, environmental diversity or potable water resources.³⁵⁵ Rather, the diversion of the Danube River into the power canal was necessary in order to prevent an ecological disaster occasioned by the decrease in the bed-level of the Danube and the increasing possibility of flooding.³⁵⁶ Slovakia, moreover, argued Hungary was barred from invoking the state of necessity doctrine as it did not in fact hold a good faith belief the construction of the Gabčíkovo–Nagymaros project would cause imminent harm to the environmental interests of Hungary. Instead, Hungary sought to extricate itself from the 1977 Agreement because of Hungary's severe financial difficulties, Hungary's reduced need for additional energy resources and Hungary's inability to resist irrational political pressure to abandon the project.³⁵⁷

Slovakia also contended in the event Hungary was able to establish it held a good faith belief in an actual state of necessity, it could not rely upon the doctrine to excuse itself from the consequences of suspending its performance under the treaty as Hungary's actions seriously impaired the essential interests of Slovakia in preventing repeated flooding, pursuing development in a sustainable and environmentally responsible manner and in generating clean energy.³⁵⁸ Furthermore, by delaying the construction of works on Hungarian territory, Hungary contributed to the occurrence of any state of necessity.³⁵⁹

In reaching a decision on the issue of whether Hungary could invoke the doctrine of a state of necessity to justify its suspension and abandonment of the Nagymaros works and its participation in the Gabčíkovo works,³⁶⁰ the Court chose not to address the issue of the applicability of the Vienna Convention as raised by the parties, but rather noted some of the rules set forth in the Convention, including in many respects Articles 60 to 62, accurately reflected customary international law.³⁶¹ The Court also chose not to explore the issue of the relationship between the law of treaties and the law of state responsibility, noting merely the law of treaties addressed whether or not a convention was in force and whether it had been properly suspended or denounced, while the law of state responsibility addressed the responsibility of a state for any suspension or denunciation which was not lawful.³⁶²

The Court, however, did address the issue of whether the suspension and abandonment of the Nagymaros works and the Hungarian share of the Gabčíkovo works amounted to a suspension and subsequent rejection of the 1977 Agreement. Here the Court rejected Hungary's argument the suspension and abandonment of the works did not amount to a suspension and rejection of the Agreement, on the basis by suspending and abandoning the works, Hungary rendered impossible the performance of the Agreement, which called for a jointly constructed and operated project.³⁶³ The Court further noted by invoking the doctrine of state responsibility, Hungary implicitly acknowledged in the absence of the justification of a state of necessity, its conduct would be deemed unlawful. As such, the existence of a state of necessity would not permit the conclusion Hungary acted pursuant to its obligations when suspending and then abandoning performance of the Agreement, or the Agreement had ceased to be binding, rather it would serve to shield Hungary from international responsibility for committing the wrongful act of suspension and abandonment of its treaty obligations.³⁶⁴

With respect to whether or not a state of necessity actually existed, the Court found in order for a state to rely upon such a claim, it must meet the criteria established by the ILC in draft Article 33 of the Draft Articles on State Responsibility, which it believed would be applicable in only exceptional cases.³⁶⁵

The Court then found although Hungary's concern for its natural environment represented an essential interest of Hungary, Hungary had not established the objective existence of a 'peril' associated with the Nagymaros works or the Gabčíkovo works, nor had Hungary established if such a peril did exist it was imminent. Moreover, the Court determined Hungary could realistically have pursued other means of avoiding any perceived perils associated with the Project. At the time of concluding the 1977 Agreement, Hungary was presumably aware of the environmental risks when it consented to the Agreement, and at various times since adopting the 1977 Agreement, depending on its interests, Hungary had sought either to delay or accelerate the construction of the Gabčíkovo–Nagymaros Project. The Court, therefore, determined had the Gabčíkovo–Nagymaros Project constituted an unavoidable grave and imminent peril, Hungary nevertheless would have been prohibited from relying upon the claim of a state of necessity as it had helped, by act or omission, to bring about the state of necessity.³⁶⁶

The Court thus found it unnecessary to address the questions of whether Hungary's actions impaired an essential interest of Czechoslovakia, whether Czechoslovakia had furthered the alleged state of necessity by breaching Articles 15 and 19 of the Agreement or whether Hungary had violated Article 27 of the Agreement by taking unilateral measures without first seeking to resolve the dispute within the mechanisms provided by the Agreement.³⁶⁷

In finding Hungary's concern for its natural environment represented an essential interest of Hungary, the Court referenced the ILC's commentary to draft Article 33, which indicated an essential interest should not be reduced to a matter only of the existence of a state and the question was to be evaluated on a case by case basis. The Court further recognized the Commission believed state practice over the last twenty years indicated a grave danger to the ecological preservation of all or some of a state's territory may constitute a state of necessity.[368] The Court also took the opportunity to recall in the *Legality of the Threat or Use of Nuclear Weapons Advisory Opinion*, the Court had stressed 'the great significance that it attaches to respect for the environment, not only for states, but also for the whole of mankind'.[369]

In considering whether the threat to Hungary's natural environment could be considered an imminent peril, the Court reasoned although the use of the word 'peril' connoted an element of risk, a state of necessity could not be considered to exist unless a 'peril' was 'duly established at the relevant point in time', as the 'mere apprehension of a possible peril' was not sufficient.[370] Similarly, the word imminent connoted an element of immediacy or proximity of the peril, which established a higher threshold than mere 'possibility'.[371] The Court further noted while the 'peril' had to be established as imminent, the harm could be long term, if the occurrence of such harm was established at the time the state of necessity claim was invoked. Hungary was thus obligated to prove when it sought to suspend the 1977 Agreement in 1989, a grave and imminent peril did in fact exist and the suspension of works was the only possible means of responding to the peril.

After evaluating the scientific evidence presented by the parties, the Court found with respect to the upstream reservoir associated with the Nagymaros works the long term dangers to drinking water resources were uncertain and in the event the damages were certain, they were not imminent since the final operating procedures of the Gabčíkovo power plant had not yet been determined.[372] With respect to the effect of lowering the river bed downstream of the Nagymaros works, the Court found at various times prior to 1989 the river bed had reached a depth envisioned by the Agreement, and thus the peril, which in fact had materialized, could not in 1989 be solely associated with the implementation of the Project. Moreover, the Court found ample alternative means existed to respond to the peril, including the regular discharge of gravel into the river or the purification of the river water prior to consumption.

Concerning the Gabčíkovo works, the Court found the feared long term damage remained uncertain and would occur as the result of a 'relatively slow natural process, the effects of which could not be easily assessed'.[373] The Court also noted the 23 June 1989 report of the Committee of Hungarian Academy of Sciences did not 'express any awareness of an authenticated peril', but rather determined it was not possible to evaluate adequately the environmental impacts of the project, and thus called for the implementation of an extensive monitoring program.[374] Moreover, the Court found ample alternative means existed to respond to the peril, including controlling the distribution of the water into the old bed of the Danube and the side arms, which was explicitly provided for in Article 14 of the 1977 Agreement, or the continuation of negotiations which might have led to a review and modification of the Project.[375]

In finding Hungary's suspension and abandonment of the project works amounted to a suspension and subsequent rejection of the 1977 Agreement, and Hungary was not entitled to invoke the claim of necessity to justify this rejection of the Agreement, the Court narrowed the options available to the parties. In particular, the Court validated Slovakia's claim of Hungary's illegitimate activity, which provided Slovakia with a certain degree of moral authority, and laid the foundation for a subsequent determination by the Court Hungary might be liable for damages suffered by Slovakia during the period of time leading up to the termination of the Agreement.

In making the legal finding Hungary's concern for the environment could constitute an essential interest of a state, coupled with the factual findings the Nagymaros works did not constitute a certain and unavoidable threat to Hungarian drinking water supplies, and the threat of long term damage from the Gabčíkovo works remained uncertain, the Court refined and restricted the negotiating positions of the parties. Specifically, the Court furthered a resolution of the dispute by affirming Hungary's right to insist in subsequent negotiations the environmental implications of the project must be considered, but denied Hungary the license to insist the current assessment of environmental consequences justified a termination of the Project.

The Court furthered the role of international law by clarifying that the law of treaties relates to whether a state may lawfully suspend or terminate a treaty, while the law of state responsibility relates to whether or not a state may be held responsible for the unlawful suspension or termination of a treaty. The Court also furthered the role of law by clearly articulating the criteria for invoking a state of necessity as a basis for eluding state responsibility for an unlawful act, and by indicating the high threshold of grave and imminent peril which must be reached in order to establish a state of necessity.

The Court also potentially furthered the resolution of other CEE disputes by validating a state's concern for its environment as an essential interest, and by reaffirming its view in the *Legality of the Threat or Use of*

Nuclear Weapons Advisory Opinion great significance is attached to respect for the environment. Moreover, the Court established that a state's concern for its natural environment represented an essential interest, and threats to a state's ecological preservation may give rise to a state of necessity. Notably though, the Court clarified these views by emphasizing the threat to that interest must be imminent and there must exist no other means of abating the threat. These findings are of particular relevance to the Temelín and Mochovce nuclear power plant disputes given their alleged threat to the ecological integrity of Austria. The findings are also of relevance to the Cernavoda and Belene nuclear power plant disputes given the respective claims of ecological aggression articulated by Romania and Bulgaria and the apparent willingness of both Romania and Bulgaria to engage in unilateral acts designed to protect their ecological interests.

B. Whether Hungary was lawfully entitled, in 1992, to terminate the 1977 Agreement

To justify its termination of the 1977 Agreement, Hungary relied upon two separate but related bodies of international law, the law of treaties and state responsibility. Slovakia contended Hungary did not meet the necessary conditions under the law of treaties, and objected to the invocation of the legal basis of state responsibility.[376] The Court dealt separately with the law of treaties and the law of state responsibility as grounds for terminating the 1977 Agreement.[377]

1. The law of treaties

To support its termination of the 1977 Agreement under the law of treaties, Hungary argued there had been an unforeseen fundamental change of circumstances which radically altered the obligations to be performed under the Agreement, Czechoslovakia had materially breached the Agreement, and new binding norms of international environmental law had evolved which precluded Hungary's compliance with the Agreement.[378]

a. A change of circumstances as a justification for treaty termination. The customary international law doctrine of a fundamental change of circumstances allows a party unilaterally to terminate or suspend the operation of a treaty on the ground 'there has been a fundamental change of circumstances from those which existed at the time of the conclusion of the treaty'.[379] The fundamental change of circumstances doctrine requires a balancing between the effect of the changed circumstances and the doctrine of *pacta sunt servanda* to determine whether the continuation of the treaty obligations in light of the changed circumstances would be unreasonable and unfair.

Immediately prior to World War I, many states invoked the changed circumstances doctrine to 'escape inconvenient treaty obligations',[380] when states felt the changed circumstances merely rendered the performance inconvenient.[381] Given the potential for abuse of the changed circumstances doctrine, the drafting states of the Vienna Convention sought to limit its application by requiring the party seeking withdrawal to prove the alleged changed circumstances: (1) were fundamental to the operation of the treaty, (2) did not exist at the time of treaty negotiation, (3) were not foreseeable by the parties, (4) would have affected the consent of the parties to the treaty, (5) radically transformed the nature of the obligations to be performed by the parties, and (6) were not the result of a breach by the party invoking the justification.[382]

Although the Court has generally proven reluctant to approve the unilateral termination of a treaty on the grounds of changed circumstances, the Court has accepted the doctrine as a principle of international law.[383] In the *Fisheries Jurisdiction Case*, for instance, the Court accepted 'a fundamental change in the circumstances which determined the parties to accept a treaty, if it has resulted in a radical transformation of the extent of the obligations imposed by it, may, under certain conditions, afford the party affected a ground for invoking the termination or suspension of the treaty'.[384] In admitting to the existence of the changed circumstances doctrine the Court sought, however, to limit its applicability by confirming in order for a change of circumstances to give rise to a sufficient ground for terminating a treaty, the changed circumstances must have 'resulted in a radical transformation of the extent of the obligations still to be performed' by the parties, and the 'change must have increased the burden of the obligations to be executed to the extent of rendering the performance something essentially different from that originally undertaken'.[385]

Both Slovakia and Hungary agreed before the Court that Article 62 of the Vienna Convention generally reflects customary international law relating to the justification of a fundamental change of circumstances.[386] Slovakia and Hungary disagreed, however, as to how Article 62 related to the three specific issues of: (1) whether the changed circumstances required are extraordinary or singular in character, with Hungary arguing a change could be 'cumulative and [could] result from the concurrence of a number of factors',[387] and Slovakia arguing the necessity of a singular changed circumstance; (2) whether a perceived increase in risk of a certain set of circumstances occurring was sufficient to warrant invocation of the doctrine, as argued by Hungary, or whether the changes must have actually occurred, as argued by Slovakia; and (3) whether a state could have relied upon the changed circumstances doctrine if its own conduct had contributed in a tangible but not substantial manner to the change, with Hungary arguing such a state could have invoked the doctrine, and Slovakia arguing any breach by the invoking state rendered the justification inapplicable.[388]

With these three disagreements in mind, Hungary argued the purposes of the 1977 Agreement included socialist integration, a single and indivisible operational system for the generation of peak power production, an economically beneficial joint investment, a framework treaty subject to and requiring revision and a treaty consistent with environmental protection.[389] These purposes were then frustrated by a series of unforeseeable changed circumstances, which had they existed in 1977 would have dissuaded Hungary from signing the Agreement and which in 1992 radically transformed Hungary's obligations under the Agreement. The changed circumstances included political changes, economic changes and changes in environmental knowledge and law.[390]

Changed political circumstances presented by Hungary included: the defeat of communism; the dissolution of the Warsaw Pact; the change in government from a totalitarian regime to a democracy with an ensuing increase in public participation in decision making and increased governmental accountability;[391] and the introduction of privately owned and uncensored media.[392] Additional changed circumstances included the detachment of Slovakia and Hungary from the Soviet Union's sphere of influence;[393] the increased environmental awareness and activism of the Hungarian population from the late 1980s and early 1990s;[394] and the decreasing ability of the totalitarian regime either to silence or ignore domestic opposition to the Project.[395]

Changed economic circumstances presented by Hungary included: the transformation from a socialist command economy to a market economy; the dissolution of the former communist bloc trade arrangements; the end of subsidized state owned enterprises, the end of guaranteed full employment; and the introduction of cost–benefit analyses.[396] Additional changed economic circumstances included the substantial improvements in energy conservation technologies developed since 1977, which rendered unnecessary the energy to be produced by the Project;[397] the shift from economic reliance on heavy industrial production to light manufacturing, which substantially reduced the need for increased energy production; the shift in CEE trade away from the former Soviet Union and toward Western Europe, which lessened the need for improved navigability of the Danube River in this region; and the recession of the early 1980s, which fundamentally affected the ability of Hungary to construct its share of the Project.

Changed environmental circumstances presented by Hungary included: the increased understanding of the relationship between large scale development and its impact on the environment; increased public environmental awareness; the development of procedures for assessing environmental impacts; and the evolution of a new environmental ethos, which transformed the occurrence of environmental harm from an unquantifiable, disregarded externality to a factor which significantly increased the parties' perceptions of the costs and risks associated with the Project.[398]

Slovakia argued the primary aim of the 1977 Agreement related to the establishment of a joint investment program for the purpose of producing energy, controlling floods and improving the navigability of the Danube River.[399] Slovakia then contended although Hungary was able to cite a number of circumstances that had changed, Hungary did not specifically indicate how these changed circumstances impeded the achievement of these purposes, how they radically altered the obligations to be performed under the Agreement, or how they would have altered the consent of Hungary if they had existed in 1977. Slovakia also noted, as Hungary had reconfirmed its intent to proceed with the project in 1989 and had in fact requested an acceleration of the Project, Hungary could only rely on circumstances that had changed since that time.[400] Slovakia then generally admonished the Court its decision on this matter would be watched closely by many states and any indication that states could terminate a treaty after meeting a relatively low threshold could cause significant uncertainty in treaty relationships.[401]

The Court, citing the *Fisheries Jurisdiction Case*, affirmed Article 62 of the Vienna Convention represented a codification of existing customary international law with respect to treaty termination on the basis of a fundamental change of circumstances. The Court further noted consistent with the negative conditional wording of Article 62, the international community's interest in stability of treaty relations required only in exceptional cases could a change of circumstances be permitted to serve as a basis for treaty termination.[402]

The Court then found although the prevailing political situation in 1977 was relevant for the conclusion of the 1977 Agreement, the Agreement was primarily concerned with establishing a joint investment program for the purpose of producing energy, controlling floods and improving the navigability of the Danube River. In the Court's view, the prevalent political conditions of 1977 were thus not 'so closely linked to the object and purpose of the Treaty that they constituted an essential basis of the consent of the parties and, in changing, radically altered the extent of the obligations still to be performed'.[403]

With respect to Hungary's claim of changed economic circumstances, the Court found these circumstances too were not so closely linked to the object and purpose as to constitute an essential basis of the Agreement and the diminished profitability of the Project did not radically transform the treaty obligations of Hungary and Slovakia. Concerning the claim of changed environmental circumstances, the Court noted it 'did not consider new developments in the state of environmental knowledge and of environmental law can be said to have been completely unforeseen', and, in fact, Articles 15, 19 and 20 of the Agreement were designed to serve as mechanisms for accommodating such change.[404]

In rejecting Hungary's claim that certain political, economic and environmental changes constituted a fundamental change in circumstances justifying a termination of the Agreement, the Court promoted the resolution of the dispute by more clearly defining for the parties the object and purposes of the agreement. In particular, the Court built on the finding above, that the object was still attainable, narrowly defining the object of the Agreement as a joint investment program for the purpose of producing energy, controlling floods and improving the navigability of the Danube River. The Court also reaffirmed the parties were obligated to achieve this object in a manner that sufficiently protected the environment, pursuant to the relevant articles of the Agreement.

The Court potentially furthered the resolution of other CEE disputes by indicating its reluctance to accept the 1989 change in political and economic regimes as a basis for terminating treaties, and indicated states would unlikely be able to cite newly developed norms of international environmental law as changed circumstances given such a change in norms is generally foreseeable. Presumably, a change in other sets of legal norms would also generally be foreseeable. CEE states will thus likely consider themselves less able to justify a purported termination of an agreement on the basis of a fundamental change of circumstances, and therefore be more inclined to use existing treaties as a basis for negotiating the resolution of their disputes. In particular, this finding may be relevant for CEE states contemplating the possibility of terminating treaties with the former Soviet Union relating to their waiver of claims for contamination caused by Soviet troops. This finding may also be modestly relevant to state parties to the Baltic Sea and Black Sea conventions who might be contemplating rationales for their failure to comply completely with their obligations under these agreements.

b. The breach of a treaty as justification for treaty termination. Article 60 of the Vienna Convention provides if one party materially breaches a bilateral agreement, the other party is entitled to terminate the treaty. A material breach is defined by Article 60 as '(a) a repudiation of the treaty not sanctioned by the present Convention; or (b) the violation of a provision essential to the accomplishment of the object or purpose of the treaty'.[405] Both Hungary and Slovakia accepted Article 60 as an accurate reflection of customary international law.[406]

To sustain the justification of breach as a basis for its termination of the 1977 Agreement, Hungary argued Czechoslovakia had breached the Agreement by failing to comply with Articles 15, 19 and 20 of the 1977 Agreement,[407] and related conventions and general rules of international law,[408] by failing to ensure the construction and operation of the Project did not cause significant damage to water quality and the general environment and by failing to carry out a joint international environmental impact assessment.[409] These breaches were, in the view of Hungary, aggravated by Czechoslovakia's refusal to engage in good faith efforts to resolve disagreements relating to the causes, extent and means for remediating the environmental impacts of the project as identified by Hungary.[410] Hungary's primary justification for invoking material breach, however, was the decision of Czechoslovakia to plan and construct the Provisional Solution, which Hungary contended amounted to a clear repudiation of the Agreement as it constituted a 'damaging unilateral diversion of the Danube'.[411]

Slovakia rejected Hungary's contention it could invoke alleged non-compliance with related conventions and rules of general international law as a basis for terminating the 1977 Agreement, as Slovakia was not in breach of those conventions or rules, and if it was, Article 60 of the Vienna Convention required a material breach of the treaty, not of ancillary conventions or international rules. Slovakia also argued its conduct did not amount to a material breach of Articles 15, 19 and 20 of the 1977 Agreement as it had not impaired the environmental integrity of the region, but had in fact enhanced the quality,[412] and it was at all times prepared to conduct good faith negotiations with Hungary to develop means for mitigating any environmental damage Hungary was able to identify.[413]

Slovakia further argued the implementation of the reversible Provisional Solution could not be considered a material breach of the Agreement, because it was implemented for the specific purpose of accomplishing the objects and purposes of the treaty relating to flood control, energy production and improved navigability, as necessitated by Hungary's material breach of its obligations to construct the Nagymaros works and participate in the construction of the Dunakiliti works.[414]

The Court first found Hungary could not invoke a material breach of general conventions or rules of general international law to justify the termination of a treaty as it is only a material breach of the treaty itself by a state party that permits another state party to terminate the treaty. According to the Court, 'the violation of other treaty rules or of rules of general international law may justify the taking of certain measures, including countermeasures, by the injured state, but it does not constitute a ground for termination under the law of treaties'.[415]

The Court then found Hungary could not properly rely upon the argument that Czechoslovakia's failure to enter into good faith negotiations as required by Articles 15, 19 and 20 of the Agreement to amend the Joint Contractual Plan in light of scientific and legal developments relating to the environment amounted to a material breach justifying its termination of the 1977 Agreement. According to the Court, there was insufficient evidence to indicate Czechoslovakia 'had consistently refused to consult with Hungary about the desirability or necessity of measures for the preservation of the environment', and thus Czechoslovakia could not be deemed solely responsible for any breaches of Articles 15 and 19 of the Agreement. Specifically, the Court noted, by suspending construction of the Nagymaros and Dunakiliti structures, Hungary had contributed to 'the creation of a situation which was not conducive to the conduct of fruitful negotiations',[416] presumably necessary to take action under Articles 15 and 19 of the Agreement to protect the environment.

The Court also rejected Hungary's argument that the implementation of the Provisional Solution constituted a material breach as Czechoslovakia had not yet implemented the Provisional Solution at the time Hungary terminated the Agreement. According to the Court, as Czechoslovakia did not violate the Agreement until the moment it diverted the Danube River into the power canal, Hungary's termination was premature and could have no legal effect.[417] As noted below, the Court did not consider Czechoslovakia's preparation of the Provisional Solution to constitute a breach of the Agreement, rather only the actual implementation of the Provisional Solution could constitute a breach.[418]

The Court further noted the wrongful act of implementing the Provisional Solution was the result of Hungary's own prior wrongful conduct associated with its decision to cease construction of the Nagymaros and Dunakiliti structures.[419] Thus, even if Czechoslovakia had been in breach of the Agreement at the time of Hungary's termination, Hungary would have had no valid basis to terminate the Agreement as it was the unlawful acts of Hungary that necessitated the Czechoslovak breach.[420]

In rejecting Hungary's argument of material breach, the Court promoted a resolution of the dispute by creating a hierarchy of legal principles governing the negotiated resolution of the dispute and by highlighting certain applicable procedural principles and obligations. More specifically, the Court established that negotiations must first be concerned with alleged or potential breaches of the 1977 Agreement itself, and any alleged or potential breaches of other obligations might permit the taking of countermeasures, but would not have a direct impact on the continuing validity of the Agreement, or the obligations of the parties contained therein.

Moreover, the Court's pronouncements on the principle of good faith signaled to the parties that in subsequent negotiations relating to the Project, the parties were bound by the obligation to negotiate in good faith, and the taking of unlawful acts designed to prejudice the negotiations would deprive that party of the right to object to certain unlawful acts taken by the other party. The Court also provided Slovakia with a certain degree of moral authority by finding there was insufficient evidence to conclude Czechoslovakia had failed to negotiate in good faith, and by pointedly criticizing Hungary for creating a condition unsuitable for productive negotiations.

The Court potentially furthered the resolution of other CEE disputes, in particular the Romanian–Bulgarian Nuclear Corridor, Danube River Gauntlet and Black Sea disputes, by indicating the significance of the duty to negotiate in good faith. However, by requiring Hungary to meet a fairly high evidentiary threshold in order to prove Czechoslovakia failed to act in good faith, the Court at the same time may have constrained the functionality of the principle of good faith. Similarly, in recognizing the role of ancillary treaties and norms of international law in a treaty-based dispute, the Court emphasized the need of the parties to consider all relevant norms and rules of international law, but by finding only a material breach of the treaty itself could give rise to a right to terminate that agreement, it minimized the role of these ancillary laws.

c. *Conflict between the 1977 Agreement and subsequent obligations under general international law as a justification for treaty termination.*
Relying upon the maxim *lex posterior derogat legi priori* (subsequent law abrogates the preceding law), Hungary asserted it was entitled to terminate the 1977 Agreement on the basis its implementation conflicted with a number of newly developed norms of international law.[421] Hungary identified these newly developed norms of international law as including,

> the principle of cooperation in order to protect the environment, especially in transboundary relations, the principles of prevention and of precaution, the duties to perform thorough environmental impact assessments and to conserve biological diversity, the protection of human rights against the exercise of countermeasures, the right to life and the duty of the state to protect it, as well as the right to a healthy and ecologically sound environment. (footnotes omitted)[422]

Hungary further asserted Articles 15 and 19 of the Agreement were capable of being interpreted in such a way as to provide the incorporation of newly developed norms of international law, and thus Hungary was entitled to terminate the Agreement on either the grounds the planning and construction of the Provisional Solution was a violation of these newly developed norms or a violation of Articles 15 and 19.[423] Finally, Hungary argued in the event the Court found the 1977 Agreement to be *lex specialis*, thereby excluding the consideration of the development of new norms of international law, the Court should determine these newly developed norms have displaced those provisions of the Agreement with which they are inconsistent.[424]

Slovakia rejected Hungary's argument on the following grounds: (1) it was in fact an argument relating to the effect of supervening peremptory or *jus cogens* norms, which Hungary was casting as an argument of supervening newly developed norms because it could not establish these principles were peremptory norms; (2) Hungary was similarly unable to establish these norms developed between February 1989 when Hungary reaffirmed its desire to implement certain provisions of the Agreement and May 1989 when it terminated the Agreement; (3) Hungary could not seek grounds to justify its attempted termination of the Agreement outside the law of treaties;[425] (4) newly developed rules of general international law which were contrary to treaty rules did not preempt those treaty rules; (5) the newly developed norms did not constitute a *lex specialis* and were generally considered to be soft law; (6) Slovakia's activities under the 1977 Agreement were in accordance with the intent of these principles;[426] and (7) Articles 15, 19 and 20 provided relevant subsequent norms could be incorporated into the Agreement.[427]

As Hungary had not argued new peremptory norms of international environmental law had emerged since the adoption of the 1977 Agreement, the Court found it was not required to address the applicability of Article 64 of the Vienna Convention to the dispute.[428] The Court did find newly developed norms of international environmental law were relevant for the implementation of the Agreement. But rather than conflicting with the Agreement, as argued by Hungary, the Court found the norms could be incorporated into the Agreement via Articles 15, 19 and 20. Although these articles did not contain specific obligations, they required the parties to ensure water quality, nature protection and fisheries when agreeing upon the measures necessary to implement the Joint Contractual Plan.[429] The Court reasoned by including these provisions in the Agreement, the parties had intended to permit the adaptation of the Project to comply with any norms of international environmental law that might emerge.[430]

The Court then determined the responsibility to implement new environmental norms was jointly held by the parties through a process of consultation and negotiation in which the parties were required to discuss in good faith actual and potential environmental risks.[431] The Court also recalled its *Legality of the Threat or Use of Nuclear Weapons Advisory Opinion*, wherein it declared 'the environment is not an abstraction but represents the living space, the quality of life and the very health of human beings, including generations unborn'.[432]

With this in mind, the Court asserted that as the awareness of the vulnerability of the environment and the recognition of the continual need to assess environmental risks had increased significantly since the adoption of the 1977 Agreement, Articles 15, 19 and 20 possessed enhanced relevance.[433] Finally, the Court recognized the parties agreed on the need to 'take environmental concerns seriously and to take the required precautionary measures', but they disagreed as to how this affected the operation of the Project. The Court thus recommended Slovakia and Hungary consider the merits of third party involvement in developing a plan for operating the Project consistent with new environmental norms.[434]

In finding that newly developed norms of international environmental law were incorporated into the 1977 Agreement via Articles 15, 19 and

20, and the parties jointly held the responsibility to implement these articles, the Court furthered the resolution of the dispute by providing the parties with an enhanced set of environmental principles to use in determining how best to operate the Project. The intent of the Court in so doing was to balance Slovakia's interest in the continuation of the Agreement, with Hungary's interest in applying contemporary principles of international law and ensuring the operation of the project did not infringe upon its essential interest in protecting the environment. Although Hungary would have preferred for the Court to find the operation of the Project violated these norms and thus Hungary validly terminated the Agreement, the Court's opinion strengthened Hungary's subsequent negotiating position by removing any doubt as to the applicability of contemporary norms of international environmental law, and even implicitly envisions the possibility the parties may find it is impossible to operate the Project consistent with these norms, and thus the Project must be terminated.

In recommending the parties seek the assistance of a third party in developing a plan for operating the Project consistent with new environmental norms, the Court furthered the resolution of the dispute by providing necessary political cover if the parties wish to involve a third party. Although the Court accepts it cannot require the involvement of a third party, the Court recognized the essential role often played by third parties in the implementation phase of a dispute, and acted to provide the parties with a justification for inviting a third party, in the event either or both the parties felt they would suffer domestic criticism for seeking the continued assistance of a third party.

Moreover, the Court furthered the role of international law in resolving other CEE disputes by affirming the status and relevance of newly developed norms of international law, and by indicating international law would permit these norms to be incorporated into existing treaty arrangements based only on the most minimal reference within the text of the treaty to a concern for the environment. This finding is particularly relevant for the Baltic Sea and Black Sea disputes as the applicable agreements make frequent reference to environmental norms. In addition, the Court potentially furthered the resolution of CEE disputes by indicating the necessity of considering the involvement of a third party, and indicating third parties frequently play a valid role in environmental dispute resolution. This finding is of particular importance for the Black Triangle and Silesian Coal Basin disputes, and possibly for the Romanian–Bulgarian Nuclear Corridor and Danube River Gauntlet disputes.

2. State responsibility and ecological necessity

Whereas the law on treaties provides a basis for declaring a treaty invalid or for rendering a treaty subject to termination, the international legal norms relating to state responsibility provide that a state may avoid international responsibility for the commission of an unlawful act by establishing a valid claim of necessity. As such, Hungary argued it could not be held responsible for the termination of the 1977 Agreement as it had no choice but to terminate the Agreement given the state of necessity occasioned by the ecological harm Hungary and Czechoslovakia would have suffered if the Agreement were implemented.[435]

Slovakia contested the validity of the doctrine of necessity as a separate basis for justifying Hungary's termination of the 1977 Agreement,[436] and argued that in the event the Court determined Hungary could properly seek to justify its termination of the Agreement on the basis of necessity, Hungary had not in fact met the requirements of that doctrine as no state of necessity existed.[437] Slovakia further argued contrary to causing an environmental catastrophe, the diversion of the Danube River into the power canal was necessary in order to prevent an ecological disaster occasioned by circumstances unrelated to the Agreement,[438] and by terminating the Agreement, Hungary seriously impaired Slovakia's essential interests in preventing repeated flooding, pursuing development in a sustainable and environmentally responsible manner and generating clean energy.[439]

As noted above, the Court found a state of necessity did not exist at the time Hungary suspended and abandoned its work on the Nagymaros and Dunakiliti structures and thus presumably not at the time Hungary notified Slovakia of its termination of the Agreement. The Court then observed if a state of necessity had existed it would not have served as a basis for terminating the Agreement, but would have merely permitted a party in noncompliance with the Agreement to escape responsibility for such noncompliance. Once the state of necessity had ended, however, the parties would be obligated to comply with terms of the Agreement, which would continue in force.[440]

In finding a state of necessity did not exist, and if it did it would not serve to terminate the 1977 Agreement, the Court promoted the resolution of the dispute by clearly signaling to the parties the sanctity of the Agreement and by reaffirming the parties were required to consider jointly the environmental consequences of the Project and to implement measures to mitigate the environmental effects of the Project. Moreover, this finding, coupled with the findings immediately above, removed from the subsequent negotiating agenda any claim Hungary may have had to legitimately terminate the Agreement. In addition, these findings established the basis for the Court's subsequent finding below that the Agreement continues in force.

C. Whether in response to Hungary's suspension and abandonment of work on the Gabčíkovo–Nagymaros Project and subsequent termination of the 1977 Agreement Czechoslovakia was lawfully entitled to construct and implement the provisional solution in an attempt to secure the objectives of the agreement

Hungary objected to the construction and implementation of the Provisional Solution on the grounds it violated a number of treaty based principles and general rules of international law establishing procedural and substantive obligations. Slovakia argued the construction and implementation of the Provisional Solution was a legitimate attempt to secure the objectives of the 1977 Agreement,[441] and the Provisional Solution did not violate any of Slovakia's treaty based obligations, nor the principles of general international law cited by Hungary.[442]

1. The distinction between a lawful entitlement to construct the Provisional Solution and a lawful entitlement to implement the Provisional Solution

Although neither Hungary nor Slovakia sought to make a clear distinction in their written pleadings before the Court as to the act of planning and constructing the Provisional Solution and the act of implementing the Provisional Solution, the Court highlighted the fact in the Special Agreement the parties sought separate determinations as to (1) whether Slovakia was entitled to proceed in November 1991 with the construction of the Provisional Solution, and (2) whether Slovakia was entitled to put the Provisional Solution into operation from October 1992.[443]

With respect to the first question, the Court found from November 1991 to October 1992, the construction of the structures necessary for the Provisional Solution occurred solely on Czechoslovakian territory and up until the moment Czechoslovakia diverted the Danube River, the Project was reversible and did not infringe upon any of Hungary's international rights.[444] The Court supported this finding by noting in international and domestic law a wrongful act is frequently 'preceded by preparatory actions which are not to be confused with the act or offense itself', and which do not themselves constitute wrongful acts.[445] A number of Judges dissented on this point, arguing the construction of the Provisional Solution was 'inseparable from its being put into operation'.[446]

In finding the preparation of the Provisional Solution did not constitute a wrongful act, the Court marginally promoted a resolution of the dispute by removing from the negotiations any claims by Hungary of damages resulting from the preparation of the Provisional Solution. More importantly, the Court created a precedent whereby states involved in other CEE transboundary environmental disputes would be restricted from arguing the construction of a potentially harmful project, such as the Temelín, Mochovce, Cernavoda or Belene nuclear power plants, is a wrongful act, and would be confined to asserting legal rights relating to the obligation to consult, to share information, and possibly to conduct a joint EIA.

To answer the question of whether the actual operation of the Provisional Solution constituted a wrongful act, the Court examined Hungary and Czechoslovakia's treaty based obligations and Slovakia's contention the Provisional Solution was a lawful countermeasure.

2. *Czechoslovakia's treaty obligations as a justification for the Provisional Solution*

With respect to Czechoslovakia's treaty based obligations, Hungary argued the Provisional Solution violated provisions of the 1977 Agreement, the 1976 Boundary Waters Convention, the 1976 Joint Contractual Plan, the 1948 Danube River Convention and the 1958 Danube Fisheries Agreement, which required the parties to operate a joint project, carry out all water management activities by mutual agreement, share relevant information on environmental impacts of such projects and consult with the affected party in an effort to avoid any negative environmental or other impacts.[447] Hungary also asserted the Provisional Solution violated Czechoslovakia's obligations under Articles 15, 19 and 20 of the 1977 Agreement, which required the parties to protect the quality of the Danube River water, conduct necessary studies to ensure the protection of the natural environment,[448] and protect fishing interests.[449] Hungary further contended the Provisional Solution changed the character of the Slovak–Hungarian border in a manner not authorized by the 1977 Agreement and contrary to the requirements of the 1976 Boundary Waters Convention.[450]

Slovakia countered with the assertion the construction and operation of the Provisional Solution was, in fact, consistent with Czechoslovakia's obligations to consult with Hungary and engage in water management projects by mutual consent on the basis that the 1977 Agreement envisioned the construction of a system of works nearly identical to the Provisional Solution. Slovakia contended as the Agreement continued in force, it not only reflected years of consultation and sharing of information, but also embodied Hungary's consent to the project. Moreover, Slovakia argued since the Provisional Solution complied with Articles 15, 19 and 20 of the Agreement, not only were Hungary's environmental concerns unfounded,[451] but in fact, the Provisional Solution had a positive impact on the water quality, natural environment and fishing resources of the Danube River basin as predicted by the studies carried out by Czechoslovakia and other international entities.[452] Slovakia also noted the Provisional Solution could not be considered to violate the territorial integrity of Hungary as the 1977 Agreement explicitly provided the boundary line would remain unchanged despite the diversion of the River.[453]

Invoking general international law, Hungary then asserted the Provisional Solution violated Czechoslovakia's obligations not to cause damage to the environment beyond one's borders by causing substantial environmental damage to Hungary in the Szigetköz region and by changing the underground water level and water regime and threatening the water supply of Budapest;[454] to notify and consult with a potentially affected state by excluding Hungary from any discussions concerning the construction and implementation of the Provisional Solution and by providing Hungary with inadequate information about its operation and effects;[455] to respect the principle of nondiscrimination by failing to take into account the harm caused to Hungary by the Provisional Solution;[456] to equitably utilize shared natural resources by depriving Hungary of its share of water quantity, water quality and power potential;[457] and to respect the permanent sovereignty of other states over their natural resources by threatening the quality of Hungary's water resources and thus the ability to utilize its soil resources.[458]

Slovakia asserted the Provisional Solution did not violate the norms of international law cited by Hungary since: (1) the Provisional Solution improved environmental quality in the adjacent region of the Danube River by improving the quality of surface and ground water, promoting the development of flora and fauna and enhancing the productivity of agricultural and forestry resources; (2) as a *lex specialis* treaty, the 1977 Agreement contained all necessary standards and obligations relating to environmental protection, which could not be modified by subsequent rules of international law; (3) certain principles cited by Hungary such as the precautionary principle and sustainable development were of the status of soft law, while other principles, such as equitable utilization, the no harm rule and the human right to a healthy environment were selectively applied or were not as articulated by Hungary; and (4) the

Provisional Solution was designed so that upon construction of the Nagymaros works it could be modified to meet the original design specifications of the 1977 Agreement and Joint Contractual Plan.[459]

The Court found whereas the Agreement provided for the construction of a 'joint investment constituting a single and indivisible operational system of works', which would be jointly owned and operated by the parties, the Provisional Solution constituted a unilateral act, providing no means for the joint participation of Hungary, and which appropriated to the sole benefit of Slovakia 80–90 per cent of the flow of the Danube River. In the view of the Court, the egregiousness of Slovakia's sole appropriation of the Danube was heightened by the fact the river was not only a shared international watercourse, but also an international boundary river.[460]

The Court also rejected Slovakia's contention the Provisional Solution was lawful as it represented a structure as similar as possible to the original Project consented to by Hungary, finding rather Hungary's express consent only encompassed a joint project, and Slovakia could not rely upon such consent in order to operate a unilateral project, which would deny Hungary its right to an equitable and reasonable share of the resources of a transboundary watercourse. The Court thus determined regardless of whether Hungary had legitimately or illegitimately withdrawn its original consent for the joint project, Slovakia could not rely upon that consent to construct a Project different from the Project envisioned by the 1977 Agreement.[461]

The Court noted because it held the operation of the Provisional Solution was an internationally wrongful act, violating the express provisions of the 1977 Agreement, it was not necessary for the Court to address the questions of material breach and international environmental law raised by Hungary.[462]

In finding the Provisional Solution constituted a unilateral act and therefore not authorized by the 1977 Agreement, absent Hungary's consent, the Court promoted the resolution of the dispute by removing from subsequent negotiations Slovakia's contention the 1977 Agreement embodied Hungary's consent for a unilateral project, and the Provisional Solution was broadly consistent with the specifications of the Agreement. If Slovakia thus wished to continue operation of the Project, it must agree to modify the Provisional Solution so as to comport with the original intention of the parties to create a single, indivisible, jointly owned and operated system of works.

In basing its finding on the unilateral nature of the Project and the lack of consent, the Court passed an opportunity to render a determination as to whether the activities undertaken by Slovakia with respect to environmental assessment and precaution were sufficient to comply with its obligations under Articles 15, 19 and 20 of the Agreement. The Court, however, may have rightly concluded it lacked the necessary expertise to make a determination as to the sufficiency of Slovakia's activities, and this assessment was best left to the parties in subsequent negotiations, with the possible assistance of a third party, which possessed the necessary expertise.

Notably, the Court did further the potential role of law in other CEE transboundary water disputes by confirming activities relating to shared watercourses require the consent of all other parties concerned, and should be undertaken consistent with the requirements of the principle of equitable utilization. In particular this finding is relevant for the resolution of the Danube River Gauntlet dispute, and certain aspects of the Black Triangle dispute.

3. *The doctrine of justifiable countermeasures as a justification for the Provisional Solution*

Slovakia argued that in the event the Provisional Solution was to be considered an unlawful act, it would be permissible as a justifiable countermeasure. In the view of Slovakia, the Provisional Solution was justified as a countermeasure since: (1) Hungary had committed a prior illicit act by breaching its treaty obligations to construct the Nagymaros works and participate in the completion of the Dunakiliti and Gabčíkovo works;[463] (2) the Provisional Solution was an appropriate response to Hungary's breach and subsequent illegal attempt to terminate the 1977 Agreement;[464] and (3) the Provisional Solution was a proportionate response to Hungary's illicit acts as it only sought to achieve the object and purpose of the Agreement and no more.[465]

Hungary countered with the argument the Provisional Solution did not constitute a lawful countermeasure to Hungary's termination of the 1977 Agreement, as Hungary's termination was lawful and therefore not subject to countermeasures.[466] In the event Hungary's termination was considered unlawful, Hungary argued the Provisional Solution would not represent a lawful countermeasure as: (1) Czechoslovakia had threatened to implement the Provisional Solution as early as 1989; (2) Czechoslovakia refused to provide Hungary appropriate information about the details of the Provisional Solution; (3) the consequences of the Provisional Solution were disproportionate to Hungary's suspension and abandonment of the Nagymaros works and its share of the Gabčíkovo works as it threatened Hungary's water supply and thus the right to life of many Hungarians; and (4) the Provisional Solution was not a necessary response to Hungary's actions, especially in light of the fact it impaired Hungary's essential interests.[467]

In addressing the question of whether the operation of the Provisional Solution could be considered a lawful countermeasure on the part of Czechoslovakia, the Court noted in order to constitute a lawful countermeasure, the action taken must be in response to a previous international wrongful act of another state, must be directed against that state, must occur after the injured state has called upon the offending state to cease its wrongful conduct, must be proportionate to the injury suffered and must be intended to induce the offending state to comply with its international obligations.[468]

The Court found although the Provisional Solution was in response to the international wrongful act of Hungary, was directed against Hungary and was implemented only after Czechoslovakia had called upon Hungary to implement its treaty obligations, the implementation of the Provisional Solution was not a proportional response to Hungary's breach of its treaty obligations.[469] Specifically, the Court found Hungary possessed a right to equitably use and share the natural resources of the Danube,[470] and by denying Hungary this right, through its diversion of the Danube River, Czechoslovakia caused harm to the ecological interests of Hungary in the riparian areas of the Szigetköz, which were disproportionate to the harm suffered by Czechoslovakia as a result of Hungary's breach of its treaty obligations.[471]

In finding the operation of the Provisional Solution did not constitute a legitimate countermeasure, the Court furthered the resolution of the dispute by reaffirming the wrongful nature of Hungary's noncompliance with the 1977 Agreement, and by reaffirming the wrongful nature of Czechoslovakia's unilateral implementation of the Provisional Solution, thus requiring the parties to further negotiate to reach an agreement on the operation of a series of works consistent with the objectives of the Agreement. Moreover, the Court, in rejecting Slovakia's countermeasure argument, highlighted Hungary's right to equitably use and share the benefits of the Danube River, and legitimized Hungary's claim of environmental harm suffered in the riparian areas of the Szigetköz. The Court thus reaffirmed in negotiating the exact nature of the series of works to be implemented, Hungary may validly seek the protection of its environmental interests.

The Court also furthered the role of international law in promoting the resolution of other CEE transboundary environmental disputes by clearly setting forth the standard applicable to the claim of legitimate countermeasures, and by accepting that in certain circumstances environmental harm may be disproportionate to the harm suffered as a result of the original wrongful act, thus delegitimizing the countermeasure.

D. The legal consequences arising from the determination of the legal issues before the Court

Slovakia and Hungary agreed in the first phase of litigation that the Court would not be called upon to quantify any reparation or compensation due, but rather to decide the rights and legal obligations of each party and then permit the parties to attempt to reach an agreed upon set of modalities for executing the Court's Judgment. Both parties did, however, set forth the legal basis upon which the Court could establish the international responsibility of the other party and principles governing the amount of reparation reasonably due. Slovakia, moreover, provided the Court with a preview of the types and amount of damages it would seek from Hungary.

Hungary, in its initial pleading, sought to have the Court declare Hungary's suspension and abandonment of the Nagymaros works and its participation in the Gabčíkovo works legitimate, the 1977 Agreement properly terminated – thus releasing Hungary from the duty to perform any of the obligations set forth in the Agreement,[472] and Slovakia responsible for the harm caused to Hungary by the construction and operation of the Provisional Solution.[473] Hungary also sought to have the Court declare Slovakia was responsible for the wrongful conduct of the former Czechoslovakia as it sought to adopt Czechoslovakia's rights and obligations under the 1977 Agreement;[474] while Hungary might be obligated to settle accounts with Slovakia for work completed under the 1977 Agreement, it would not be obligated to compensate Slovakia for any work relating to the Provisional Solution; and as the Dunakiliti dam, by-pass canal and Gabčíkovo series of locks were jointly owned by Hungary and Slovakia, which were unlawfully appropriated by Slovakia to implement the Provisional Solution, the parties would have to reach agreement on compensation.[475] Finally, Hungary noted subsequent to the decision of the Court any settlement would have to take into account damages suffered by Hungary as a result of Czechoslovak breaches of the Agreement prior to 31 December 1992, damages suffered by Hungary as a result of the operation of the Provisional Solution, and damages suffered as a result of Slovakia's breach of Article 4 of the Special Agreement requiring the implementation of a temporary water management regime.[476]

In subsequent pleadings, Hungary sought to have the Court order the cessation of the operation of the Provisional Solution, the restoration of the Danube River and the affected region to its original condition, and the reparation to Hungary by Slovakia of any damages associated with an inability to restore the river and region to its original condition.[477] In calculating appropriate reparation, Hungary urged the Court to take account of moral damage associated with the uncertain conditions of health and livelihood faced by current and future Hungarian generations and to develop a methodology and criteria to be used by the parties to calculate compensation associated with future risk and to measure nonuse environmental values.[478] Hungary also sought to persuade the Court its determination of legal consequences should be guided by the evolution of international environmental law, especially the principles of sustainable development, precautionary action, environmental assessment and the protection of groundwater resources.[479]

Slovakia sought to have the Court declare the 1977 Agreement remained in force and Hungary acted illegitimately when it suspended and subsequently abandoned construction of the Nagymaros works and participation in the Gabčíkovo works and when it purported to terminate the Agreement.[480] Specifically, Slovakia sought to have the Court declare Hungary breached its international legal obligations to comply with the provisions of the Agreement, and in order to 'wipe out all the consequences of the illegal act',[481] Hungary was obligated to cease its illegal conduct immediately, in particular, its preparations for dismantling the coffer dam constructed at Nagymaros and to make restitution to Slovakia by performing its obligations under the Agreement.[482] To cover damages not remedied by restitution, Slovakia urged the Court to instruct Hungary to pay compensation for costs incurred for the interim protection of the Gabčíkovo–Nagymaros Project structures, costs of maintaining the old Danube River bed in navigable condition from 1990–92, losses associated with the inability to use the bypass canal from 1990–92, the construction costs of the Provisional Solution and losses in the areas of navigation, flood protection and electricity production caused by the failure to construct the Nagymaros works.[483]

Slovakia also asked the Court to reject Hungary's calls for ordering the cessation of the operation of the Provisional Solution and for providing restitution on the bases that the acts taken by Slovakia were consistent with the 1977 Agreement and with general principles of international law, and as the successor to the former Czechoslovakia with respect to the 1977 Agreement, Slovakia was entitled to act according to the provisions of the Agreement.[484] Slovakia also requested the Court disregard

Hungary's claims associated with the Temporary Water Management Regimes provided for in the Special Agreement, as this issue was not directly related to the 1977 Agreement and therefore not properly before the Court.[485]

Recognizing the question of the legal consequences of its above determinations represented an opportunity to render a prescriptive declaration, the Court sought to frame its answer to this question in a manner most likely to promote a resolution of the dispute consistent with those determinations. In this regard, the Court first noted as the 1977 Agreement remained in force as a treaty *lex specialis*, it was the primary source of law governing the future relationship between Slovakia and Hungary, with the secondary sources being other relevant conventions to which Slovakia and Hungary were parties and the rules of general international law, in particular, rules of state responsibility.[486] The Court then determined as the Agreement had not been fully implemented, and as the acts and omissions of the parties had contributed to creating a factual situation different from that envisioned by the Agreement, the current factual circumstances should be 'placed within the context of the preserved and developing treaty relationship, in order to achieve its object and purpose in so far as that is feasible'.[487]

The Court then determined as the 1977 Agreement provided for the creation of a joint regime of ownership and operation of the Gabčíkovo–Nagymaros Project, Slovakia and Hungary should exercise a joint ownership interest in the Project and should share equally in its operational expenses and its benefits, unless they agreed otherwise.[488] Noting Article 10 paragraph 2 of the Agreement provides works located on the territory of one of the contracting parties shall be owned and operated by that party, the Court indicated in making this determination it was seeking to transform the Provisional Solution from its *de facto* status into a treaty based regime.[489] The Court further supported this determination by noting it appeared from the evidence submitted to the Court the Provisional Solution could be operated in a manner that would accommodate both the economic and environmental needs of Slovakia and Hungary. Such operation would comply with Article 9 of the Agreement, which provides for the equal participation of the parties in the Project and is consistent with the international norm of equitable and reasonable utilization of shared water resources as articulated in the Convention on the Law of the Non-Navigational Uses of International Watercourses.[490]

In calling for a joint operation of the project, the Court reminded the parties in addition to the production of energy, the Project was also intended to improve navigability of the Danube River, increase flood control, regulate the level of ice-discharge and protect the natural environment. In the Court's view, none of these objectives holds absolute priority over the other objectives, and in order to achieve these objectives, the parties must fulfill their various 'obligations of conduct, obligations of performance, and obligations of result'. In particular, the Court noted the parties have a continuing and evolving obligation to take into consideration current standards relating to environmental protection, as prescribed by Articles 15 and 19 of the 1977 Agreement.[491] The Court then reminded the parties in fulfilling their obligation to protect the environment, they must consider the new norms and standards developed in the past twenty years, which seek to balance economic development and environmental protection as expressed in the concept of sustainable development. As such, the Court determined Slovakia and Hungary were required to reconsider the environmental effects of the operation of the Gabčíkovo plant and to settle on an agreed amount of water to be released into the old river bed and into the respective side arms.

The Court then reminded Slovakia and Hungary it could not be called upon to set an operating regime for the Project, rather the parties themselves must engage in meaningful negotiations consistent with the rule of *pacta sunt servanda* to 'find an agreed solution within the cooperative context of the Treaty'.[492] In attempting to reach an agreed solution, the parties are bound by an obligation to negotiate in good faith for the purpose of realizing the objectives of the 1977 Agreement.[493] The Court then noted during the course of the dispute the parties at various times called upon the assistance of the European Commission and although this assistance did not resolve the dispute – in large part because of the intransigence of the parties, the involvement of the Commission in negotiations seeking to give effect to the Judgment of the Court would be evidence of their good faith approach to those negotiations.[494]

Turning to the question of reparations associated with the internationally wrongful acts committed by Slovakia and Hungary, the Court noted the governing standard set forth in the *Factory at Chorzów* case which required reparation must be made with the intent of 'as far as possible, wip[ing] out all the consequences of the illegal act and reestablish[ing] the situation which would, in all probability, have existed if that act had not been committed'.[495] In the view of the Court, such reparation would properly take the form of cooperation between Slovakia and Hungary for the purpose of equitably and reasonably utilizing and protecting the shared water resources of the Danube River according to the multi-purpose single unit regime established by the 1977 Agreement. To achieve this cooperation, the Court recognized the parties might agree to maintain the Čunovo works, with some modification of its operating regime, and agree not to build the Nagymaros works.[496]

Specifically addressing the question of damages, the Court found as the sole successor to the rights and obligations of the former Czechoslovakia arising from the 1977 Agreement, Slovakia was obligated to pay compensation not only for its own wrongful conduct, but also for the wrongful conduct of the former Czechoslovakia. Slovakia was also entitled to be compensated for any damages owed to the former Czechoslovakia as a result of Hungary's unlawful acts.[497]

The Court then recalled while the parties had not requested the Court to determine the quantum of damages due, they had asked the Court to indicate upon what basis such damages should be paid. The Court observed as both parties committed intersecting internationally wrongful acts, 'the issue of compensation could satisfactorily be resolved in the framework of an overall settlement if each of the parties were to renounce or cancel all financial claims and counter-claims'.[498] The Court also observed the separate settlement of accounts for Hungary's failure to contribute to the final construction of the Gabčíkovo works was a separate issue from compensation for Hungary's internationally wrongful acts and was to be resolved in accordance with the 1977 Agreement. Moreover, the Court noted as Hungary was to share in the operation and benefits of the Čunovo works, it must compensate Slovakia for a proportionate share of the construction and operation costs.[499]

In reaffirming the 1977 Agreement remained in force and the relationship between the parties was primarily governed by the Agreement, and secondarily governed by the conventions to which Slovakia and Hungary were parties, as well as customary international law, the Court furthered the resolution of the dispute by establishing a clear hierarchy with respect to the principles, norms and rules to be used to guide the subsequent negotiations in the implementation phase.

In placing the works of the Provisional Solution within the framework of the treaty, and by granting to Hungary an equal share of those works, so long as Hungary made a contribution equal to its share of the construction costs, the Court furthered the resolution of the dispute by legitimizing the Provisional Solution, and thereby transforming it from a wrongful act into a lawful physical structure. By so doing, the Court removed from the subsequent negotiations any claims by Hungary as to the illegitimacy of the Provisional Solution, and established a clear basis for negotiating the physical construct of a joint system of works – that of the current works of the Provisional Solution.

In characterizing the 1977 Agreement as an evolving treaty, capable of incorporating the physical works of the Provisional Solution as well as the legal developments relating to the protection of the environment, the Court furthered the resolution of the dispute by preserving the treaty as the legal basis for establishing the nature of the regime governing the operation of the project. Moreover, the incorporation of new environmental norms into the Agreement established a legal basis from which Hungary could request modifications in the current works of the Provisional Solution as necessary to protect its environmental interests.

In declaring parity among the various objectives of the Agreement, the Court furthered the resolution of the dispute by denying the parties a license to debate which objectives should be considered paramount in designing an operating regime for the project, or a modification of the existing works. The Court thereby preserved the integrity of objectives that might be perceived as second tier objectives, in particular the objective of environmental protection. Notably, the Court identified the role of the notion of sustainable development in assisting the parties in negotiating a balance between the objectives of economic development and environmental protection, but interestingly recognized it only as a concept, apparently declining to characterize it as possessing the status of a rule or principle of international law.

By informing Slovakia and Hungary it could not be called upon in subsequent proceedings to establish an operating regime for the parties, the Court sought to promote a resolution of the dispute by signaling the role of adjudication had been maximized, and the parties had to negotiate in good faith to reach a final solution within the parameters of the decision of the Court. If the parties were in disagreement as to specific facts, or as to how best apply the decision of the Court, the parties were better served by employing the assistance of a third party, rather than an arbitral or adjudicatory body. The Court, of course, recognized if the parties disagreed as to the substance or applicability of certain norms then the Court could properly be called upon to adjudicate these issues.

Finally, in finding each party was responsible for a degree of reparation and damages to the other party, and recommending the parties settle these claims by agreeing to cooperate in the development of a joint operating regime for the project, and renounce their claims and counterclaims, the Court sought to further the settlement of the dispute by removing from the subsequent negotiations the issue of damages. In particular, the Court likely felt any negotiations would be more fruitful if the parties focused on the need to modify the current works, if deemed necessary, and develop a joint operating regime, and negotiations over the amount of reparations due and damages owed would detract from the efforts of the parties to agree upon a sustainable solution.

In seeking to promote a resolution of the dispute by making the above determinations, the decision of the Court will potentially affect the resolution of other CEE transboundary environmental disputes. The determination that a treaty can continue to exist despite its avowed termination by one of the parties, may cause some CEE states in future disputes to reconsider their arguments that certain treaties are to be considered terminated, and may prompt them to use these treaties as a basis for negotiating a resolution of the dispute. The effect of this determination may be limited, however, given it is a somewhat unconventional interpretation of the animate nature of treaties.

The determination treaties possess a capacity to evolve and incorporate factual situations occurring outside the purview of the treaty, as well as newly developed norms of international environmental law, may promote the resolution of some CEE disputes by providing the disputing parties with a clear indication they should seek to negotiate a resolution of the dispute on the basis of the treaty, as modified by relevant norms, and they should incorporate the deviant factual situations into the treaty. The role of this determination may be inhibited, however, by the novelty of the Court's decision, which will likely be criticized in subsequent writings of publicists.

Notably, the role of the principle of sustainable development has been highlighted by the Court and will likely play a more prominent role in CEE disputes, in particular the Black Triangle, Silesian Coal Basin and Danube River Gauntlet disputes. Unfortunately, the Court declined to provide a precise definition of the concept of sustainable development, a point of much contention among some CEE states.[500]

Finally, the Court's somewhat unusual suggestion that Slovakia and Hungary involve the European Commission in the implementing phase of the dispute should help CEE states focus on the proper role of the various dispute resolution mechanisms and signal the importance of relying upon more than one mechanism as the dispute progresses through the dispute resolution process.

6
The Implementation Phase: Back to the Negotiating Table and Possibly Back to the Court

The decision of the ICJ signaled the end of the resolution phase of the Gabčíkovo–Nagymaros Project dispute, with the implementation phase beginning with the efforts of Slovakia and Hungary to negotiate an operating regime consistent with the parameters contained within the order of the Court. As will be discussed in this chapter, when the parties eventually agree upon an operating regime, the implementation phase will further entail the development of mechanisms for the enforcement and verification of that regime, and for any necessary modification of the regime for environmental or other reasons. The efforts of the parties to negotiate a joint operating regime and develop mechanisms for enforcing, verifying and modifying that regime will be affected by the interests of sub-state actors and interested third parties, which may in some instances offer their good offices or serve as mediators or conciliators. The efforts of the parties to implement the agreement will also be influenced by the same situational circumstances that influenced the dispute formation, pre-resolution and resolution phases of the dispute.

I. The preliminary efforts of Slovakia and Hungary to negotiate a joint operating regime

Immediately following the Court's decision, Slovakia and Hungary entered into negotiations in an effort to agree upon how best to implement the decision. Hungary acted on the broad intent of the decision of the Court to propose initially a return of the river to its natural watercourse, and a zero option plan with each state's claims for compensation canceling the other's.[501] Slovakia, invoking the specific findings of the Court, rejected this proposal and insisted damages and construction costs should be treated separately and Hungary should pay the costs of the work carried out by Slovakia after Hungary's breach of the 1977 Agreement.[502]

Slovakia then proposed on 10 November 1997 that Hungary construct the Nagymaros works or compensate Slovakia for its inability to generate peak performance at the Gabčíkovo plant. On 24 November, Hungary rejected this proposal and countered with an offer to construct two small hydroelectric plants in place of the Nagymaros works and the agreement to respectively waive claims for damages. Hungary also insisted the negotiations address navigation, energy production and sewage purification as well as environmental issues.[503]

When the Slovak Prime Minister refused to attend meetings scheduled in Budapest, in protest at Hungarian statements relating to the Hungarian minority in Slovakia, Austria offered its good offices and hosted a meeting between the Slovak and Hungarian Prime Ministers in Vienna on 15 December 1997, where they agreed in principle to reach a solution by 15 March 1998, and avoid the necessity of resubmitting the case to the ICJ.[504]

Subsequent to a meeting between the heads of delegations in Bratislava on 12 January, the head of the Hungarian delegation indicated Slovakia had withdrawn its demand the Nagymaros works be constructed and appeared willing to settle for an arrangement whereby Dunakiliti would become operational and where Hungary would build a second hydroelectric plant between Szob and Budapest, possibly at Pilismarot.[505] On 14 January 1998, however, the head of the Hungarian delegation informed the parliamentary Environment Commission Hungary would not consent to the construction of a second hydroelectric plant until sufficient environmental and economic studies were conducted.[506] Then on 19 January 1998, the Slovak and Hungarian delegations entered into a self-described 'theoretical agreement' that Hungary would not require Slovakia to decommission the Čunovo dam so long as Slovakia agreed to a joint operating regime that would divert a sufficient amount of water into the old river bed to protect Hungarian environmental interests and permit the creation of a resort oriented reservoir at the Dunakiliti dam site.[507] Hungary and Slovakia also established joint monitoring teams to assess the environmental effects of the continued operation of the Čunovo dam.[508]

Working under their theoretical agreement, on 27 February 1998, Slovakia and Hungary initialed a protocol stipulating prior to 25 March 1998, the states' two Prime Ministers would sign an agreement providing for the joint operation of the project, as well as the construction of an additional dam within Hungary.⁵⁰⁹ After signing this agreement, the Hungarian government came under significant pressure by elected officials, environmental groups and the general public not to construct a second dam, and to complete an environmental study of any action contemplated under the joint operating regime. As a result of this pressure, the Hungarian government informed the Slovakian government it may take as many as four years to work out a final resolution of the dispute and it might take another four years to implement that solution.⁵¹⁰

In response, Slovakia demanded Hungary comply with the decision of the Court and threatened to return the dispute to the Court by the 25 March 1998 deadline.⁵¹¹ Hungary replied it would not sign the agreement until after it had conducted adequate environmental studies, which it envisioned would take until the end of 1998, but Hungary was willing to continue to negotiate on other aspects of the regime for joint operation.⁵¹² Slovakia then adopted a Parliamentary resolution on 1 April 1998, urging Hungary to comply with the decision of the Court and threatening to return the dispute to the Court if the protocol was not signed by July 1998.⁵¹³ The resolution also called upon the European Parliament and other EU authorities to appeal to Hungary to abide by its international legal obligations and sign the protocol.⁵¹⁴ The Hungarian Cabinet then announced due to environmental reasons it would not consider the possibility of constructing the Nagymaros works, but would conduct an environmental evaluation of a possible dam at Pilismarot.⁵¹⁵

In the summer 1998, Hungary held parliamentary elections, which led to a change in government, with the new government rejecting the idea of constructing a second dam and insisting an EIA be conducted on the continued operation of the Provisional Solution.⁵¹⁶ Frustrated by this change of course, on 3 September 1998, Slovakia exercised its option under the Special Agreement,⁵¹⁷ and filed a submission with the ICJ requesting it to direct Hungary to comply with the Court's initial decision. Specifically, Slovakia requested the Court to find Hungary bore responsibility for the failure of the parties to agree on the modalities for executing the first Judgment, to direct Hungary to negotiate in good faith, and to direct Hungary to conclude a Framework Agreement with Slovakia, which would enable the parties to achieve the objectives of the 1977 Agreement.⁵¹⁸ As of the time of publication, the Court had set a schedule for receiving the written pleadings for the parties, but had not rendered a decision on the matter. Since the time of filing, however, Slovakia too conducted parliamentary elections, wherein the coalition of opposition parties took control of the government and appointed Mikulas Dzurinda of the Slovak Democratic Coalition as Prime Minister.[519] The new government and Hungary have also restarted negotiations.[520]

II. Situational circumstances affecting the negotiation and implementation of a joint operating regime

In considering whether to abandon the Project or to adopt a Framework Agreement providing for the operation of the Project in a manner that will provide Hungary a share of the energy entitlement and encompass measures necessary to protect the environment, Slovakia will likely calculate whether the costs of environmental measures and transfer of energy revenue to Hungary outweigh the economic and political benefits of continued operation of the Project.

This calculation may involve ecological determinations relating to the actual ability and cost to mitigate environmental harm caused by operation of the Project; the economic benefit to be derived from the generated electricity, minus Hungary's share, and whether Slovakia's transforming economy requires this additional electricity. The calculation would as well include an assessment of the national political benefits to be derived from the operation of the Project versus the national and international political costs arising from domestic and international opposition to its continued operation.[521] The continuing political cost of the Project's impact on the Hungarian minority population resident in the region will also play a role – in this case, if the Project is perceived to bring economic development to the region, its termination will be assigned a negative value, whereas if it is perceived to curtail economic development, then its termination will be assigned a positive value.

In conducting its own cost–benefit analysis, Hungary will likely seek to determine whether the possibility of Slovakia actually being able to mitigate environmental harm and the economic benefit derived from its energy entitlement are sufficient to warrant discontinuing alternative political avenues for persuading Slovakia to abandon operation of the Project.

This calculation may involve ecological determinations as to whether it is actually possible to mitigate the negative impacts on the environment, and whether Hungary is genuinely concerned with those impacts. Hungary will also likely consider the economic value of its energy entitlement, whether it can use that energy to offset its settlement with Austria, or whether it requires the energy for domestic consumption, and whether it is possible for Hungary to compensate Slovakia for agreeing to cease operation of the Project. Hungary too will balance the political cost of domestic opposition to the continued operation of the Project and the international political cost of continuing to raise the issue on the international stage once it has been addressed by the ICJ. Finally, Hungary will be sensitive to the human rights concerns of whether the continued operation is beneficial or detrimental to the national Hungarian minority living in the Gabčíkovo region.

The ultimate ability of Slovakia and Hungary to agree upon a joint operating regime consistent with the Court's decision and to implement that compromise will in large part be affected by the relationship of this issue with other issues of importance to Slovakia and Hungary, including the conclusion of a number of planned economic cooperation agreements, the construction of a major highway between Bratislava and Györ, and the implementation of the Slovak–Hungarian friendship treaty.[522] The resolution of the dispute will also be affected by the domestic democratic transformations within Slovakia and Hungary, in particular whether the liberal coalitions elected to power in the summer and autumn of 1998 will remain in power,[523] the relationship between the Hungarian government and Hungarian environmental and human rights NGOs, and the political influence of the local Hungarian population affected by the project. Finally, the dispute will be affected by the political feasibility of constructing additional works and the extent to which the EU maintains its involvement and brings constructive pressure upon the parties to resolve their dispute within the parameters set by the ICJ.

III. Sub-state actors and interested third parties influencing the negotiation and implementation of a joint operating regime

In addition to the above mentioned circumstances, the negotiation of a regime for joint operation or possible termination of the Project will be influenced by a variety of sub-state actors and interested third parties. Most importantly, the continued operation of the Project will be influenced by the recent change in national government in Slovakia. Throughout the phases of the dispute resolution process, the Slovak government was led by officials supportive of the Gabčíkovo–Nagymaros Project. The primary political influence during this time has been Vladimír Mečiar, who aside from a brief time in opposition, served as Slovakia's Prime Minister since its independence in 1993 until elections in the autumn of 1998. It is expected that the new liberal democratic government of Slovakia will distance itself from the interests associated with the completion of the Project, but will not wholly abandon it.

The Hungarian national minority living in Slovakia will also play a significant role in that it is organized and vocal and has attracted the attention of important international actors such as the EU. If the Hungarian national minority is significantly agitated by the continued operation of the Project, or by its treatment by the Slovak government, it could indirectly impede the negotiation process, such as when the September 1997 meeting between the Slovak and Hungarian Foreign Ministers was canceled by Slovakia because of Hungarian complaints about the mistreatment of the Hungarian minority in Slovakia.[524] Similarly, the Slovak water and construction interests are likely to continue to exercise significant political authority and influence, and various development oriented Ministries are likely to keep control over key decisions affecting the environment, such as the Ministry of Agriculture's continued control over the administration of water resources.

Moreover, environmental NGOs have played an irregular role in influencing Slovak policy as a result of internal fragmentation and a lack of expertise, coupled with a tendency to react to specific projects rather than to affect the development and implementation of legislation.[525] Until recently the NGOs have also lacked support from the general public, which was slow to recognize the causes and consequences of environmental pollution.[526] Hungary will thus find that although the decision of the Court requires the consideration of environmental interests in the development of an operating regime, it will not yet be able to significantly rely upon environmental actors in Slovakia to influence the Slovak government. The area where environmental actors may be most successful, though, is in encouraging Slovakia to engage in additional environmental monitoring prior to negotiating a joint operating regime.[527]

In Hungary, the political pendulum has been swinging back and forth between the liberal democrats who attained power in the early 1990s, the reformed communists who held power in the mid 1990s, bringing Gyula Horn back as the Prime Minister, and now as of the summer of 1998 returning power to the liberal democrats and Prime Minister Premier Victor Orban. While in power, the reformed communists – who had adopted the 6 February 1989 Protocol, which accelerated the construction of the Nagymaros and Gabčíkovo works by 15 months, sought to use the decision of the Court to reach an agreement with Slovakia which would implement as much of the Project as possible, with little attention to environmental concerns. This was confirmed by the Hungarian government's willingness to initial the Protocol calling for the construction of a second dam, although the decision of the Court clearly did not require Hungary to construct a second dam.[528]

The power of the reformed communists was constrained on this matter, however, by public opinion and the ability of the other political parties to express effectively their concerns through the Hungarian democratic process, and to eventually remove the reformed communists from power. In this regard, on 28 February 1998, 30,000 people protested in downtown Budapest against the construction of a new dam on the Danube,[529] while on 14 March 1998 another 1,000 environmentalists organized a similar demonstration.[530] These demonstrations coupled with an active role by opposition politicians have swayed Hungarian public opinion, such that according to a Gallup poll, 62 per cent of the Hungarian population is opposed to the construction of a new dam on the Danube River, while 49 per cent believe the construction of such a dam would serve the interests of Slovakia, with only 15 per cent believing it would serve the interests of Hungary.[531] Notably, some opponents of the second dam have argued it would cost an additional one to three billion dollars.[532]

Reflecting public sentiment, the Hungarian Parliament adopted a resolution requiring the former government not to adopt the Protocol until it has been ratified by the Parliament.[533] The Parliament further forced the Prime Minister to agree to establish a cross party committee of all six Hungarian parliamentary parties to set a strategy for the negotiations with Slovakia.[534] This committee also includes the Ministers of the Environment, Transportation, and Water Management.[535]

The new government has expressly declared its opposition to the construction of a new dam and has publicly committed itself to reaching a resolution of the matter in a manner which ensures the protection of the environment.[536] Moreover, the National Election Committee determined in April 1998 the Hungarian population could hold a nonbinding referendum on the question of 'Do you agree with the building of a dam on the Hungarian part of the Danube?' if 200,000 signatures were collected by July 1998.[537]

Concerning the role of interested third parties, in 1998 Hungary was invited to begin negotiations toward joining both the European Union and became a member of NATO in March 1999 while Slovakia was invited to join neither organization. Given the high political and economic stakes associated with EU and NATO membership, Hungary will likely seek to avoid having any dispute concerning the implementation of the Court's decision detract from membership negotiations. Hungary will also most likely implement any agreed operating regime in good faith. Slovakia, while less concerned with the perception of the EU and NATO member states in the short term given its rejection, will be concerned with their perception in the long term as it will likely continue to seek EU and NATO membership. Thus although Slovakia will be under less third party pressure to compromise on establishing an operating regime, it will likely seek to implement such a regime in good faith.[538]

If the EU or NATO member states become involved in the implementation phase of the dispute, beyond the EU's role as potential mediator, their primary interests will be to promote stability in the region,[539] with their secondary interest being environmental protection. Interestingly, although the Slovak Parliament called upon the European Parliament and European Commission to urge Hungary to comply with the decision of the Court, Slovakia has to date rejected a request by Hungary that the European Commission be invited to assist with the negotiations relating to the Protocol.[540]

IV. Prospects for the enforcement and verification of a joint operating regime

When Slovakia and Hungary eventually settle their intersecting claims for compensation and damages and agree upon an operating regime for the Project, they will need to establish a series of mechanisms for enforcing and verifying the implementation of that regime. Given the sensitive environmental nature of the region and the prominent role it played in the arguments of the parties and the decision of the Court, the parties may also have to develop a mechanism for modifying the regime to respond to any threats it may pose to the environment.

As the decision of the Court provides for the joint operation of all Project related facilities, including the Čunovo and Gabčíkovo works, Hungary ultimately will be in a favorable position to verify the amount of water released into the old river bed and any special dispensations for environmental reasons. Hungary will also share in the physical control of the water diversion mechanisms and thus be able to ensure the enforcement of the agreed upon regime. Similarly, if for some reason Hungary should seek to divert additional water into the old river bed, Slovakia would be able to ensure the protection of its interests in maintaining the specified flow in the power canal.

While implementing the joint operating regime, the parties will likely be faced with such issues as the precise composition of personnel charged with operating the Project, the legal arrangements for joint ownership of the Project, the creation and operation of an oversight body, the involvement of third parties in the operation and oversight of the regime, the establishment of intervals for the reevaluation of the regime and means for incorporating environmental information and requirements into the regime. The parties will also have to determine the extent to which the regime should be modified to respond to reasonably perceived threats to environmental health and safety occasioned by the operation of the Project.

In seeking to answer these questions, the parties may be guided by their international legal obligations to give domestic effect to their bilateral arrangements, to share relevant information, to provide notice of activity that may affect the implementation of the agreement, to consult on disagreements concerning the implementation of the agreement and to resolve peacefully any such disputes. The parties may also call upon various international environmental law principles, as articulated by the Court, to ascertain whether certain reasonably perceived threats to the environment require a modification of the operational regime.

Conclusion to Part II

As noted in the introduction, the purpose of an extensive examination of the Gabčíkovo–Nagymaros Project dispute was to further the understanding of the role of international law and the legal processes of inquiry, mediation and adjudication within the regime of transboundary environmental protection. In particular, the case study demonstrates how specific legal principles and various aspects of the legal process operate to address the multifarious elements of a transboundary environmental dispute in such a manner so as to allocate the respective rights and responsibilities of each party as necessary to produce a formal decision that can form the basis for the settlement of the dispute. This case study also furthered the understanding of the influence of state and sub-state actors and interested third parties on the dispute resolution process and the role of law within that process.

Interestingly, the role of international law throughout the process of resolving the dispute was affected by the fact the dispute has its origins in a number of socio-political and economic circumstances that have evolved substantially in the last twenty years. Since the conclusion of the 1977 Agreement to alter the course of the Danube River, both Slovakia and Hungary have undergone a transformation from totalitarian regimes with command economies to market-based democratic regimes. The former Soviet Union, Warsaw Pact and CMEA, all three of which played a role in the decision to conclude the Agreement, have dissolved, and there are substantial questions as to whether the hydroelectric energy and improved navigation resulting from the construction of the Gabčíkovo–Nagymaros Project are still of significant value to Slovakia and Hungary. Slovakia asserts that not withstanding these dramatic changes there is still a need for the Project, while Hungary asserts the Project is a relic of another time and should be abandoned. But for the socio-political and economic transformations, this dispute would not have occurred and certainly would not have become the subject of international discourse with the involvement of the EU and ICJ.

Ecological circumstances, as highlighted by a number of sub-state actors, have also played an important part in the resolution of this dispute. The simple evolution of environmental technology coupled with an emerging public environmental ethos raised substantial questions as to the effect of the Project on the regional ecosystem and the ability of Slovakia and Hungary to mitigate any adverse consequences. In the end, because the threats to the environment were so severe, the public objection so vociferous, the bilateral relations between Slovakia and Hungary so important and the desire to participate fully in the international community so pervasive, Slovakia and Hungary agreed to reach a political resolution of this dispute guided by the principles of international law.

In the initial stages of the dispute, international law obligated the parties to seek to resolve their dispute peacefully, to consult with each other in good faith, and to share relevant information. As a result of the perceived noncompliance with these obligations, the parties sought to establish a means of inquiry, and subsequently conciliation, with the assistance of the European Commission. Although unable to bring about a resolution of the dispute, the mechanisms of inquiry and conciliation assisted the parties in articulating the exact nature and parameters of their dispute and in agreeing to submit the dispute to arbitration or adjudication. In particular, the EU strongly influenced the parties to continue to rely upon the principles of international law as they sought to resolve this transboundary environmental dispute and to make use of a formal dispute resolution mechanism. The international legal process of adjudication was also used in part because of the general evolution of the Slovak and Hungarian domestic political systems towards the rule of law, the growing importance of domestic environmental law to resolve domestic resource disputes and the perception of the ICJ as an adequate dispute resolution mechanism.

Once submitted to the ICJ, the question of the capacity of international law to promote a resolution of the dispute became relevant. In the case of the Gabčíkovo–Nagymaros Project dispute, the Court was able to employ principles of international law relating to treaty termination, environmental protection, state responsibility and state liability in a manner that narrowed the dispute from eleven possible outcomes, with each party advocating the most extreme outcomes, to a single outcome – that of preserving the fundamentals of the treaty regime while ensuring environmental protection. The role of the Court in narrowing the possible outcomes of the dispute cannot be underestimated, as throughout the time the case was before the Court, the parties continued to negotiate intermittently, but were unable to reach any consensus as to which regime was most consistent with their political and economic interests as well as international law.

Notably, the Court was also able to interpret the law in a manner so as to provide each party with certain essential elements necessary for them to reach an agreement, which were not forthcoming from the other party. For instance, the decision of the Court provides Hungary will share in the operation of the Čunovo dam and Slovakia will be entitled to operate the Provisional Solution. Had Slovakia offered to permit Hungary to share in the operation of the Čunovo dam, which controls the amount of water diverted into the power canal, Hungary would likely have been willing to agree to some form of operation of the Provisional Solution. Similarly, had Hungary agreed in principle to the operation of the Provisional Solution, Slovakia would likely have been willing to permit its joint operation.

Although international law and the legal process were effectively able to guide the parties toward a structured and peaceful means of resolving the dispute, and eventually to narrow the options available to the parties, the principles of international law and the legal process were not able to, and should not reasonably be expected to, dictate the exact nature of the operating regime best suited to balancing economic and environmental interests. The exact nature of the joint operating regime to be established by Slovakia and Hungary will be properly dictated by the nonlegal factors that have influenced the dispute throughout its four phases.

Relying on these observations and the observations relating to the secondary case studies examined in Part I, Part III will deduce a number of general conclusions regarding the role of international law and the legal process within the international regime of transboundary environmental protection among CEE states, and more particularly those aspects of the regime relating to transboundary environmental dispute resolution. The purpose of Part III will be to promote an understanding of how international law and the legal process might better operate to promote CEE transboundary environmental dispute resolution and thereby enhance CEE transboundary environmental protection by systematically delineating the context within which international law operates, and by identifying the circumstances and factors that influence the manner in which states use international law.

Part III

Understanding the Role of International Law

7
The Regime of International Law: Its Nature and Function

To structure the articulation of general conclusions concerning the role of international law and the legal process within the international regime of transboundary environmental protection among CEE states, this chapter first explains the nature and operation of international regimes and the role of law as a constituent element of the regime of transboundary environmental protection, including its relation to other constituent elements such as social norms, international organizations, state actors and state policies, and the identifiable interests of sub-state actors and sub-state actor transnational networks. Particular attention is paid to situating this discussion within the current intellectual debate among social science scholars as to the way in which the power relationships among states are regulated and whether international law plays a role in shaping the manner in which a state exercises its political and economic power. This chapter then combines the observations derived from the secondary and primary case studies with the select work of regional specialists, political and social scientists, and legal scholars to identify and articulate the functions served by international law during the different phases of the dispute resolution process.

I. The nature of international law as a constituent element of the international regime of transboundary environmental protection

To address the question of the nature of international law as a constituent element of the international regime of transboundary environmental protection, this section briefly defines the concept of an international regime, its mode of operation and its constituent elements. This section then argues that international law should be considered a constituent element of an international regime.[541]

A. International regimes

International regimes are interrelated sets of shared 'implicit or explicit principles, norms, rules, and decision making procedures around which actors' expectations converge in a given area of international relations',[542] and which affect the behavior of states by constraining and shaping their use of power.[543] Principles are defined as 'beliefs of fact, causation and rectitude', norms as 'standards of behavior defined in terms of rights and obligations', rules as 'specific prescriptions and proscriptions for action', and decision making procedures as 'prevailing practices for making and implementing collective choice'.[544]

A specific set of interrelated principles, norms, rules and procedures is generally considered to possess a 'characteristic set of features', including 'the political process by which it is created and maintained, the regime's substance, compliance effects on participating countries (effectiveness) and institutional learning'.[545] These regimes achieve their objective of constraining and shaping the exercise of state power by structuring the utilization of the collective interdependence of sovereign states to manage and resolve international conflicts in particular issue areas and thus promote cooperative behavior.[546] International regimes accomplish this purpose by setting forth norms and social institutions which serve to reduce the transaction cost of establishing the parameters of interdependent power relationships.[547]

The norms and social institutions of a regime then influence state behavior because states desire to be accepted into and operate within the world community,[548] and because states generally believe it is 'right' to respect international norms.[549] Once a state is accepted into the world community, its behavior is constrained both by its desire to maintain good will with other community members, and by the reality that if a state breaches community protocol on a particular issue with a fellow community member, then it may suffer reciprocal offense by that or other community members in related areas of the community relationship.[550] These consequences may include diplomatic protests, public boycotts, denunciations before and possibly by international organizations, domestic based lawsuits by sub-state actors, economic sanctions, state based claims for compensation, the possible use of force and the general deterioration of bilateral and multilateral relations.

The strength of an international regime may be measured by reference to its effectiveness in constraining state behavior and its durability, particularly in situations where the political relations between states have deteriorated.[551] In some instances, states may create pseudo regimes, which are in fact not intended to constrain state behavior in particular issue areas, but rather serve merely symbolic purposes.

International regimes are comprised of constituent elements, which are classified either as components of the social structure of the regime or components of the agency structure. Social structure associated constituent elements include the principles, norms and rules forming the prescriptive substance of the regime, as well as state policies and the identifiable interests of sub-state actors. Agency associated constituent elements refer to the states themselves, international organizations and sub-state actor transnational networks.[552]

B. International law as a constituent element of an international regime

To date, scholars studying international regimes have not clearly defined international law as a constituent element of a regime.[553] Some scholars have argued that law performs the limited function of serving as a source of rules,[554] and whereas international regimes function as animate social institutions, international treaties are inanimate stipulations of rights and duties, and in and of themselves do not constrain state behavior.[555] Other scholars have, however, acknowledged that although the norms and social institutions associated with regimes can be both formal and informal, they are more often than not reflected in customary international law, or specifically set forth in bilateral and multilateral treaties.[556]

Legal scholars have also tended, until recently, to neglect reciprocally the applicability of regime analysis to the political bargaining process leading to the development of new laws, the role of power in inter-state negotiations, and the political circumstances influencing whether and when states comply with the norms and rules contained within a particular body of international law.[557] One commentator explains this reciprocal neglect as a consequence of the focus of political scientists on 'how groups of human beings organize themselves and interact with one another', and the focus of international legal scholars on 'determining the existence, meaning, scope of application and effect of legal rules, and not so much with understanding the processes through which those rules are created'.[558] Recently, however, legal scholars have begun to investigate the relationship between international law and international regimes,[559] with at least one commentator concluding international law operates as a social institution.[560]

Notably, many legal scholars have used terms to describe international law that are strikingly similar to those used by international relations scholars to describe regimes. For instance, most legal scholars who grapple with the question of the nature of international law would agree with Georg Schwarzenberger international law is 'a body of mutually advantageous rules of social conduct',[561] and with Hans Kelsen this body of rules of social conduct is broadly considered to constitute 'a system of norms which prescribe or permit a certain conduct for states'.[562] And, according to Louis Henkin this system of norms then 'affords a framework, a pattern, a fabric for international society, grown out of relations in turn', which then serves to guide and limit state action, but seldom controls their behavior.[563] This framework then takes on a binding force, according to Brierly, as states desire to create an ordered framework of rules designed to promote peaceful interaction with other states.[564]

Although some might conclude from this descriptive symmetry that law itself operates as an international regime, this conclusion is at the same time over inclusive and under inclusive. Equating law with a regime assumes within the framework of law social norms that affect state behavior, but which lawyers would not consider hard law, and excludes from the framework of law political dispute resolution mechanisms and informal enforcement mechanisms, which although lacking a legal basis, may significantly affect the operation of the legal process.

If law is considered to make up a constituent element of an international regime, then its role in promoting dispute resolution may be considered in an integrated fashion with the other elements of the regime of transboundary environmental protection. As a constituent element, international law embodies certain norms, rules and decision making procedures classified as part of the social structure, as well as law based processes such as inquiry, good offices, mediation, conciliation, arbitration and adjudication, which would be classified as part of a regime's agency structure. Notably, the constituent element of law does not include the principles of a regime, which relate to beliefs of fact and causation, but does include those relating to rectitude. The constituent element of law also excludes social norms which may have some effect on state behavior, but would not be considered legally binding, nor decision making procedures not dictated by binding precepts, and processes which are solely political.

II. The functions served by international law during the four phases of the transboundary environmental dispute resolution process

As a constituent element of the international regime of transboundary environmental protection, international law and legal mechanisms may operate to serve various functions during the transboundary environmental dispute resolution process. To facilitate the isolation and identification of different social and political factors affecting the negotiation process – defined as the 'process whereby two or more parties attempt to settle what each shall give and take, or perform and receive, in a transaction between them',[565] social scientists divide the process into the three phases of pre-negotiation, negotiation and post-negotiation.[566] Similarly, as noted in the introduction to this book, to facilitate the examination of what role international law may serve during the course of the resolution of a transboundary environmental dispute, the environmental dispute resolution process may be considered to consist of the four phases of dispute formation, pre-resolution, resolution and implementation.[567]

A. The environmental dispute formation phase

During the environmental dispute formation phase, one or more of the parties begin to experience, become aware of or fear the effect of environmental degradation on their territory and/or population.[568] The environmental degradation is frequently the result of the actual or planned utilization of domestic or transboundary natural resources and may be related to the over-utilization or mismanagement of those resources. The affected state then makes some claim upon the state it perceives as responsible for the environmental degradation. In most cases, the latter state then either declares it is not literally or legally responsible for the harm caused to the claimant state and/or makes a counter claim for damages as a result of an activity of the claimant state. At this point a transboundary environmental dispute is considered to exist.

International law operates at this stage to assist a party in framing its claim in the context of established international rights and obligations, such as the obligation to equitably utilize a transboundary watercourse or the duty not to cause significant injury to other states. International law may further operate to assist a party in indicating it is willing to work within a legal framework to resolve the dispute, as in the case of the Gabčíkovo–Nagymaros Project dispute.[569]

The substantive international legal obligation of a state to provide notice of activities that may cause harm to other states may also be the catalyst for the formation of a dispute if the notified state would not otherwise have become aware of the potential harm, as in the case of Romania's notice to Bulgaria that it intended to bring on line the Cernavoda nuclear power plant. Similarly, the legal obligation of a state to consult with other states that may be harmed by that state's activity often constitutes the basis for an initial exchange of views as to whether a state's potentially harmful activity will in fact give rise to the formation of a transboundary dispute, and if so which dispute resolution mechanisms the parties may be willing to employ to assist in the avoidance or resolution of the dispute, such as inquiry, good offices or mediation.[570]

Employing a mechanism of inquiry, as in the case of the EU sponsored tri-party secretariat for the Black Triangle, might assist the parties in establishing the facts essential to the formation of the dispute. An inquiry might for instance identify the types and quantities of pollutants imported or potentially imported into a recipient state and the source of those pollutants, both in terms of which state or states are emitting the pollutants and which industrial or other enterprises are contributing to that emission. The process of inquiry may also assist the parties in identifying real and potential harms to human health and the natural environment, as well as possibly identifying the legal rules applicable to the avoidance or resolution of the dispute and potential dispute resolution mechanisms.[571] It should be noted, however, parties must be cautious in engaging a mechanism of inquiry as members of the inquiry team may be permitted to advocate on behalf of one of the parties in a subsequent arbitration or adjudication, as in the Gabčíkovo–Nagymaros Project dispute, where a Dutch member of a PHARE funded project advocated on behalf of Slovakia.[572]

The involvement of a third party through the mechanisms of good offices or mediation might facilitate the exchange of information between the parties,[573] and create an atmosphere conducive to extended consultations aimed at reaching an early resolution of the dispute. If the dispute appears to be unavoidable, good offices or mediation may facilitate the ability of the parties to identify more formal dispute resolution mechanisms involving an enhanced role for third party participants, as in the case of the Gabčíkovo–Nagymaros Project dispute. Similarly, by involving a third party in the dispute resolution process at the early dispute formation phase through the use of good offices or mediation, the disputing parties may be able to attract the attention and resources of the international or regional community and thus increase the leverage that may be employed to promote the resolution of the dispute. The involvement of third parties may also enhance the ability of the parties to become aware of and use as models similar disputes that have been peacefully resolved.

Third party efforts to mediate a dispute may be inhibited where the parties are not yet willing to admit a dispute exists or where they have made unconditional promises to domestic constituencies, which cannot be overridden by the authority of a mediator.[574] In these circumstances, the parties may seek to employ more formal legal mechanisms for initiating the dispute resolution process.

B. The pre-resolution phase

During the pre-resolution phase, states engage in further contact with one another and with their respective sub-state actors to ascertain the parameters of the dispute, their ability to reduce the causes and consequences of environmental degradation and the possible options for dispute resolution.[575] States may also pursue contacts with third parties to determine their interest in and ability to facilitate dispute resolution, such as through the provision of financial support or the application of pressure on other parties to the dispute. Sub-state actors, in some circumstances, will establish contact with sub-state actors across the international border, either for the purpose of forming alliances (as in the case of United States and Canadian sub-state actors forming transboundary alliances both to support or oppose an acid rain agreement) or for the purpose of previewing demands they expect their state to make in the subsequent resolution phase.[576] The creation of Euroregions in CEE is only one example of the institutionalization of these transboundary contacts.

During the pre-resolution phase, states also engage in strategic positioning, both with respect to other states, setting expectations or inflating bargaining positions, and with respect to their own sub-state actors, assembling support for the state's negotiating position, lowering sub-state actors' expectations and building a negotiating team.

Since the primary function of international law is to promote stability and order between nation-states,[577] international law operates in the pre-resolution phase to provide the 'status, rights, responsibilities, and obligations' of states participating in the dispute resolution process and an 'infrastructure of assumption, practices, commitments, and expectations upon which states may rely' during this process.[578] International law may also promote the formation of a regime for transboundary environmental protection or compel a state to bring itself within a pre-existing regime.

More specifically, international law requires states to provide information concerning the extent of potential or occurring environmental harm, to cooperate in abating environmental degradation and to settle the emerging dispute by peaceful means. These obligations are of particular relevance in the Danube River Gauntlet dispute. In some instances, international law may require states to participate in a joint EIA or direct their attention to various means for resolving the dispute, including conciliation, negotiation, arbitration and even the pre-negotiation submission of the dispute to an adjudicatory body, such as the ICJ.[579]

In the case of conciliation, parties generally enlist the formal participation of a third party for the purpose of identifying possible compromise solutions to the dispute or possibly establishing the parameters of a formal party-to-party negotiating process, which may bring the parties to an understanding of how best to resolve the dispute. During the process of conciliation, the third party may also assist in sharpening the parties' understanding of the causes of the environmental degradation giving rise to the dispute and to the means by which that environmental degradation may be abated or its consequences mitigated. The conciliator may also indicate it would be willing to provide financial incentives to resolve the dispute or impose sanctions in the event of a failure to resolve the dispute, and may identify other third party states or institutions capable of providing similar incentives and/or assistance with the implementation of any agreement designed to resolve the dispute.[580]

In the preliminary stages of negotiation, the parties may directly meet without the formal involvement of a third party and exchange views as to the causes and consequences of the environmental degradation giving rise to the dispute as well as justifications for any actions they may have taken in response to the conduct of the other party. During the initial stages of negotiation, the parties will likely articulate the conditions they believe must be met in order to resolve the dispute and any technical means of implementing a resolution consistent with those demands. When states involve sub-state actors in their negotiations, states have an additional opportunity to legitimize their claims by permitting the sub-state actors to present their views of the consequences of, or the justifications for, the activities causing the environmental degradation giving rise to the dispute. As the negotiations or mediation/conciliation efforts develop to include a discussion of the actual terms of an agreement designed to resolve the dispute, the process moves from the pre-resolution to the resolution phase.

Notably, where the parties are unable to reach an agreement and permit the dispute to remain active, or where the parties break off negotiations or otherwise indicate a desire to cease their attempts to resolve the dispute, the dispute is then characterized as slipping back into the pre-resolution stage. Similarly, where states indicate a desire to resolve the dispute but do not engage in diplomatic activity designed to reach agreement with the other party or parties to the dispute, the process may be characterized as the pre-resolution phase.

C. The resolution phase

Parties move to the resolution phase when they perceive it to be in their self-interest to reach a formal agreement as to a specific course of action that will resolve the transboundary environmental dispute and possibly eliminate the circumstances causing the dispute. Parties may determine their self-interest is served by reaching an agreement if they have been able to persuade the other party to accept many of their demands, such that the benefits to be derived from an agreement would outweigh the harms associated with maintaining the *status quo*, as in the Black Sea dispute, or if third parties have offered incentives or threatened sanctions which when considered by the party lead it to conclude it will obtain a net benefit by agreeing to a resolution of the dispute.

During the resolution phase, which is usually marked by negotiations between the parties but can include mediation or conciliation, the parties clearly identify their conflicting interests at the base of the dispute and their common interests that might form the foundation for resolving the dispute. After evaluating the various feasible options, which may include an assessment as to which options are most consistent with what they perceive as their international rights and obligations, the parties commit to a specific course of action.[581] The successive attempts of the Baltic Sea Basin states to negotiate an effective convention for controlling and reducing the level of emissions into the Baltic Sea exemplifies these aspects of the resolution phase.

During the resolution phase states may utilize substantive international law to support their arguments, evaluate possible procedural obligations under which the negotiations might continue and identify solutions for resolving the dispute. International law may also serve as a vehicle for expressing moral authority, for signaling firmness on a particular position, as with Bulgaria asserting the Kozloduy nuclear power plant meets international safety standards, or for the purpose of engendering the support of third parties by positioning a state as law-abiding, as with Hungary's initial unilateral submission of the Gabčíkovo–Nagymaros Project dispute to the ICJ.[582] If during the resolution phase there is a need to create additional norms, international law also provides a process for reaching agreement and a basis for the supposition that states are obliged to fulfill obligations contained in the agreement.[583]

In their attempts to utilize international law for these purposes, states frequently, intentionally or unintentionally, misrepresent both the facts and the substance of international law or disagree as to which facts or international laws are relevant to the dispute.[584] In these circumstances states might agree to submit certain factual and legal questions, or the entire dispute itself, to a third party dispute resolution mechanism. Importantly, the continuation of negotiations does not prevent the submission of certain aspects of a dispute to a judicial dispute resolution mechanism.[585] The parties may also agree alternatively to submit the dispute to a nonjudicial dispute resolution mechanism or to submit nonlegal aspects of the dispute to such a mechanism, and this decision will not prohibit the judicial mechanism from ruling on the legal aspects of the case.[586]

States may also resort to a dispute resolution mechanism when the negotiations reach a political impasse, when one or both states perceive it to be in their strategic interests or when third parties associated with the dispute perceive a value in recourse to such a mechanism. Once the dispute resolution mechanism runs its course, states may return to negotiating a mutually agreeable resolution of the dispute within the parameters set by the dispute resolution mechanism, as with the Gabčíkovo–Nagymaros Project dispute. If the parties seek to use a dispute resolution mechanism during the resolution phase, they will most likely select arbitration, or adjudication by a regional court or by the ICJ.

If the parties submit their dispute to the ICJ, the Court will limit its rulings to legal findings and will avoid pronouncing upon nonlegal issues.[587] The legal inquiries submitted to the Court may include a request for a determination as to the status of obligations under a pre-existing agreement,[588] identification of the principles and rules to be applied during negotiations,[589] and a determination as to specific legal rights and obligations.[590] As noted by Lord McNair, the Court's role is to resolve legal disputes or answer legal questions relevant to the dispute, not to resolve the social and political aspects of the dispute.[591] Importantly, even after parties agree to submit a dispute to the Court, they frequently disagree as to the actual role of the Court in resolving the dispute, as in the Gabčíkovo–Nagymaros Project dispute where Hungary sought a broad role for the Court in rendering a decision that would include determinations on disputed questions of fact and opinion, while Slovakia sought a more narrow role limited to the legal aspects of the three questions submitted to the Court.[592]

The existence of a legal dispute within a larger political context, however, will not preclude the ICJ from resolving the legal aspects of a case.[593] In fact, it is frequently the resolution of those legal questions that creates the foundation for a subsequent political settlement by facilitating the resolution of the social and political aspects of the dispute.[594]

Parties are generally aware it is not appropriate for the ICJ to serve as a legislative mechanism responsible for crafting a final solution to the dispute.[595] However, when states fail to juxtapose legal and political avenues of dispute resolution, or to submit their dispute to a hybrid mechanism, they run the risk of condensing social, political and ecological disputes into legal arguments in an attempt to achieve a legal resolution of the dispute.[596] Alternatively, legal arguments may be presented to a political mechanism unqualified to make useful legal determinations. In the most successful cases of dispute resolution, political dispute resolution mechanisms work in tandem with legal dispute resolution mechanisms to resolve the scientific and technical issues related to the dispute.

Where the parties submit some aspect of their dispute to international arbitration each party generally selects one of the arbitrators, with those arbitrators agreeing upon the third if it is a three person panel. Similarly, the parties generally establish the rules governing the arbitration, including whether they wish the panel to consider nonlegal as well as legal issues in making its determination. The advantages of arbitration lie in the fact it is generally less costly and more efficient than adjudication. Arbitration, however, provides less political cover for unfavorable decisions than adjudication. In many instances arbitration panels find it expedient to reach conclusions that provide some benefit to each party, which may or may not be desirable depending on whether a party wishes to achieve an adjusted compromise, or whether it is seeking a clear determination its proposed or actual course of action is legitimate.[597]

States may also consider taking unilateral action when they reach an impasse in the resolution phase,[598] as did Czechoslovakia when it implemented the Provisional Solution in October 1992, after it was unable to reach agreement with Hungary on the implementation and/or modification of the Gabčíkovo–Nagymaros Project. The disadvantages with unilateral action are it usually violates a state's international legal obligations to negotiate and to act in good faith, it engenders mistrust and significant animosity with the other party or parties to the dispute, and exposes the state to censure by the international community. The advantages of unilateral action are it might create a *fait accompli*, which may serve the interests of that state in the event the dispute slips back to the pre-resolution phase, or in the event the dispute is subsequently submitted to arbitration or adjudication.

D. The implementation phase

The implementation phase involves the formal adoption and implementation of any agreement resulting from the resolution phase and attempts by states to provide domestic effect to the negotiated settlement, as well as any agreements necessary to implement that settlement.[599] In order to accomplish this task, states may attempt to persuade their sub-state actors of the value of the agreement, institutionalize any incentives accruing to those sub-state actors, fulfill any domestic ratification procedures, establish a regime for carrying out the agreement's provisions and monitoring compliance and provide a process for negotiating minor changes to the agreement. Correspondingly, the act of regime formation, wherein states give effect to new legal rules created in the dispute resolution phase, occurs during the implementation phase.[600] States may also begin to lay the foundation for the future modification of the agreement or for additional steps necessary to achieve a long term remediation of the environmental harm causing the dispute.

International law operates in the implementation phase primarily to ensure states fulfill their agreements, including the obligation to take all necessary domestic measures to give the agreement domestic effect, by providing a variety of mechanisms and procedures for readjusting that state's behavior.[601] International law also obligates states to cooperate in the implementation of the agreement, and many procedural principles relevant in the pre-resolution stage relating to sharing information and providing notice of activity affecting the implementation of the agreement will again become relevant. States may also establish joint legal mechanisms, such as HELCOM, for the purpose of monitoring and possibly enforcing the agreement. In some instances they may encounter a dispute as to the interpretation of the agreement, which may then become subject to negotiation or may be submitted to arbitration or adjudication.[602] In many instances, the agreement itself will indicate which options are available to the parties to resolve a dispute relating to the implementation of the agreement, as with the Danube River Convention.

8
Influencing the Utilization of International Law: Sub-State Actors, Interested Third Parties, Situational Circumstances and Factors of Functionality

To continue the articulation of general conclusions concerning the role of international law and the legal process within the international regime of transboundary environmental protection among CEE states, this chapter examines the groups of relevant sub-state actors and third parties capable of influencing transboundary environmental dispute resolution and the role of law within that process. This chapter then examines the categories of situational circumstances structuring the ability of states, sub-state actors and third parties to promote the resolution of disputes and the use of international law. Finally, in order to establish a basis from which to predict whether CEE states may be more inclined to rely upon international law, this chapter identifies factors that promote the functionality of international law within the regime of transboundary environmental protection, and thereby enhance the likelihood CEE states will rely upon international law to assist in the resolution of their transboundary environmental disputes.

I. Sub-state actors

Although states are the primary actors within the regime of transboundary environmental protection, and are thus both subject to and capable of using international law, the manner in which states use international law to achieve their objectives is significantly affected by the interests and capabilities of a collection of sub-state actors. A state may thus be considered both an entity with a general and predictable pattern of interests and behavior, and an entity that reflects an aggregated collection of interests, which may vary given the nature of the dispute in which it is involved.

Support for the proposition sub-state actors must be studied in order to fully understand the actions of states can be found among contemporary game theorists,[603] who argue in pluralist societies a number of interests, pursued by a variety of sub-state actors, coalesce to form the national interest of a state – defined as a 'consistent social preference order'.[604] Similarly, some social scientists assert that in order for states to balance the equities affected by a transboundary environmental dispute or to maximize the value for one state while minimizing the cost to the other state, it is necessary to identify those sub-state actors who might benefit, or conversely experience a detriment, from the resolution of the dispute in a particular manner.[605] It is also necessary to identify those sub-state actors who possess the expertise to assist in crafting a transboundary environmental regime necessary to resolve a transboundary environmental dispute, or who are capable of exercising political influence over the development or implementation of a regime.[606]

The identification of various sub-state actors contributing to the aggregation of interests reflected by a state is therefore necessary not only to determine the extent to which a state may be inclined to resolve a dispute and to seek the assistance of international law, but also to determine a state's particular capability to negotiate effectively an agreed resolution of a dispute and to implement that agreement. It should be noted sub-state actors are frequently intentionally excluded from direct participation in negotiations or decision making processes, and thus they exercise their influence primarily by affecting the aggregation of the state's national interest and seldom through separate channels of direct communication with other states.

The primary sub-state actors with an interest in the resolution of transboundary environmental disputes include government officials, economic actors, environmental actors and the general public.

Government officials: The government of a state is formally charged with reflecting a state's aggregated national interest, and thus the nature of that government will affect how accurately or effectively it aggregates and reflects the nation's interest in using international law to resolve transboundary environmental disputes, as well as how effectively a state makes use of the varying capabilities of its constituent sub-state actors. A state's government will also have its own interests in reaching a particular settlement of a transboundary environmental dispute,[607] as evidenced by the fact from 1988 to 1999 the successive Czechoslovak, Slovak and Hungarian governments approached the Gabčíkovo–

Nagymaros Project dispute with different perspectives from those of their predecessor governments.[608]

Similarly, since state governments are interested in both their own long term survival and the long term advantage of the state, they will be interested in not only resolving the dispute at hand in such a way that it serves their interest, but will also consider the effect of their behavior on their domestic political position and on other strategic and economic concerns in which the issue is nested, such as potential EU membership, as evidenced by the Czech Republic's and Poland's interest in working with the EU to resolve the Black Triangle dispute.[609]

In addition, individual bureaucrats will exercise differing degrees of influence on and control over a state's negotiating positions and behavior and may provide support and legitimacy to other individuals or sub-state actors seeking to influence whether a state will seek to use international law to resolve a transboundary environmental dispute.[610] A particularly illustrative example is the case of Bulgarian nuclear power officials taking the public lead in defending the operation of the Kozloduy plant and arguing its operation is consistent with international law. Government officials also need to consider the agendas and influence of any utilities or state owned entities exercising significant 'bureaucratic' power, and assorted government ministries, such as the Ministry of the Environment, Ministry of Foreign Affairs and Ministry of Industry. For instance, the Hungarian Ministry of Water Management continued construction of the Nagymaros works, under the pretense of maintenance work, after the Hungarian government formally postponed all such work in order to consider termination of the project.[611]

Economic actors: In heavily industrialized societies, such as those in CEE, economic actors invariably play a significant role in the aggregation and reflection of a state's national interest relating to environmental matters, and it is frequently the environmental degradation associated with the activity of these economic actors that gives rise to a transboundary environmental dispute.[612] In particular, because environmental agreements often require restrictions on industrial activity affecting the economic interests of a state, the agreements prompt economic actors, such as the numerous industrial enterprises operating along the Danube River Gauntlet, to intervene in the resolution and implementation phases.[613] In some instances, economic actors may, however, play a positive role in resolving a dispute where environmental pollution generated by other sub-state actors negatively affects their interests. Such economic actors would include, for instance, Polish

138 *Understanding the Role of International Law*

industrial enterprises who cannot use the Oder River for industrial purposes because it corrodes their machinery.[614] Multinational economic actors may also play a role. Western energy companies, for instances, have sought to ensure CEE states continue operation of nuclear power plants so they may be able to market Western safety technology to these plants, and as such perceive of the 41 nuclear reactors in CEE as 'business opportunities'.[615]

The ability of CEE states to acquire funding from the economic actors making up the private investment community in order to construct projects or enterprises which may affect the environment is also of relevant concern. For instance, Czechoslovakia was able to persuade the J. P. Morgan Bank group to arrange funding for the Provisional Solution.[616] In other circumstances, private investors may be persuaded to provide funding for projects designed to promote environmental remediation.

Environmental actors: In many instances, environmental actors in the form of NGOs and local governments play an important role.[617] Political science research indicates in the pre-resolution phase, NGOs may provide relevant information and raise public pressure on governments to enter into negotiations.[618] In the dispute formation phase, NGOs often seek to initiate the negotiation process before the environmental harm becomes irreversible.[619] NGOs also seek to shift public attitudes toward environmental issues, ensure environmental protection remains on the public agenda, publicize the nature and seriousness of environmental problems, and serve as conduits for the dissemination of scientific research.[620] During the resolution phase, NGOs are in some instances permitted to participate in transboundary environmental dispute resolution negotiations and in almost all instances perform an important role by independently monitoring the implementation of agreements.[621] The Czech based Rainbow Movement, for instance, in 1997 started operating six independent radiation monitoring stations around the Dukovnay nuclear plant in the Czech Republic.[622]

Local governments often adopt the position of environmental actors because these governments frequently bear the financial costs of transboundary environmental pollution and must deal with fewer competing international considerations than central governments. In some instances, however, local governments adopt the position of economic actors, in particular when they own and operate large industrial enterprises, as is the case with the municipal government of the town of Most in Northern Bohemia where 80 per cent of the residents are directly or indirectly employed by the mining industry.

General public: As CEE states continue their democratic transition, the role of the general public in influencing the national interest of states will likely continue to grow.[623] In particular, public participation motivates otherwise indifferent policy makers during the dispute formation and pre-resolution phases and serves as a conduit for broader local concerns during the resolution phase.[624] During the initial stages of the Gabčíkovo–Nagymaros Project dispute, for example, 100,000 Hungarian citizens focused the attention of the Hungarian government, and actors in the international community, by joining together to form a 100-mile human chain to protest the project.[625]

In some instances, transborder domestic political coalitions may be created that support international trade-offs for the protection of the environment.[626] These groups then create, from within the state, circumstances where 'good will' is generated or where, conversely, 'ill will' is engendered by a failure to act upon the requests of the internationally linked domestic coalitions to adopt a legal means for resolving the dispute. States may build upon these internationally linked domestic coalitions by permitting nationals of neighboring states to seek local remedies within their state.[627]

II. Interested third parties

Separate from states directly involved in a dispute, certain third parties may have an interest in influencing the operation of the regime of transboundary environmental protection in such a way as to promote the use of international law to resolve transboundary environmental disputes, as with the EU and the Gabčíkovo–Nagymaros Project dispute. The concern of third parties may arise from a direct impact of the dispute on their territory or national interest, a general desire to promote the rule of law or a desire to promote environmental protection.

In many instances, interested third parties may themselves be states. In the case of CEE, as will be discussed below, the interested third parties include the EU and its member states, the Scandinavian and Nordic states, the United States, IFIs (World Bank, IMF, EBRD and EIB) and international human rights and environmental organizations.

The *European Union and its member states* have an interest in promoting the decrease of transboundary environmental pollution and the resolution of CEE transboundary environmental disputes, as many of its member states experience the harmful effects of this pollution. The disputes also inhibit intra-CEE economic cooperation, and the EU possesses an interest in not importing these disputes into the EU when it admits certain CEE states into the Union. The EU may influence dispute resolution by offering specific financial assistance for the resolution of certain disputes, threatening to withhold general financial assistance if certain disputes are not resolved, and modulating the speed of EU association and potential membership based on the extent to which CEE states pursue good faith efforts to resolve the disputes.

Scandinavian and Nordic states, which possess both symbiotic environmental relationships and expanding trade relationships with some CEE states,[628] are particularly active in providing assistance to CEE states responsible for polluting the Baltic Sea and for generating SO_2 emissions, which fall as acid rain in the more northern states.[629] Thus, in May 1990, Sweden agreed to provide Poland with $45 million for funding the implementation of pollution abatement projects, and also in May of 1990, Finland agreed to forgive $28 million in debt owed by Poland in exchange for an agreement Poland would invest an equal amount in environmental protection measures.[630] The Scandinavian and Nordic states also actively participate in the Baltic Sea Joint Comprehensive Environmental Action Program, and are co-parties to the Convention on the Protection of the Marine Environment of the Baltic Sea. Notably, Scandinavian and Nordic states tend to invest in states where the reduction of pollution will have a positive effect on environmental circumstances in the Scandinavian and Nordic regions, and in many instances, these states select specific projects based on this predominant criterion.

The United States possesses an interest in promoting bilateral relations with individual CEE states, promoting the export of United States technology, improving the global environmental condition, and promoting the general rule of law. To promote a resolution of CEE transboundary environmental disputes, the United States frequently relies on the provision of financial resources and technical expertise. The United States may also exercise more general diplomatic pressure to encourage the resolution of transboundary environmental disputes, but retains little interest in actually brokering the resolution of any particular dispute.

NGOs operating in CEE states are establishing strong links with Western NGOs and receive substantial amounts of funding and information from their counterparts to facilitate their efforts to promote environmental protection, and in limited instances, seek to assist with the resolution of transboundary environmental disputes.[631] In addition, international NGOs and United Nations organizations, such as the United Nations Environmental Programme, are interested in the protection of human rights and the environment, and thus are concerned with CEE transboundary environmental disputes to the extent their resolution affects human rights or promotes the abatement of environmental pollution. The degree to which these various groups can affect the resolution of CEE transboundary environmental disputes rests with their level of influence on CEE governments, their ability to provide useful information or expertise to domestic actors, and their ability to influence the behavior of other third parties, such as the EU and IFIs.

The *World Bank, International Monetary Fund, European Bank for Reconstruction and Development*, and *European Investment Bank* hold a primary interest in promoting the economic transformation of CEE states from command to market economies, ensuring the debt obligations of CEE states are serviced, and in some instances promoting environmental protection.[632] The capacity of the IFIs to promote a resolution of CEE transboundary environmental disputes lies primarily in their ability to provide funding for the amelioration of environmental pollution in one state which gives rise to a transboundary environmental dispute, and to condition certain loans on either the reduction of pollution or good faith efforts to resolve outstanding transboundary environmental disputes.[633] To enhance their ability to promote environmental protection the IFIs participate in the operation of the GEF Fund, which draws on $2 billion made available from the World Bank, national governments and UN agencies to support international environmental management programs and the transfer of environmentally benign technologies to reduce global warming, promote bio-diversity, protect natural habitats, and reduce the pollution of international waters and ozone depletion.[634]

Institutions such as the OECD and NATO are engaged in a number of development and assistance projects with the EBRD, World Bank and IMF. As a result of their desire to integrate into these Western institutions, CEE states are potentially responsive to efforts encouraging environmental protection and transboundary environmental dispute resolution. Beyond the continued operation of the Committee on the Challenges on Modern Society, which now includes, as observers, states participating in the NATO Partnership for Peace program, NATO places no noticeable emphasis on the general protection of the environment or the resolution of transboundary environmental disputes.[635] NATO, however, has begun to take an interest in the resolution of disputes relating specifically to the environmental hazards existing on former Soviet military bases. The OECD is more active and has in some instances conditioned membership of certain CEE states on the adoption of specific environmental programs.[636] Hungary's admission to the OECD in

March 1996, for example, was conditional upon Hungary adopting specific changes to its environmental regulations.

In general, interested third parties are quite capable of exercising power and authority to influence the four phases of the transboundary environmental dispute resolution process. In exercising their power and authority, third parties may act individually or collectively and may at times pursue cross purposes with other interested third parties, as with the Austrian objection to and the American funding of the Temelín nuclear power plant, and the Austrian objection to and French funding of the Slovak Mochovce nuclear power plant.

Similar to the primary state actors, interested third parties may be composed of sub-state actors with varying interests and capabilities. In some instances, the interests of these sub-state actors may conflict with the articulated interests of the third party. Thus for example, while EU Foreign Affairs Commissioner Hans van den Broek was calling on Slovakia to end its nuclear program, the Slovak government signed co-operation agreements with German and French companies relating to the construction of the Mochovce power plant.[637] In other instances the pursuit of economic interests by third party sub-state actors might bring about a more safe and efficient operation of industrial enterprises responsible for much of the transboundary pollution.[638] Third party states may also be subject to political pressure by domestic NGOs or political entities, such as when the Austrian Green Party sought to persuade the Austrian Government to disallow the $360 million claim by Donaukraftwerke for its work on the Nagymaros dam.[639]

Third parties may frequently have a positive affect on the dispute resolution process by providing an incentive to the disputing parties to comply with international rules of good behavior, and to invoke international dispute resolution mechanisms.[640] For instance, although the EU does not yet directly link potential EU membership with the resolution of transboundary environmental disputes, it frequently makes CEE states aware their cooperation on these disputes will be one of the factors considered when assessing their qualifications for full membership.[641] Two particularly vivid examples of EU influence relate to the successful pressure applied to both Slovakia and Hungary to submit the Gabčíkovo–Nagymaros Project dispute to the ICJ, and warnings by the French European Affairs Minister, on 18 December 1995, to the Bulgarian Foreign Minister that continued operation of the Kozloduy nuclear power plant jeopardized Bulgaria's chances of joining the EU.[642]

On a more general level, the EU includes in its association agreements with CEE states, environmental provisions relating to efforts to reduce the emission of pollutants and the harmonization of their environmental legislation and institutions with EU standards.⁶⁴³ The EU is also attempting to influence the environmental behavior of CEE states by providing assistance for environmental remediation projects, and in some case conditioning this assistance on the resolution of transboundary environmental disputes.⁶⁴⁴ The EU, for instance, has informed Romania and Bulgaria that future PHARE funding for projects in the Danube River basin area is contingent upon their resolution of the Giurgiu–Ruse dispute.⁶⁴⁵ The EU has also provided assistance for projects designed to specifically resolve such a dispute, as with the EU grant of $13 million to support the Working Group for Neighborly Cooperation on Environmental Issues, which brings together officials from the Ministries of the Environment in Poland, the Czech Republic and Germany in an attempt to structure a resolution of the Black Triangle dispute by harmonizing environmental policies and developing an environmental recovery plan for the region.⁶⁴⁶

Interested third parties may also frequently provide the funds necessary to implement means agreed upon by the parties to resolve the dispute, as in the case of the Baltic Sea dispute, where the World Bank has invested $240 million to promote environmental protection and plans on investing up to $1 billion over the next 20 years.⁶⁴⁷ Similarly, in the case of Poland, the United States, France and Switzerland established a debt-for-nature swap arrangement, which created the Polish EcoFund by forgiving 5 per cent of Polish debt in exchange for an agreement that an equivalent amount of local currency funding would be allocated to environmental projects.⁶⁴⁸

The efforts of some third parties to promote environmental protection in CEE might be limited by the fact many of their activities are driven by a profitability quotient, leading them to decline support for certain environmental projects that might not be profitable, or to finance profitable projects that have the potential to cause significant harm to the environment.⁶⁴⁹ The EBRD, for instance, recently invested in a heavily polluting aluminum plant in Zaire nad Hormone, Slovakia that consumes substantial amounts of energy from the Bohunice nuclear power plant and produces aluminum sold at below cost and world market prices.⁶⁵⁰ The IFIs efforts also suffer as these international financial institutions are still in the process themselves of recognizing the value of environmental protection,⁶⁵¹ and as such, the EBRD was recently criticized for its willingness to disregard its own environmental procedures, restricting public participation, and its failure to create a plan for promoting an economic structure in CEE based upon the principle of sustainable development.⁶⁵²

144 *Understanding the Role of International Law*

In some instances, third parties also may diminish their credibility by failing to abide by the rule of law in their own dealings, or by permitting the abuse of their incentive programs, such as where EU member states abuse EU assistance programs by using them as cover for exporting or transferring products to CEE banned in the EU, such as certain harmful pesticides or technologies considered less environmentally desirable.[653] In certain cases, interested third party states attempt to minimize the damage caused by such activity, as when in 1993, Germany provided over $2.5 million for the return to Germany of 425 tons of poisonous and hazardous waste illegally exported to Romania from 1991 to 1993.[654] Yet in other cases, these same states actively participate in the export of unsafe material, such as when, in February 1996, Germany shipped 235 used nuclear fuel elements to Hungary's Paks nuclear plant. These fuel elements originated from the Greifswald/Lubmin power station in the former East Germany, which was closed for safety reasons. Although Hungary apparently saved the cost of $21 million for new fuel rods as it can use the fuel rods for one year, Hungary is responsible for their ultimate disposal.[655]

Some interested third party states may serve the unique function of acting as a bridge between the disputing states and institutional third parties which have the capacity to assist in promoting a resolution of the dispute. The most relevant examples in CEE include Germany and Austria which are a party to a number of the disputes, including the Baltic Sea, Black Triangle, and Temelín and Mochovce nuclear power plant disputes, and to a certain degree the disputes relating to the pollution of the Danube River. Because of their membership in the EU they may be able to facilitate EU involvement in and funding of means to resolve the dispute. Efforts by Germany, for instance, to reduce environmental degradation in CEE states that afflicts Germany include assisting CEE states in the drafting of environmental legislation, engaging in joint monitoring operations, and financing the purchase of pollution abatement technology.[656] In an effort to improve the quality of the Elbe River, for example, the German Government provides funding to a number of the Czech municipalities located along the River to assist in their construction of sewage treatment plants.[657]

Finally, it should be noted an added dynamic to the role of interested third parties is the reemergence of competing regions of influence traditionally held by Western states in CEE. France, for instance, maintains a traditional relationship with both Slovakia and Romania, Germany a relationship with Bulgaria, the Nordic states with the Baltic states, Austria with Hungary, Slovakia and the Czech Republic, and the United States with Poland, Slovakia and Bulgaria. This creates a natural tension among Western states as they try to both promote intra-CEE fraternity and pursue and encourage their special bilateral relationships with CEE states.[658]

III. Situational circumstances affecting the role of international law in the transboundary environmental dispute resolution process

Since international law operates within the regime of transboundary environmental protection, which is inherently a socio-political regime, the ability of the various states, sub-state actors and interested third parties to affect how international law is used to resolve transboundary environmental disputes is influenced by ecological, economic, domestic political and international political, and national minority cirumstances.[659]

A. Ecological circumstances

Because any transboundary environmental dispute is at its core a dispute about an ecological balance or imbalance, ecological circumstances influence both the nature of the dispute and the dispute resolution process itself. Transboundary environmental disputes customarily grow out of the over utilization of natural resources by a single state, the competing utilization of those resources by two or more states, as in the Danube River Gauntlet dispute, and/or the degradation of regional and local ecosystems, causing environmental, economic and health consequences, as in the Black Triangle and Silesian Coal Basin disputes.[660] These consequences vary with respect to the upstream (upwind)–downstream (downwind) locations of the respective states, the relative effect of pollution on the respective states and the ability of local ecosystems to absorb pollution. Moreover, some social scientists argue in the case of environmental disputes, negotiations are most successful when there is symmetry between the parties in terms of exposure to environmental harm and situational bargaining power.[661] Depending on the type and extent of harm caused by the over utilization of natural resources, particular state and sub-state actors will naturally possess different interests in resolving the dispute and will call upon certain bodies of international law to assist in the resolution of that dispute.

Relevant ecological circumstances influencing how international law may be used to assist in the resolution of a transboundary environmental dispute may be considered to include: the geography and ecosystem in the area of the dispute; the nature and type of pollution, including the mobility of pollutants, the types of harm caused by pollution and the ecosystem's absorption capacity (critical load factors); the existence of alternative exploitable natural resources; the existing technical ability to abate or control such pollution;[662] the ability to identify sources of pollution; the difference between the ecological nature of the transboundary environmental dispute and other domestic resource disputes; and the ability to assign responsibility accurately for environmental degradation in the case of two or more sources of pollution.

B. Economic circumstances

Economic circumstances define the extent to which state and sub-state actors, particularly, but not exclusively, economic actors, may seek to use international law to assist in the resolution of transboundary environmental disputes. Economic circumstances also define the relative authority or position of these actors; promote or discourage the integration of the interests and behavior of the various actors participating in or affected by the transboundary environmental dispute; promote or discourage the cross border association of sub-state actors with similar interests; and affect the interest and ability of concerned third parties to promote the use of international law.

Political science research indicates societies with a higher quality of life tend to be more concerned with environmental quality and tend to initiate the pre-resolution phase more readily for the purpose of adopting a legal regime to resolve a transboundary dispute.[663] Economic actors, however, often delay the commencement of environmental negotiations and subsequent regulation as they perceive themselves as likely to bear the costs of any agreed upon remediation measures or production restraints.[664] Those economic actors which do support the commencement of negotiations frequently are those which have already reduced their emissions or installed the necessary abatement technology, or those which export such technology.[665]

In the implementation phase, as public expenditure on the environment increases, there is a corresponding increase in state efforts to verify and enforce environmental protection agreements in order to ensure the effect of the public investment is not diluted by a neighboring state's noncompliance with the agreement.[666] Certain international relations scholars further argue if there is a significant density of economic transactions between two or more states, there is a high probability those states will have created transboundary environmental degradation, requiring a regulatory regime for its management and a pattern of trust and cooperation which will facilitate the establishment of that regime.[667]

In the CEE transboundary environmental dispute context, the extensive trade between Germany and Poland and the Czech Republic represents an example of how such structural symmetry might both require and promote the creation of a regulatory regime.⁶⁶⁸ Because of the substantial environmental degradation caused by the extensive back-ups of diesel trucks at limited border crossing points, such as Görlitz, when on most days, the line of trucks waiting to cross into Germany stretches at least ten kilometers back into Poland, all three states have recognized the need for greater environmental cooperation and are willing to initiate efforts to establish a preliminary regulatory regime as part of a broader regime designed to structure their significantly increasing economic relations.

Although most CEE states currently hold relative parity in their economic power, the next ten to fifteen years will be an open economic playing field, and these states may attempt to increase their economic power and experiment with using this power to influence their relations with other CEE states.⁶⁶⁹ Yet, as CEE states continue their economic transformation from a command to a free market economy, and as they shift from an environmental policy based on 'react and cure' to a policy based on 'anticipate and prevent', domestic environmental regimes might serve to integrate economic and environmental policy with specific sector policies such as industry, land use, tourism and energy development.⁶⁷⁰ Notably, Poland, the Czech Republic and Hungary are generally considered the CEE states most likely to succeed with their economic transformation in the near future, with this having a positive effect on the Baltic Sea, Black Triangle and Silesian Coal Basin disputes.⁶⁷¹

Since a state's capacity to abate environmental degradation and the existence of adequate abatement technology also influence that state's propensity to seek international environmental regulation, it is important to consider the condition of the state's economic and technical capabilities.⁶⁷² From the realist perspective on international environmental relations it would be argued that states will pursue efforts to resolve transboundary environmental disputes if they perceive it to be either in their short term or long term interest to do so, with this assessment including an examination of a state's desire to protect its population and natural resources from harm, as weighed against any economic or other costs that might accrue.⁶⁷³

The economic cost–benefit analysis associated with the reduction of pollution resulting from the resolution of a transboundary environmental dispute is also important.⁶⁷⁴ On the domestic level, the cost–benefit analysis frequently includes considerations relating to increased environmental protection such as reduced employment and social dislocation as compared to reduced health costs and increased tourism receipts.[675] In deciding whether or not to commit the necessary resources to environmental protection, CEE states must consider it is estimated to require $50 billion to bring CEE water quality up to EU standards, and $30–35 billion to bring CEE thermal power plants up to EU standards.[676] However, in April 1995, a group of American and European economists estimated economic costs associated with pollution were as high as 10 per cent of each state's GDP, compared to 1–2 per cent in EU,[677] and meeting EU standards for environmental pollution could yield benefits for CEE states equal to 5–11 per cent of their GDP.[678]

On the international level a cost–benefit analysis might determine a state's willingness to pursue agreements requiring greater environmental protection. For instance, according to two social science scholars, in order to reduce long range SO_2 emissions by 30 per cent, it would cost the Czech and Slovak Republics 0.16 per cent of their GDP and Hungary 0.32 per cent. Since these states would substantially benefit from reduced emissions, and the associated cost is relatively low, they would be identified as 'pushers' for an agreement requiring a 30 per cent reduction in SO_2 emissions. Since Poland and Romania would have relatively high respective costs of 0.69 per cent and 2.42 per cent, but would also experience relatively high benefits from such reduction, they would have an interest in serving as 'intermediaries' in negotiating a reduction. Bulgaria, however, would experience a high cost of 1.81 per cent and a low benefit, and would thus act as a 'dragger', with the intent of obstructing such an agreement.[679] This may explain the recent efforts to resolve the Black Triangle dispute, and the perceived negative inertia associated with the air pollution aspects of the Danube River Gauntlet dispute.

Relevant economic circumstances influencing how international law may be used to assist in the resolution of a transboundary environmental dispute may be considered to include: a state's energy and other natural resources;[680] the composition of its industrial base; its GNP and living standards; the domestic consumption or export demand for products produced by polluting industries; the relative integration of environmental and economic policies; symmetry between economic/environmental costs and benefits; the costs associated with protecting the environment either through industry closure or the implementation of abatement technologies; the economic cost of environmental degradation; and the access to capital for investment in new technologies.[681]

C. Political circumstances

Political circumstances, which are divided here into domestic and international circumstances, affect the aggregation of a state's national interest in using international law and influence a state's willingness and ability to pursue a legal means for resolving a transboundary environmental dispute. Similar to economic circumstances, political circumstances define the interests of a state's constituent actors and the relative influence of those actors on the dispute resolution process. Scholars subscribing to the policy-oriented approach to legal theory are particularly interested in examining the individuals or entities influencing the decision making process, the factors affecting the perceptions or values of these various sub-state actors, the policy goals of these actors and the contexts in which such decisions are made in order to ascertain the nature of international law.[682] Moreover, regime theorists would further emphasize the need to understand how these domestic political circumstances promote or discourage the correlation of the interests and behavior of various sub-state actors with international political circumstances which promote or discourage the correlation of cross border interests and affect the behavior of concerned third parties.

1. Domestic political circumstances

In assessing the interest and ability of sub-state actors to further the use of international law to assist in the resolution of transboundary environmental disputes, it is necessary to identify circumstances promoting the integration of public opinion into the decision making process.[683] This proposition is supported by some game theorists who argue that in accounting for the sub-systematic aggregation of a state's national interest it is necessary to focus on the need in democratic states for voter support of state behavior during all stages of the dispute resolution process. More specifically, some game theorists note voters possess different degrees of influence at various stages of the dispute resolution process because of the different roles they may play in the formation or ratification of policy, with special interests having more influence at the formation stage and the general public having more influence at the ratification stage.[684]

Similarly, scholars subscribing to the liberal approach to legal theory assert dispute resolution outcomes are determined by the varying and relative strengths of their constituent domestic interests and their interactions at the sub-governmental level.[685] Moreover, social science research indicates integration of public opinion with the domestic environmental decision making process increases information available to decision makers and enhances the public's willingness to support and implement ensuing environmental policy,[686] and public involvement facilitates consensus building and provides an opportunity for government authorities to educate the local population about environmental and economic trade-offs.[687]

There is a further need to examine the domestic factors shaping a state's position in international environmental negotiations.[688] For instance, a state attempting to balance the costs of pollution abatement with a desire to limit its ecological vulnerability must consider its citizens are primarily interested in enlarging their own wealth and power, as opposed to regional wealth and power. States may therefore come under domestic political pressure to abate transboundary pollution only when the economic and social costs of doing so are low and where there is direct benefit to the local population, as exemplified by the Russian government's unwillingness to enter into negotiations for resolving the disputes arising from environmental degradation caused by the stationing of Soviet forces in CEE. As such, some social science commentators argue that if public concern is focused on international environmental issues, then it is more likely a state will enter into and comply with binding international agreements. If public concern, however, is focused on the local environment then there is less of a chance a state will enter into binding international agreements, or if it enters into one, then there is less of a chance it will actually comply with the agreement.[689]

As CEE states become more democratic, they may be more willing to abide by international law and to respect other states' municipal laws, as according to scholars subscribing to the liberal approach to legal theory, the political and economic tenets of a liberal democratic state are more consistent with the operation of an independent body of law. States failing to democratize, however, may be less inclined to use international law to resolve their disputes.[690] Neighboring liberal democratic states also may be more willing to facilitate the development of transboundary political links between like-minded sub-state actors and to provide an opportunity for actors of one state to participate as individuals or NGOs in the rule making or judicial processes of another state.

The identification of domestic factors, such as the degree of public participation and access to environmental remediation technologies, also indicates whether symmetry might exist between the disputant states, which would increase their desire and ability to agree upon and implement legal mechanisms for reducing and remediating transboundary environmental degradation.⁶⁹¹ In particular, political scientists subscribing to the structural-symmetry theory assert environmental policy is developed and depends in part on the respective state's socio-economic structures, such as their degree of industrialization, urbanization, economic wealth and technical capabilities, and the greater the degree of structural symmetry the greater the degree of potential co-operation.⁶⁹²

Moreover, according to some social scientists, the type of government may not be as important as the fact there has been a change in government. Studies indicate states are generally more likely to consider participating in or creating regimes after change in the form and character of domestic political power relationships, as evidenced by the post 1989 escalation in efforts to resolve the various nuclear power disputes and the Black Sea dispute, as well as the post November 1997 escalation in efforts to resolve the Danube River Gauntlet dispute and Romanian–Bulgarian Nuclear Corridor dispute.⁶⁹³

Relevant domestic political circumstances influencing how international law may be used to assist in the resolution of a transboundary environmental dispute may be considered to include: popular environmental awareness and activism; the existence of a process for public participation in the development of environmental policies; the possibility and level of citizen participation in the judicial process; the perceived need for industrial growth as compared to environmental protection; the ideological composition and varying influence of government and bureaucratic actors; and the degree of democratic transition.⁶⁹⁴

2. *International political circumstances*

Without a central authority to establish or coordinate environmental or economic regulation, it is distinctly possible states will adopt different approaches to environmental protection and economic development.⁶⁹⁵ It is therefore important to focus attention on the questions of whether any transboundary authorities are emerging which might promulgate or harmonize environmental and economic regulations, and whether states are adopting divergent or compatible environmental and economic regulatory policies.

In assessing the current normative structure governing state relations, it is necessary to discern whether states have created any regional norms, in particular through bilateral or multilateral treaties, which might promote a resolution of transboundary environmental disputes, such as the Baltic Sea and Black Sea disputes,⁶⁹⁶ and to examine whether institutions, such as the EU, have sought to extend the influence of

152 *Understanding the Role of International Law*

their norms through their inclusion in EU association agreements or through assistance with harmonizing domestic legislation. For instance, in 1995, Poland prepared and adopted a Framework Act on the Environment, which was designed to reform Poland's sectoral approach to environmental law and incorporate relevant legal principles contained in EU legislation, as required by its 1991 Agreement of Association with the EU, and in 1998, the Czech Republic amended its hazardous waste to comply with EU regulations.[697] In addition, the EU has indicated that over the next decade it is willing to allocate a total of $55 billion from Structural Funds and the Cohesion Fund to assist CEE states with infrastructure and environmental improvements in order to meet EU standards.[698]

Game theorists would further assert there is a need to ascertain within which other international games the transboundary environmental dispute is nested and whether states ascribe additional positive or negative secondary values to the benefit or detriment that might be caused to neighboring states by a resolution of the transboundary environmental dispute.[699] It is particularly necessary to identify clearly which game states are playing and the relationship of that game to other games given that, although most states may initially be perceived as playing the game of transboundary environmental dispute resolution for the purpose of promoting environmental protection, it is possible they are playing a different game, such as strategic economic positioning, or the game of transboundary environmental dispute resolution is a subsidiary game of a separate primary game, such as EU integration. For instance, in the Black Triangle some Polish government officials fear Germany is playing the game of strategic economic positioning under the guise of the game of regional environmental protection. While in the same dispute, the Czech Republic may only be playing the game of regional environmental protection since it facilitates its chances of success in what it perceives as the more important game of EU integration.

The social environment in which international law functions affects its role as well, with the social environment of CEE characterized by the equal strength of various states and the law of reciprocity.[700] In order to confirm this perception, it is important to explore the degree to which CEE states are interdependent, their relative degrees of power and the extent they may seek to capitalize on any disparate power relationships. Relevant questions include the functionality and durability of current regional cooperation mechanisms and external (for example, the EU) pressure upon CEE states to craft and act within interdependent relationships.

Relevant international political circumstances influencing how international law may be used to assist in the resolution of a transboundary environmental dispute may be considered to include: the general power relationships and normative structures for cooperation between states; pressure and/or support from IFIs or other international organizations to settle disputes; the provision of financial resources necessary to construct projects that will generate or aggravate disputes; pressure from organizations, such as the EU, with which states wish to become associated or to join; and the availability of nonjudicial international mediation or arbitration opportunities.

D. National minority circumstances

As domestic political theorists would acknowledge, national minority groups constitute a domestic actor who may exercise unique influence on both the domestic and international political processes relating to transboundary environmental dispute resolution. National minorities, for instance, play a special role in the domestic process of the aggregation of a state's national interests as, under international law, they are provided with equal rights to participate in the public policy process, yet they are frequently denied the fair consideration of their interests.[701] National minorities also possess specific rights with respect to the protection of their culture, which may include their natural environs, yet again, these rights are frequently under threat by national majorities.[702]

Since national minority populations are generally resident in border regions, they frequently maintain a unique interest in the resolution of transboundary environmental disputes. With respect to a state whose national majority population exists also as a minority population in a neighboring state, there is the likelihood the former state will seek to exert influence over the behavior of the latter state as it relates to domestic or international decisions affecting the well-being of the ethnic minority population with which the former state identifies.

To assess accurately the effect of national minority circumstances on the dispute resolution process, an inquiry should be made as to the existence and status of national minorities in CEE states, and an assessment of which specific rights to public participation and cultural protection they are entitled within the relevant human rights normative structure. Consideration must also be given to the social institutions existing to protect and promote those human rights and whether those institutions have directed their attention to the circumstances of national minority populations affected by transboundary environmental degradation.

Relevant national minority circumstances influencing the transboundary environmental dispute resolution process thus include: the existence of national minorities within a state's territory; the existence of national minorities in an area of environmental degradation in neighboring states who identify with the national majority of the polluting state; rights and mechanisms for the public participation of national minorities in economic and environmental decision making; and the willingness of international human rights organizations or bodies to apply pressure for dispute resolution and to offer mediation services or good offices.[703]

IV. Factors promoting the functionality of international law within the regime of transboundary environmental protection

International law does not always operate as a constant, and it is therefore necessary to determine which factors may indicate a willingness of states to use or rely upon international law, and may indicate whether international law will adequately function within the regime of transboundary environmental protection. Notably, these factors relate both to a willingness of states to use international law and the legal process, and an ability of international law and the legal process to enhance the efforts of a state to resolve a transboundary environmental dispute.[704]

A. The practice of applying international law to resolve international disputes

States with a practice of applying international law to resolve disputes with other states may be more likely to use and more adept at using international law to promote the resolution of their transboundary environmental disputes. According to some international relations scholars, states which have participated in international regimes have been exposed to their value and will be more likely to create or use additional regimes to resolve outstanding disputes.[705] States with a past practice of using regimes will also be more proficient at using the rules of law contained within general regimes and may use them as models for creating nuanced sets of laws and processes for the regulation of specific behavior.[706] When states have used a set of rules to resolve a dispute, they are inclined to use those rules to resolve future disputes or if they are insufficient, to investigate the possibility of using additional rules of a similar nature.[707] Conversely, when a state has unsuccessfully attempted to use international law to resolve its international disputes, it will tend to perceive law as an inadequate tool for promoting a resolution of that state's transboundary environmental disputes.

If two or more states have a practice of successfully using international law to resolve disputes among themselves, they also will likely have developed patterns of mutual trust and may, therefore, be more inclined to experiment with using international law to resolve their mutual transboundary environmental disputes.[708] These states may also be more familiar with the substance of international law applicable to their dispute and possess a mutual understanding of the benefits, as well as the risks, of applying that law.

B. The practice of applying municipal environmental law to resolve domestic resource disputes

States with a practice of applying municipal environmental law to resolve their domestic resource disputes are likely accustomed to the use, operation and function of law, are familiar with the utilization of law making and law implementing institutions and are thereby more likely to use international law to promote the resolution of their transboundary environmental disputes.[709]

Some political scientists have concluded that management of local common resources often indicates the willingness and ability of nations to engage in international cooperation to manage transboundary resources, because states are familiar with the practice of using laws to manage conflict over resources and their value for producing an optimum and enforceable allocation of those resources.[710] States that utilize municipal environmental law will also have experience in translating normative legal principles into a functional framework for generating options for the resolution of environmental disputes.[711] Similarly, individual domestic political interests which have a practice of using domestic law to resolve resource conflicts may be more adept at prompting their states to use international law to influence the allocation of international resources. Individual policy makers may also develop a certain level of expertise with resource management regulations, which they may be able to transfer from the sub-systematic to the systematic level.[712]

The active use of domestic environmental law might also engender the creation of institutions which, with some modification, could be used by environmental interests to promote the reduction of transboundary environmental pollution. The usefulness of these institutions might be enhanced to the extent they are modified to permit citizens of neighboring states to participate in the rule making, administrative and enforcement processes, as well as the judicial process.[713] Similarly, the greater the degree of congruence between neighboring states' municipal laws, the greater the chance citizens of those states may utilize each others' domestic resource allocation processes, as they will be familiar with the relevant laws and mechanisms.[714]

C. The correlation of municipal and international environmental law

The approaches of certain international relations scholars emphasizing the sub-systematic factors affecting the creation of regimes, in particular the cognitive, sub-systematic and normative–institutional approaches, would suggest states may also be more likely to use international law to assist in the resolution of transboundary environmental disputes when the municipal environmental laws of a state are parallel to international environmental laws and when international law is given full force and effect in municipal law.[715] Where a state's own municipal environmental legal norms parallel international environmental legal norms, a state may have a greater understanding of the relevant norms of international law, and those norms will likely be more credible for that state. If both parties to a dispute possess municipal legislation that is parallel to international environmental norms, then they may also be more likely to reach agreement to use international law in resolving their dispute, since they have both played a similar game by similar rules. Alternatively, the 'internal legal orders' reflected in the municipal environmental laws in a particular region might serve as a source from which to deduce regional principles of international environmental law applicable to the resolution of transboundary environmental disputes.[716]

Some international law scholars posit that states use international law more frequently when they adopt domestic legislation giving effect to international law, and when international law is embraced in the national life and institutions of a particular state such that the state has a tradition of abiding by the rule of law.[717] Scholars subscribing to the liberal approach to legal theory further assert democratic states are more likely to use international law, since they may be freely criticized by public advocacy groups for violating international law, and because the domestic checks and balances system of most democracies affords an opportunity for different bureaucratic or political actors to invoke international law in support of their policy positions.[718] Some international environmental law scholars add that in the area of international environmental law, it is particularly important for a state to possess domestic institutions capable of giving effect to international environmental obligations, particularly domestic verification mechanisms able both to verify compliance with international obligations and to increase a state's familiarity with the general operation of a verification process.[719]

D. The accessibility of international dispute resolution mechanisms

Where states involved in a transboundary environmental dispute perceive there to be an objective institution capable of competently applying the principles and rules of international law to their dispute, those states may be more inclined to use international law to promote a resolution of that dispute.[720] Although the existence of an adequate international dispute resolution mechanism is not necessarily sufficient to lead to the resolution of a transboundary environmental dispute, its existence substantially enhances the utilization of international law, and its absence lessens the chances of the proper application of international law.

According to some international law scholars, the application of international law is improved where international institutions and mechanisms exist for the purposes of law making, law determination and law enforcement.[721] However, unlike municipal legal systems where these three functions are customarily performed by a central political body, in the international system the authority to perform these functions is either fragmented or absent.[722] Some international environmental law scholars have further determined that institutions developed to perform these functions will be more successful if they are permanent bodies with a mandate to measure state performance and to carry out regulatory and supervisory functions, rather than to operate as a formal dispute resolution mechanism designed to assess liability.[723]

A dispute resolution mechanism may take a variety of forms, such as the ICJ, which may be called upon to clarify and apply certain rules of international law to the dispute, or a more interventionist entity with a juridical capacity, which is mandated to mediate the dispute and proffer suggested options for its resolution.[724] In either case, the dispute resolution mechanism will likely recognize the objective nature of international environmental law and the value of its application, may be familiar with the specific principles of international law relevant to the circumstances of the dispute and may be capable of applying to the dispute international principles that might be too indistinct to be applied directly by the disputant states.

Decisions of dispute resolution mechanisms also promote the implementation of international law, as they articulate legal options available to the parties, and they provide a state with political cover to implement

158 *Understanding the Role of International Law*

an unpopular but necessary decision.[725] The availability of political-based regional environmental dispute resolution mechanisms may also improve the likelihood of transboundary environmental dispute resolution by providing the parties a forum within which to negotiate and an opportunity for third parties to assist in structuring possible solutions.

E. The incentive to use international law to assist in the resolution of transboundary environmental disputes

If states perceive there to be value associated with the use of international law and if this value is not outweighed by the costs of using international law, then states will naturally be more likely to use international law to promote the resolution of their transboundary disputes.[726] This principle is relevant for two types of value: first, the endemic value associated with maximizing a state's net political and economic gains with respect to a particular dispute and second, the external value created by third parties.

Some international relations scholars posit that when states are faced with disputes resolvable through coordination, such as in the Baltic and Black Sea disputes, states will pursue a maximin strategy, to reach a nash equilibrium, which will lead to the formation of an international regime since such a regime will serve to perpetuate the nash equilibrium.[727] In dilemma situations, such as the Gabčíkovo–Nagymaros Project and Black Triangle disputes, where the pursuit of a maximin strategy produces a stable sub-pareto-optimal result,[728] states might seek to create an international regime in order to build mutual trust and a pattern of behavior that could support the achievement of a pareto-optimal result.[729] Since the incentive to defect makes it inherently difficult to perpetuate a pareto-optimal result in a dilemma situation, states will seek to form an international regime to maintain the pareto-optimal result by creating disincentives to defect.

In deadlock situations, such as the Soviet Environmental Legacy or the Temelín Nuclear Power Plant dispute, where it is perpetually in the net interest of one state to defect from cooperation (although this produces a sub-pareto-optimal result), that state will avoid the creation of any international regime, while the victim state will seek to link the deadlock situation with other circumstantial situations in order to create an incentive for the former state to submit to an international regime, such as potential EU membership.[730]

In addition to the endemic value created by using a legal regime to maximize political and economic gains, value may be created externally when international organizations, such as the EU, indicate their relationship with those states will be evaluated in part based on the degree to which they seek to resolve their transboundary environmental disputes and the extent they use international law to assist in the resolution of those disputes.[731] Similarly, external value may be created when donor states or organizations indicate a continued financial relationship is in part based upon the use of international law to resolve transboundary disputes.[732] Conversely, there may be an absence of value associated with the use of international law when donor states or organizations propose and/or support projects incompatible with general principles of environmental law or when these states or organizations do not require procedures, such as EIAs, which are hallmarks of certain emerging principles of international environmental law. External value may also be created when certain sub-state actors pressure a state to use and abide by international law in its international relations.

F. The perceived capacity of international law to promote the resolution of transboundary environmental disputes

Where states perceive international law to possess the substantive and procedural capacity to promote the resolution of a transboundary environmental dispute, they will be more inclined to employ international law. Where states perceive international law to be vague, hortatory and ineffectual, they will naturally be disinclined to rely upon it as an instrument for promoting dispute resolution.

The capacity of international law to assist in the resolution of transboundary environmental disputes revolves around the attempt of international law to balance the rights and obligations of a sovereign state with the rights and obligations of the international community of states.[733] The value of state sovereignty is then balanced endogenously between a state's interest and right in seeking to exercise sole control over the activities occurring within its border, yet also seeking to protect resources within its territory from degradation caused by activities occurring within the sovereign sphere of other states.[734]

This characteristic of international law is particularly important in the context of CEE transboundary environmental dispute resolution, as prior to 1989, CEE states subscribed to socialist international law, which did not generally seek to establish this equilibrium of sovereignty, but rather placed primary emphasis on the principle of noninterference in the internal affairs of a state.[735] As CEE states attempt to adapt to the application and use of general international law principles, they must also grapple with the contemporary modification of the nature and scope of sovereignty, which entails the evolution of authority over many natural resource decisions to international organizations and the sharing of decision making authority with other states.[736]

Given the complexities of CEE states' relationship with general international law, CEE states are unlikely to apply and use international law effectively, unless the existing law is specifically relevant to the needs of CEE states and is inherently and operationally functional.[737] In order for international laws to affect the behavior of states and the structure of an international regime designed to regulate state behavior, these laws must exist and be cognizable by those subject to their authority. Similarly, these laws must be of such a quality as to ensure states applying them can perceive how they should modify their behavior to comply with the substantive intent embodied within the laws.

Despite their obvious nature, these criteria are important in the context of international environmental law since there is frequently a great deal of debate between states and among academics as to which 'laws' are binding hard law and which are prescriptive soft law.[738] Similarly, there is frequently a great deal of debate as to the actual substance of certain hard and soft laws. Therefore, in order to determine accurately whether states perceive international environmental law as having the capacity to assist them in the resolution of transboundary environmental disputes, it is necessary to ascertain whether, from their perspective, specific laws relating to all four phases of the dispute exist, are relevant and functional. The capacity of international law will be determined by whether the laws positively, constructively or indeterminately exist, whether they are temporally and/or dispositionally relevant and whether they are inherently and operationally functional.

Positively existing principles exist as customary international law and/or within multi and bilateral treaties. Constructively existing principles are generally emerging principles of customary international law contractually brought into force by specific multi and bilateral treaties and thus binding upon their parties. Indeterminate legal principles are those, which although they may be emerging as norms of customary international law, are not contained in multi and bilateral treaties adopted by CEE states. Indeterminately existing legal principles are similar to soft law principles in that they serve as 'signposts on the way to customs and treaties',[739] and they provide states flexibility to structure dispute resolution.[740]

Relevant legal principles are those appropriate to the resolution of the subject matter of a particular dispute. Temporal relevance refers to the need for legal principles to be applicable to the specific timing or stages of a dispute, and dispositional relevance refers to the alignment of these principles with various disputant states' attitudes and approaches to international law.

Functional legal principles are those that can be applied to the circumstances of the dispute to generate specific options for resolution. Inherently functional legal principles are generally accepted principles capable of being understood by disputant states and not subject to a multitude of varying interpretations and speculative legal argument. Operationally functional legal principles are principles, which if understood and applied by the parties, yield the outcome intended by the rules.

Conclusion to Part III

The purpose of this chapter has been to draw general conclusions from the case studies discussed in Parts I and II concerning the social and political context in which international law operates and the functions it performs within this context, actors and circumstances which affect the role of law in promoting transboundary environmental dispute resolution, and factors that influence whether states are likely to rely upon international law and which influence the functionality of international law.

As such, this chapter first hypothesizes international law should be conceived of as operating as a constituent element within the international regime of transboundary environmental protection. Where international regimes consist of principles, norms, rules and decision making procedures which modulate state behavior, international law serves to provide the substance of many of those norms, rules and decision making procedures, as well as a variety of dispute resolution processes. The legal elements of the regime may then serve to assist the regime in promoting transboundary environmental dispute resolution by serving a variety of functions during the phases of the dispute resolution process.

In the dispute formation phase, for instance, international law may assist a party in framing any claims it may have in the context of established international rights and obligations, and ascertaining what obligations other states have to consult with and inform it of potential harm to its territory. The legal processes of inquiry and mediation may assist the parties in establishing the facts essential to the formation of the dispute, and might facilitate the exchange of information between the parties, while creating an atmosphere conducive to extended consultations aimed at reaching an early resolution of the dispute. In the pre-resolution stage, international law may be used to compel another party to share information and to engage in joint EIAs, while the legal process of conciliation may help the parties in sharpening the parties' understanding of the causes of the environmental degradation giving rise to the dispute and in identifying possible compromise solutions.

In the dispute resolution phase, substantive international law may assist the parties in establishing a legitimate foundation for their actions, as well as evaluating possible procedural obligations under which the negotiations might continue and identifying solutions for resolving the dispute. The legal processes of arbitration and adjudication may then be used to resolve factual disputes, identify the principles and rules applicable to the dispute, or even determine each party's specific legal rights and obligations. Finally, in the implementation phase, international law may serve to ensure states fulfill their agreements, including the obligation to take all necessary domestic measures to give the agreement domestic effect. States may also establish joint legal mechanisms for the purpose of monitoring and possibly enforcing the agreement.

The ability of states to use law to serve these functions is significantly influenced by the interests and capabilities of a variety of sub-state actors, including government officials, economic actors, environmental actors and even the general public. The use of law is also influenced by a variety of interested third parties which may include the member states of the EU and the European Commission, the United States and the IFIs. Sub-state actors may influence the use of law by defining the national interest of a state to be pursued and the role of law as an aspect of that national interest, fostering the creation of internationally linked domestic coalitions, and advancing the ability of the state to implement and enforce any agreed upon decision. Sub-state actors may also inhibit the use of international law by objecting to the creation of binding obligations or the use of impartial means of dispute resolution. Third parties may affect the use of law by creating an incentive to use law, and by identifying suitable legal processes through which the dispute can be resolved.

The ability of the various states, sub-state actors and interested third parties to affect how international law is used to resolve transboundary environmental disputes is influenced by an array of circumstances, including the ecological circumstances giving rise to the dispute, as well as the economic circumstances influencing the respective political weight of the sub-state actors and the susceptibility of the state to financial pressure from third parties. Similarly, domestic and international political circumstances define the interests of a state's constituent actors and the relative influence of those actors on the dispute resolution process, as well as that state's willingness to enter into and comply with binding international agreements necessary to resolve the dispute.

In addition to the influence of situational circumstances on the interests and behavior of states, sub-state actors and interested third parties, the use of law is affected by certain factors which may promote or inhibit the actual functionality of international law within the regime of transboundary environmental protection. The practice of states of using international law within the context of other international regimes, such as regimes regulating trade relations or the protection of human rights, will affect the functionality of international law within the regime of transboundary protection in that if a state has a past practice it will likely be familiar with the advantages and disadvantages of relying upon international law, as well as with the most effective means of using law to achieve its objectives. Similarly, if states have a past practice of using municipal law to resolve their domestic resource disputes, then they are likely to be familiar with the content and scope of many of the principles applicable to transboundary environmental disputes. When there is a correlation between a state's municipal environmental law and international environmental law, the state will be able to give sufficient effect to international law and thus maximize its utility.

Two of the most important factors would appear to be the accessibility of international dispute resolution mechanisms capable of carrying out the many functions noted above, and the perceived capacity of international law to assist in the resolution of transboundary environmental disputes. A state that perceives law to positively exist, be relevant and serve a functional purpose will likely seek to use international law as a constructive tool of dispute resolution. If, however, states perceive specific international legal principles to be vague, tangentially relevant and serve no clearly identifiable purpose, then they are unlikely to use law, or if they use law are unlikely to employ it in a manner that would actually further the resolution of the dispute.

Part IV now turns to an application of these models to evaluate the prospects for an increasing role for international law in promoting CEE transboundary environmental dispute resolution by examining the situational circumstances affecting the interest and ability of states, sub-state actors and third parties in employing international law, and assessing the perceived capacity of international law to actually assist in the resolution of these disputes.

Part IV

Prospects for an Increasing Role for International Law in Promoting Central and East European Transboundary Environmental Dispute Resolution

9
Gauging the Operability of International Law: the Evolving Circumstances

As discussed in Chapter 8, ecological, economic, political and national minority circumstances of CEE states influence the interest in and ability of CEE states, sub-state actors and interested third parties to resolve transboundary environmental disputes. This chapter explores the various circumstances affecting the behavior of states, and in particular draws conclusions relating to whether the changing ecological dynamics and the political and economic transition currently underway in CEE will enhance or detract from the ability of CEE states to use international law to assist in the resolution of their transboundary environmental disputes. The chapter also indicates, in light of these various circumstances, which elements of international law and the legal process might be most productively utilized to assist in the resolution of these disputes.

I. Ecological circumstances

Ecological circumstances in CEE influence the type of transboundary disputes afflicting CEE states, their ability to resolve these disputes through cooperative means, and the substantive areas of international law most applicable to the dispute. Ecological circumstances may also influence the ability of a state to take the necessary measures to resolve the dispute, and the level of interest a state and its sub-state actors may possess with respect to resolving a particular dispute vis-à-vis other international disputes.

The basic geography of Europe can be described as a 'conglomeration of peninsula upon peninsula', with almost all geographic areas of CEE in close proximity to a sea. This proximity to large masses of water affects the humidity, wind direction and precipitation levels of the various CEE ecosystems. The ecology of the European continent is further

168 *Prospects*

affected by human trade and industrial activity, natural and man-made transportation routes, the drainage of water basins into the various seas and sub-regional atmospheric systems. Notably, whereas many states in other geographic regions suffer from disputes relating to water quantity, the peculiar ecological circumstances in CEE dictate that most CEE water disputes relate to quality.[741]

CEE states generally have longer and slower rivers than those of the West, and as such suffer less from erosion and the accumulation of silt. These slower rivers, however, exhibit a lower rate of seepage into the groundwater system, and from the groundwater into neighboring seas, and thus most insoluble toxic substances quickly concentrate and have long term effects on the environment. The CEE regional ecosystem further suffers from a substantial loss of energy into the environment as water flows from high altitudes to lower plains, and in plain areas from a lower hydrological energy capacity and relatively low relocation of organic material, which is exacerbated by lower rates of precipitation.[742]

Because of the overlapping peninsular geography of Europe, the sub-regional atmospheric systems are of primary importance since the high humidity caused by the proximity to large bodies of water creates a solution platform capable of dissolving and carrying large amounts of sulphur compounds. Although prevailing winds flow westward, large pockets of cold air intercept the flow of warmer air masses, causing frequent depositions of sulfuric acid across CEE. This phenomenon is augmented as the recipient rivers of SO_2 precipitation recontribute the sulphur to the ecosystems through which they flow.[743]

Despite extensive monitoring and atmospheric mapping technologies, it is very difficult to determine how much SO_2 is transported between geographic regions and across state boundaries. It is similarly difficult to precisely determine the relationship between the amount of SO_2 emissions and the amount of acid rain falling in a particular area.[744] Economic actors are thus likely to resist calls for control until the efficiency of particular controls on their economic units can be verified vis-à-vis controls on alternative economic units.

Given the sheer number of enterprises emitting pollution and migration patterns of those pollutants, it is seldom possible to accurately assess which states and which enterprises are responsible for specific harms caused by transboundary pollution, and it is therefore difficult to attach liability. Moreover, because CEE air currents and water resources frequently circulate in reverse patterns, CEE states export and import substantial amounts of pollution, and no one CEE state can either claim to be the chief recipient of pollution, or be charged with being the principal exporter of pollution in the region. CEE states are thus likely to resist bilateral arrangements to resolve transboundary air and water disputes, and seek multilateral arrangements – which are significantly more difficult to negotiate and to implement.

In recent history, the ecology of CEE has been subject to three phases of influence from industrial behavior. The period from 1948 to 1960 was characterized by rapid economic development with minimal attention to its ecological consequences. From 1961 to 1970, the region experienced slower economic growth and evidenced some of the first signs of serious local ecological degradation, such as substantially lowered water quality. During this time, CEE states actively undertook some basic environmental remediation efforts, but frequently this activity, for example the construction of higher industrial emission stacks, did not serve to positively influence the regional environment.

From 1971 to 1989, CEE states experienced alternating periods of economic growth and economic stagnation, coupled with efforts to raise the standard of living, including efforts to improve environmental quality. Unfortunately, most of these efforts were symbolic in nature and had no appreciable affect on environmental quality.[745] As a result of the environmental degradation caused by these failed policies, people living in CEE now suffer disproportionate levels of over exposure to lead, acute and chronic respiratory conditions, high incidence of fluorosis, excess infant and adult lung cancer mortality, abnormal physiological development and high rates of infant chemical asphyxia. In general, these afflictions are caused by wide ranging lead contamination of soil and air, high quantities of airborne dust, SO_2 and NO_x, nitrate pollution of watercourses, and contamination of agricultural produce by lead, pesticides, hydrocarbons and chlorinated organics.[746] Because of this widespread environmental degradation, environmental concerns are not limited to certain geographic areas, but rather can be considered to affect the general public as a whole, with particular interest being held by the public who live in environmental hot spots. Environmental actors may thus be able to tap into the voting power of the general public to support domestic legislative measures designed to reduce environmental pollution.

With the transformation from a command to a market economy, many CEE states are experiencing a 'clean collapse' wherein numerous industrial enterprises simply shut down as a result of their inefficiencies or the absence of a demand for their products. Consequently, CEE states are benefiting from a substantial reduction of harmful emissions. In Slovakia for instance, from 1988 to 1992, the emission of SO_2 dropped

170   *Prospects*

from 606,000 to 374,000 tons.[747] It is estimated the clean collapse would continue throughout 1998, and thereafter as industrial production increases, the level of emissions will remain the same, or further decrease as a result of the introduction of cleaner technologies and pollution abatement mechanisms.[748] According to some regional commentators, in order to comply with EU standards, CEE states will not be able to rely upon new capital investment alone, but must also continue to retrofit and decommission a number of existing plants.[749] As these reductions occur, there will likely be a positive effect on the resolution of transboundary environmental disputes if the reductions are made in border areas, or in ways that reduce the transboundary flow of pollutants. Alternatively, there could be an increase in tension as certain states may believe they are allocating scarce resources to reduce pollution while other states, which do not allocate such resources, reap much of the benefit.

II.   Economic circumstances

The widespread environmental degradation in CEE is in large part a result of the heavy industrialization, and the unregulated nature of socialist industrial production. With the transformation to market economies, CEE states are beginning to address some of the circumstances that led to their current state of environmental affairs; however, this process tends to occur more slowly and at a higher social cost than originally anticipated. As the process continues, the evolving economic circumstances will affect the extent to which state and sub-state actors, particularly, but not exclusively, economic actors, may seek to use international law to assist in the resolution of transboundary environmental disputes by defining the relative authority or position of the relevant actors and promoting or discouraging the integration of the interests and behavior of the various actors participating in or affected by the disputes. Economic circumstances may also affect the operation of international law by promoting or discouraging the cross border association of sub-state actors with similar interests, and enhancing or limiting the interest in and ability of concerned third parties to involve themselves in the dispute.

A.   Pre-1989

The socialist ideology of the former CEE governments was based primarily upon the principle of public ownership of a state's economic and natural resources. In theory, a socialist society would have substantial influence over the regulation of production, utilization of natural resources and provisions for environmental protection, with the aim of balancing the economic and social needs of society. And because all costs to society were interrelated and reduced the socialist standard of living, a state was theoretically obligated to reduce those costs, whether they be economic or environmental.[750]

Despite the socialist ideal of economic harmony, there existed substantial tension between the utilization of resources to improve the standard of living and environmental protection arising from the fact a number of economic sectors in the socialist CEE states were substantially underdeveloped.[751] This tension was almost exclusively resolved in favor of raising living standards and ignoring environmental costs.[752] Because of their status as semi-detached economic units, state enterprises could readily treat environmental damage as an unaccountable externality.[753] In addition, economic planning did not take into consideration the environmental costs of facility siting, raw material consumption or production quotas, and as such, industrial production proceeded with little concern for environmental protection.[754]

Because of the scarcity of economic resources, CEE enterprises concentrated on short term solutions based on maximizing the utilization of resources and neglecting long term planning, the conservation of natural resources and investment in infrastructure.[755] As a result, many heavy industrial enterprises operated inefficiently and consumed excessive amounts of energy and raw materials per unit of production.[756] The cost of steel produced in the Lenin steelworks in Nowa Huta (Kraków), for example, was six times the world price.[757] The harm caused by such inefficiency was augmented by the extensive use of lignite as a power and heating source, the extensive use of low grade Russian oil, and the misallocation of natural gas resources – Bulgaria, for instance, consumed substantial amounts of natural gas, which it used to produce fertilizer at one-tenth the market price.[758]

In addition to suffering from an economic ideology denying the existence of environmental degradation and the need for its abatement, CEE states suffered from an economic reality relying upon heavily polluting industrial enterprises, and a political doctrine prohibiting the decentralization of authority necessary to establish minimal environmental safeguards.[759] These economic circumstances favored the interests of large economic actors and the central government,[760] created little desire for the resolution of transboundary disputes resulting from the pollution of industrial enterprises,[761] and allowed for almost no consideration of the use of legally grounded 'Western' transboundary economic mechanisms, such as tradable emission permits.[762]

172 *Prospects*

B. Post-1989

Since 1989, CEE states have begun the difficult economic transformation from a command to a market economy. This transformation presents CEE states with an opportunity both to develop a market economy possessing the necessary safeguards to ensure environmental protection and to develop a program of sustainable development.[763] To ensure the transformation to a market economy does not further exacerbate transboundary environmental disputes, and provides an opportunity for the innovative use of international law, CEE states must undertake substantial environmental investment and must regularly conduct a transboundary cost–benefit analysis.

1. The transformation to a market economy

One political history commentator notes after forty years of communist stewardship, in the late 1980s CEE states epitomized a 'new third world, ... stuck at a certain outdated, uneconomic, wasteful, industrial and scientific base, which lacked the resources, the human capital and know-how to modernize'.[764] Unfortunately, after forty years of forced socialization, the economic actors in CEE are naturally slow to change their patterns of behavior.[765]

Despite the environmental consequences of large scale energy and industrial production, there is substantial political support for the continued operation of industrial enterprises as they provide significant assistance for municipal infrastructure projects and generate substantial employment. In the Black Triangle region, for example, it is estimated 50,000 Germans and 100,000 Poles and Czechs are directly employed by the coal industry.[766] The economic and environmental transformation of large industrial enterprises has been further limited by the slow decline in the state subsidy of energy, and by the continued state ownership and/or operation of national and regional industrial monopolies.[767] This situation is slowly beginning to change as more industrial enterprises are privatized.[768] In the near term, however, environmentally unfriendly economic actors may thus be able to exercise significant influence on the aggregation of their respective CEE state's national interest, and will ensure their special interests are represented in transboundary negotiations.

In the long term, the change to a market economy should have a number of consequences for CEE states, including: a diversification of economic actors; a substantial increase in energy prices;[769] the imposition of hard budget constraints (with a shift from production targets to financial and efficiency concerns);[770] the replacement of out-dated equipment with modern technology; and a reorientation of production to consumer preferences and incomes. Most of these changes will have a positive affect on air quality and a correspondingly positive impact on CEE states' ability to resolve related transboundary environmental disputes. There will likely be, however, an increase in water pollution and disputes related to its transboundary consequences, caused by the shift in manufacturing to processed goods.[771]

Nevertheless, the development of a market economy does not automatically result in increased environmental protection and may pose additional threats to the environment. Without clear and enforced environmental legislation, for instance, newly privatized enterprises may continue the environmentally harmful activity of the previous state owners. Similarly, newly privatized enterprises will likely calculate the cost of environmental protection to determine whether it might undermine the market position of the enterprise, and whether it makes wise economic sense to invest scarce capital in nonproductive investments.[772] There is thus a risk these enterprises may determine it to be more financially rational to pay environmental fines than to undertake environmental investments.[773] To guard against the inherent environmental downsides of a market economy, CEE states will need to employ the polluter pays principle and impose strict regulatory limits on emissions.

2. The economics of sustainable development

The political transition, coupled with the change to a market economy, provides an opportunity for CEE states to develop an industrial program based on the principle of sustainable development and to surpass Western states both in terms of technological investment and industrial policy. Two primary barriers exist, however, to the establishment of a policy of sustainable development.

The first barrier is the underdeveloped political process in most CEE states, which is as yet fairly incapable of integrating environmental and industrial policy.[774] The interested third party of the OECD, however, is applying pressure for the integration of economic and environmental policies, and the CEE states appear to be responding to this pressure. The second barrier to the implementation of sustainable development is the interest of some Western industrial enterprises in creating a sustainable market for their nonsustainable technologies. CEE states, for instance, are faced with falling energy demand and a need for wholesale economic restructuring; however, their attention is being diverted by

Western nuclear power companies from demand-side solutions, such as energy conservation, to the supply-side solution of the continued operation or completion of nuclear power plants. To promote the continued development of nuclear power, Western companies are integrating the east–west power grids and rebuilding and finishing CEE nuclear power plants in return for electricity payments.[775]

3. The need and capacity for environmental investment

To ensure the transformation to a market economy leads to improved environmental protection, both domestic and international, CEE states must invest in and support economic policies that account for resource scarcity and facilitate the development of input-conserving technologies based on the integration of economic and environmental objectives.

In an attempt to pursue such a policy, the Czech Republic increased its expenditure on environmental protection from 0.7 per cent of its GDP in 1989 to 2.3 per cent in 1992.[776] The Czech Republic also created a National Fund for the Environment, which subsidizes local environmental protection projects, and on the whole, spent $93 million from 1994 to 1997 on environmental protection, with approximately 50 per cent of the funding allocated for projects in Northern Bohemia.[777] Similar funds exist in Poland, Slovakia, Hungary and Bulgaria.[778] To date, there is an indication that as CEE states continue to invest in environmental protection, particularly in border regions, they are increasing their interest in concluding agreements with neighboring states in an attempt to ensure those states also contribute resources to regional environmental protection.

Despite their efforts, however, it is unclear whether CEE states are financially capable of procuring the necessary capital investment, which for the entire CEE is estimated at $30–35 billion.[779] Direct foreign investment might help to meet or reduce the financial requirements of environmental investment since Western companies may be readily able to transfer more efficient production technologies and may bring with them more cost and environmentally sensitive management cadres.[780] Most foreign direct investment, however, flows into Hungary, the Czech Republic and Poland, with little investment flowing to Slovakia, Romania or Bulgaria.[781] This disparity, if it continues, will result in varying levels of industrial efficiency and improvements in environmental quality, as well as varying state ability to participate in and meet any target reductions contained in regional environmental agreements.

4. The difficulties of conducting a transboundary environmental cost–benefit analysis

Economic cost–benefit analysis assists a state in determining which actions, if any, a state should commit to taking in order to resolve a transboundary environmental dispute. In most CEE transboundary environmental disputes, this cost–benefit analysis would likely entail a balancing of the economic costs of decreased or alternative industrial and energy production with the benefits of an increase in the quality of natural resources and decreased health expenditures. The employment of a cost–benefit analysis to weigh the environmental costs of transboundary environmental disputes and the costs of various measures to reduce pollution and resolve the disputes is difficult, however, given the inherent complications with developing a general cost–benefit analysis for environmental protection, and the more specific problem that different CEE states associate different costs with various types of harm.[782]

In the Black Triangle dispute, for example, the parties have been unable to conduct a joint cost–benefit analysis because they are not able to agree upon a formula providing for a calculation of Polish concerns relating to damage caused to forestry resources, while accounting for the Czech Republic's concerns relating to the risk to human health.[783] In other cases, it might be difficult to calculate the trade-off between economic costs and environmental costs,[784] such as in the Gabčíkovo–Nagymaros Project dispute, where Slovakia calculates its costs due to lost energy production and additional construction requirements to be $178 million per year, while it is difficult for Hungary to attribute a dollar figure to the environmental costs of the project.

III. Domestic political circumstances

Prior to 1989, the CEE domestic political circumstances did not permit the inclusion of public views into the environmental decision making process. The decision making process thus suffered from a lack of diverse and insightful input, while the general public suffered from the lack of a public mechanism to utilize law to improve their environmental situation, or to even express their environmentally based frustrations. This build-up of frustration contributed to the 1989 revolutions, and since 1989, CEE states have attempted, with increasing success, to enhance the inclusion of the views of environmental actors and the general public in the environmental decision making process.

A. Pre-1989

Political circumstances were of primary importance in CEE states prior to 1989, since in order to perpetuate the existence of the communist regime, every aspect of society was deemed political and subject to ideological and functional control by the authorities.[785]

The pre-1989 communist regimes advanced their political control by suppressing information, including environmental information, and by manipulating public opinion.[786] In cases where information regarding environmental consequences of industrial projects was available, it was often simply disregarded. For instance, although the 1977 Hungarian government was aware of the potential environmental harm that might result from the Gabčíkovo–Nagymaros Project, any views questioning the rationality of the Project on environmental grounds were suppressed or ignored.[787] By suppressing information concerning environmental quality, in particular, health risks, the totalitarian regimes denied a basis for the public to become factually aware of the environmental hazards affecting their life, and denied the public the documentation to press for serious change in the means and priority for utilizing a state's natural resources.[788]

Consistent with the policy of suppressing information concerning environmental degradation, none of the CEE states' legal structures provided for meaningful public participation in the environmental rule making process. Nor did these states provide for participation of citizens of neighboring states who might be affected by the decisions of the polluting state.[789] And as environmental initiatives were centrally planned and financed, there was no substantial input into these policies by local government officials.[790] Similarly, environmental activism was discouraged as a form of social protest unacceptable in communist society and could either be silenced or ignored.[791] To prevent organized political opposition, the communist states pursued a policy of atomization intended to prevent social cohesion, with one method of atomization being to destroy historic social norms and thereby leaving the population dependent on the norms provided by the communist party. In some cases, the communist party destroyed norms for which it had no replacement counter-norms, as with ecological norms, and thus created a normative void.[792]

Prior to 1989, the general public in CEE thus lacked an awareness of the environmental consequences of heavy industrialization and an ethic of environmental protection. Where such awareness and an ethic did exist, the population lacked a means for influencing state policy relating to the environment. Local government officials similarly lacked an interest in or ability to influence environmental policy. As such, environmental actors and the general public developed little or no understanding of how to participate in an environmental decision making process, and had no reason to consider the employment of international or domestic law.

B. Post-1989

In the mid-1980s, the environmental pollution of CEE was of such an extent its existence could no longer be denied, and the environmental sentiment of the general public could no longer be suppressed. Environmental degradation thus came to play an important role in the 1989 revolutions,[793] and in part because of this role, the post-1989 CEE governments are providing an increasing opportunity for public participation in environmental decision making, and are in the early stages of devolving environmental decision making authority to local governments.

1. Increasing public participation in environmental decision making

As CEE states continue their evolution toward liberal democracies and permit greater public participation in environmental decision making, there is the increased likelihood they may seek to use international law to resolve their disputes and they may integrate economic and environmental decision making. Some commentators, however, hypothesize that because of the communist policy of atomization and destruction of social norms, it will take some time for new CEE democracies to create functioning liberal democratic societies and institutions capable of promoting environmental protection. In particular, CEE states initially suffered from the lingering distrust of state institutions, the lack of a cadre of qualified professionals and the lack of administrative experience necessary for effective environmental regulation.[794]

Since the mid-1990s, however, CEE states have begun to overcome these difficulties and to adopt extensive mechanisms for public participation in the environmental decision making process. Although effective use of these mechanisms requires a detailed understanding of the emerging political process and a period of repetition, many new NGOs are beginning to successfully navigate these obstacles and positively impact their state's environmental decision making process. Other public interest groups representing industrial interests also are making extensive use of new mechanisms for public participation. As the participation of both of these groups increases, there will likely be an increased integration of economic and environmental policies and pressure for the resolution of transboundary environmental disputes affecting their interests.

There is, however, the inherently democratic risk that short term economic interests will impede the articulation of environmental interests. In fact, some commentators note because the new political process is not well developed, and because of the weak political state of the environmental NGOs, there is the chance CEE governments will succumb to organized pressure from economic actors and adopt short term policies aimed at reducing economic hardship.[795] Similarly, the specific adoption and utilization of the principle of the rule of law is somewhat limited as CEE NGOs have yet to extensively operate on the transboundary level, and are not yet generally educated in or pre-disposed to using domestic or international law.

2. The decentralization of environmental policy formulation and implementation

Under the central planning of pre-1989 regimes, decisions regarding resource allocation tended to be subjective and politically motivated, with the decision making process characterized by soft budget constraints, rapid and negligent project design, and slow implementation. To encourage the transition to a free market and decentralized planning many interested third parties, including important donor organizations, have applied pressure to CEE states to provide more authority for local governments to formulate and implement environmental policy, and in some instances, for local governments to engage in transboundary environmental cooperation.[796] If coupled with access to local remedies by foreign nationals, the decentralization of environmental policy formulation and implementation may also serve to de-escalate transboundary disputes.[797]

The drive for decentralization is echoed by local governments since they frequently believe the harm suffered by their particular region is unique to environmental harms suffered in other regions of their state, and they are more likely to engage in efficient cross-border environmental cooperation with other local governments as they are unencumbered by the full set of international concerns burdening the central government.[798] In some instances, regional governments in West European states have developed sub-state regional transboundary environmental cooperation between themselves and local governments in CEE states. Most cooperation currently occurs between German regional governments, such as Bavaria and Saxony, and local governments in Poland and the Czech Republic. Local Czech municipalities, for instance, have established two agreements with Saxony involving nature protection.[799]

Despite the incentives, most CEE states retain central control over environmental decision making. The rationales for this approach are both practical and political, with the primary practical rationales being that local governments currently lack the capacity to engage in effective development and enforcement of environmental policy, and CEE states often lack a regional layer of government between the central government and municipalities.[800] Thus, for instance, in Hungary there are over 3,200 municipalities, none of which are sufficiently large to effectively develop or enforce environmental policy.[801] The political rationale is that CEE state governments see themselves as being in a period of transition and are concerned that the decentralization of environmental decision making, if it occurs in a national legislative void, will result in a myriad of inconsistent and unenforceable regulations.

IV. International political circumstances

International political circumstances influence the role of law by creating conditions where CEE states are able to recognize the existence of international disputes, where they come under pressure to cooperate on matters of environmental protection, and where the use of law is facilitated by formalized programs for cooperation. In instances where these programs are funded by international actors, CEE states will be more able to utilize and implement international norms and processes. Where CEE states have a practice of cooperating on matters of general interest, as well as environmental matters, there will be an opportunity for the utilization of international law beyond the obligation to cooperate.

A. Pre-1989

Prior to 1989, transboundary environmental degradation did not officially exist as a public policy issue, and therefore CEE states established few if any mechanisms to promote environmental cooperation. When states entered into agreements to pursue joint development projects, they would minimize or disregard information relating to the environmental consequences of those projects. Where a transboundary dispute did exist, this dispute was suppressed by the need to maintain a united political front as socialist states or by the influence of the Soviet Union.[802] Aside from the forced suppression of environmental disputes and subservience to Soviet hegemony, there was little substantive bilateral cooperation among CEE states. In order to ensure its political and economic domination of CEE, the Soviet Union actively discouraged intra-CEE state relations that would further regional integration or cooperation on specific issues, and instead required each CEE state to maintain its primary political and economic relations with the Soviet Union.

B. Post-1989

Since 1989, CEE states have expressed a near unanimous desire to substantially limit their international relations with Russia,[803] and have exhibited a disinterest in cooperating with other CEE states both on general issues and on environmental matters. CEE states have, however, come under pressure by the EU to cooperate on matters of environmental protection and have developed a number of formalized programs for cooperation.

1. Cooperation on matters of general interest

Even prior to the establishment of Soviet hegemony over CEE, there was no significant tradition of cooperation between CEE states. Since the demise of the Soviet control, CEE states have examined their bilateral relationships with other CEE states and have found the traditional lack of desire for cooperation remains. CEE states have also discovered a number of unresolved historic controversies have reemerged to complicate their relationships, and there are a number of contemporary controversies, which from their perspective, justify limiting cooperation.[804]

Circumstances which initially served as a disincentive for intra-CEE cooperation arose from the desire of certain states to separate themselves out from any 'package' of a state's being considered for membership in the EU or NATO. The Czech Republic, for instance, perceived its association with 'Eastern Europe' as a matter of historical chance and not destiny or geopolitical design, and it wished to rectify this historical error by accelerating its integration into Western Europe. This attitude in turn diminished the interest of other CEE states in cooperating with the Czech Republic.[805] CEE states also initially perceived themselves to be in a state of competition for resources and assistance from Western states.[806] As the EU and NATO have increased their insistence on a cooperative approach to integration, Poland, the Czech Republic and Hungary have augmented political cooperation on matters of mutual interest. There is, however, a growing political separation between the northern and southern tier CEE states.

2. Cooperation on environmental matters

Although CEE states initially exhibited only mild interest in co-operating on transboundary environmental issues, preferring to treat the issue of the environment as a win–lose game and not as an integrative win–win game, they have recently embarked on a number of programs of mutual inquiry.[807] Most of these programs are funded by or involve interested third parties. These programs include: (1) the Polish–Czech–German Black Triangle Environmental Program established in August 1991, for coordinating environmental measures in the region; (2) the Black Sea Environmental Management Program for the purpose of identifying the principle sources of pollution of the Black Sea; (3) the Environmental Program for the Danube River Basin for the purpose of developing a strategic action plan for the reduction of pollution;[808] (4) the Baltic Sea Joint Comprehensive Environmental Action Program for the purpose of addressing issues of wetland protection and reducing agricultural run-off and acid rain; and (5) the Romanian and Bulgarian Environmental Cooperation agreement designed specifically to address issues arising from air and water pollution in the Giurgiu–Ruse region.[809]

Additional technical cooperation programs include: the creation of the Association of Carpathian National Parks, various biodiversity projects sponsored by GEF, joint monitoring projects along the Elbe River Basin, the Tibrek project sponsored by Ukraine, Slovakia and Hungary relating to air and water pollution in the Ruthenian region, and joint infrastructure projects such as the German–Polish construction of sewage treatment plants along the Nisa River.

This cooperation creates a normative base from which CEE states are beginning to structure a resolution of their transboundary environmental disputes. The cooperative studies undertaken within these programs are also beginning to generate the information necessary to construct a political settlement within the parameters of international law, and at a minimum, the current levels of technical cooperation will serve to build the epistemic community necessary for the future implementation of any agreements designed to resolve transboundary environmental disputes.[810]

3. European integration

As noted above, the EU is considering all CEE states for potential membership, with CEE states expressing a strong interest in pursuing what they perceive as the final step in reorienting their political and economic ties from the East to the West. Some regional experts express optimism if CEE states are integrated into the EU, then it will be possible to establish a unified regional approach to environmental protection, which may be based on the rule of law.[811] Other commentators note the EU possesses a short window of opportunity within which to affect environmental policy, and if EU membership questions are delayed for another 5 to 7 years, then CEE states will likely adopt a variety of inconsistent approaches, which will be doubly difficult to integrate once membership decisions are made. Impending EU membership, if it is realistically imminent, may however be used as an incentive for some CEE states to resolve their disputes, or to implement decisions of third party dispute resolution mechanisms.

A related instance of European integration is the recent establishment of Euroregions. Although sponsored in large part by the EU, they bear no relationship to future EU membership, but rather relate to the integration of specific transboundary regions. The 36 Euroregions in CEE are nongovernmental entities straddling national borders and are comprised of local and national officials from their member states, but have no power to make binding decisions.[812] Most Euroregion functions relate to economic development, the promotion of tourism, cultural exchange and environmental protection. The value of Euroregions lies in their practical suggestions to central governments and the further creation of an epistemic community comprised of political and technical experts.

V. National minority circumstances

In a number of CEE border regions suffering from environmental degradation there exists substantial populations of national minorities from a neighboring state. The existence of these national minority populations gives rise to political tension, as well as an opportunity to promote transboundary cooperation.[813]

The possibility of conflict arises from the substantial 'social distance' between different nationalities in CEE, and the fact up to 60 per cent of people in CEE perceive of ethnic identity as a valid source of transboundary tension.[814] The existence of regional environmental degradation further increases this social tension.[815] The opportunity for cooperation arises from the practical circumstance that a common culture and language increases receptivity to transboundary activity, and the political circumstance that national minorities often desire some political linkage with their national brethren. The geographic distribution of minorities in a particular state and the manner in which a host state treats its national minorities affects whether the existence of national minorities in a border region will promote or inhibit transboundary environmental dispute resolution.

A. Pre-1989

The current distribution of national minority populations in border areas is largely the result of World Wars I and II. The 1920 Peace Treaty of Trianon ending World War I divided the Austro-Hungarian Empire into Austria, Hungary and Czechoslovakia, ceding much of the former Empire to neighboring states, and with it, former Austro-Hungarian nationals.[816] The Peace Treaty, for example, established the boundary between Czechoslovakia and Hungary along the Danube River's principle channel of navigation, providing Czechoslovakia with a large swath of territory populated almost exclusively by ethnic Hungarians.[817] This and other demarcations created a situation where Hungary, which once had a population of 15 million Hungarians, now possessed a population of only 10 million, with 5 million dispersed in neighboring states.[818]

In 1947, the Paris Peace Treaty following World War II attempted to create states based on ethno-linguistic boundaries that coincided with political boundaries by modifying the boundary of every state in CEE, except Austria and Albania. The readjustment of these borders, and the subsequent migration of 31 million people, had the effect of legitimizing national policies of ethnic homogeneity, but actually failed to create ethno-linguistic states, leaving significant, but more vulnerable, minority populations in all CEE states.[819] The pursuit of ethnic homogeneity, and in some instances revenge for aggression suffered during World War II, led many states to engage in population transfers with neighboring states or to simply expel members of ethnic minorities.[820] This massive relocation of ethnic populations created a social dislocation wherein traditional populations with a historical and cultural attachment to a region were replaced by 'newcomers' to the region who were attracted by economic incentives and lacked the environmental ethos of those whom they replaced.

Rather than resolving the social and political tensions between the remaining national minority populations and the majority population, the pre-1989 CEE communist regimes simply suppressed these tensions by refusing to recognize any specific minority rights and, in some instances, even the existence of a national minority.[821] This policy resulted from the ideological incompatibility of communism with the protection of minority rights, as under a communist system, class is the determinate of human behavior, while under the theory of nationalism, culture is the determinate. The communist states thus repressed national identification in order to suppress the idea that nations and governments should be formed around cultural identities rather than around the engine of the social revolution (the communist party).[822] Since recognition of minority rights would amount to recognition of nationality as a basis for classification, the communist system rejected calls for the application of minority rights.[823]

In addition to suppressing a national identity among the national minorities on their territory, the pre-1989 CEE communist regimes sought to suppress an identification of their national majority population with the interests of like national minorities in neighboring states, as the regimes wished to stifle any chance of creating a rift between fellow communist states, or the chance of creating a foundation for nationalist resentment against the Soviet Union.[824] Thus, for instance, when the construction of large industrial projects undertaken by communist regimes directly threatened the interests of national minorities, such as the Gabčíkovo–Nagymaros Project which physically isolated the Slovak-Hungarian villages of Dobrohost, Vojca and Bodiky from the rest of Slovakia and reduced their opportunities for economic development,[825] the Hungarian government refused to allow its nationals to discuss these effects of the project. As the former communist states sought only to suppress and not resolve ethnic tensions, once the power of the governments receded, the national attitudes reawakened and are to some extent stronger as a result of their temporal suppression.[826]

B. Post-1989

Since 1989, national identity has reemerged as a potent force in CEE domestic and international politics. Although CEE states have undertaken efforts to accommodate the interests of their national minorities, these efforts are lacking in effect and are not sufficient to prevent the issue of national minorities from complicating intra-CEE relations and transboundary environmental disputes.

1. The reemergence of national identity

With the fall of the communist regimes in 1989, and the fading of the notion of a monolithic and homogeneous socialist society, majority and minority ethnic identification is on the increase.[827] Because many individuals have lost their social identities provided by the former regimes, they actively seek to associate themselves with an identifiable group, and since ethnicity is easily recognizable and attachable, it provides a convenient basis for establishing a new identity. Similarly, since statehood is a building block of contemporary international society, more radical ethnic groups are pursuing the concept of an ethnically based state in order to legitimate their claim to certain resources.[828]

The concern national minorities might seek to pursue self-determination and territorial realignment leads most CEE states to act very slowly in providing for the protection of minority rights. This delay then spawns a number of ethnically based political parties and causes friction between these states and neighboring states whose national ethnic majorities share the ethnicity of the minorities in the other state.[829]

The failure to adequately and sensitively address national minority issues has delayed agreement on a number of basic friendship and reconciliation treaties, which might form the basis for negotiations to resolve transboundary environmental disputes, or strengthen the practice of using international law to lessen transboundary tension.[830] Until CEE states resolve their basic conflicts relating to the treatment of national minorities, they are unlikely to be willing to expend substantial political energy negotiating or implementing complex environmental cooperation agreements, or undertaking costly pollution abatement measures for the perceived benefit of neighboring states.

In an attempt to resolve some of the ethnic tensions, Romania, Slovakia and Hungary, in the autumn of 1995, signed a Council of Europe convention designed to protect national minorities.[831] In addition, the United States has applied pressure to both Hungary and Romania, asserting that without a basic treaty of friendship between these states, it was unlikely they could be admitted to NATO.[832] In the end, states such as Hungary are hopeful the admission of Hungary and other CEE states into the EU will lead to a noticeable increase in the protection of national minority rights.

2. *The relationship between national minorities and certain transboundary environmental disputes*

National minorities tend to have a higher level of awareness of environmental concerns than the general population because they are frequently less mobile and have a historic and cultural identification with their native landscapes. In transboundary regions, this awareness is further heightened as national minorities have a tendency to perceive the citizens of a neighboring state, or the national majority in their own state, as exploiting transboundary resources in order to increase their own wealth and in disregard of the wealth or needs of the national minorities who identify with the region as their cultural home.[833] As such, the resolution of many of the CEE transboundary environmental disputes is complicated by the presence of national minorities.

The Black Triangle dispute is impacted by the circumstance that following World War II, the former Czechoslovakia expelled 3 million Germans from the region. The dislocation of the local population originally contributed to the degradation of the Black Triangle as the Czechoslovakian government resettled large numbers of Czechs, Slovaks and Moravians into the region who did not possess a cultural attachment to the landscape, and who where primarily induced to resettle by the promise of financial rewards. To reap these financial rewards, the new settlers, with the active support of the Czechoslovakian government, embarked on a program of rapid and wasteful regional industrialization.[834] The contemporary resolution of the Black Triangle dispute is hindered by claims of the relocated Sudeten German population in Germany, and some German politicians, for an official apology and for compensation for or return of confiscated property.[835]

In the case of the Gabčíkovo–Nagymaros Project, there are three dynamic aspects of national minority circumstances affecting dispute resolution. First is the cultural, economic and environmental impact of the project on the Hungarian national minority. Second is the desire of Slovakia, which for the first time in its history has established a permanent state, to develop a sense of national pride, if not a sense of nationalism, by completing and operating such a large scale public works project. The third relates to claims by Slovakia that certain political parties in the Hungarian government called for the termination of the 1977 Agreement in large part because of their own desire to combine ecological considerations with nationalism (by focusing on the plight of Hungarians in Slovakia affected by the project) to form the platform for the 1989 revolution.

In the Silesian Coal Basin dispute, there is the positive possibility the self-identification of the local population as Silesians will serve as a basis for building stronger transboundary efforts to reduce air and water pollution. There is a similar possibility the transborder population of Ruthenians in the Carpathian mountain region stretching from Slovakia through Hungary and Ukraine to Romania, might provide a common foundation for improved transboundary environmental co-operation.

National minority concerns thus may be considered to play a particularly important role in the resolution of CEE transboundary environmental disputes. Their proper handling may promote dispute resolution while their improper handling may hinder dispute resolution. In any event, national minority actors are likely to play an important role in the resolution of transboundary environmental disputes and will invariably call upon and attempt to construct additional international legal norms that provide protection and support for their interests.

10
Predicting the Future: an Increasing Role for International Law?

Given the relatively active role of international law and the legal process in the Gabčíkovo–Nagymaros Project dispute, and the significant reliance on international law by the participants in the Baltic Sea dispute, as well as its increasing use in the Black Sea and Danube River Gauntlet disputes, this chapter seeks to gauge whether there might be an increasing use of international law by CEE states to assist in the resolution of their transboundary environmental disputes. To accomplish this task, this chapter examines the status of the six factors, identified in Chapter 8, that promote the functionality of international law within the regime of transboundary environmental protection.

I. The practice of Central and East European states applying international law to resolve international disputes

Where CEE states have a practice of utilizing international law to resolve their disputes, they will naturally be familiar with the substance and functions of international law, will recognize the value of using international law, will have demonstrated to other states their willingness to be obligated by the rule of law, and may possibly have created a functional regime for promoting transboundary dispute resolution.

A. Pre-1989 practice

Prior to the dissolution of the former Soviet Union and the evaporation of Soviet hegemony over CEE, the CEE states applied principles of socialist international law to structure their intra-CEE relationships. Socialist international law in its essence represented a merging of the principles of proletarian internationalism and socialist internationalism to encapsulate the principle of the unity of socialist states in the international class struggle between socialism and capitalism.[836] With respect to relations between CEE states, socialist international law mandated the strengthening of friendship and cooperation in all areas of political, social, economic, cultural and military activity, and as such provided a basis for the forcible assertion of Soviet hegemony, as evidenced by its invasion of Hungary in 1956 and Czechoslovakia in 1968 in order to 'protect socialist gains'.[837] This same socialist law did not, however, provide CEE states with any practical rights in relation to other CEE states as there could inherently be no contradiction between the socialist norms of CEE state foreign policy and socialist international law, and thus there was no need to create substantive reciprocal rights.[838]

The socialist states considered general international law to consist essentially of the principles of sovereignty, self-determination and non-interference in the internal affairs of a state. General international law thus primarily served to regulate the peaceful co-existence of capitalist, newly independent and socialist states, and was perceived as having little or no relevance for intra-socialist state relations. And because the former Soviet Union represented the primary guardian of the socialist mantra, it held primary responsibility for utilizing general international law to represent the interests of CEE states whenever a conflict existed between these states and the capitalist West.[839] Prior to the 1989 transformation, CEE states were thus seldom exposed to the operation of the principles of general international law.

B. Post-1989 practice

With the dissolution of the former Soviet Union and the transformation from totalitarian to democratic regimes, CEE states are actively abandoning the precepts of socialist international law and are reorienting their view of international law to coincide with the principles of general international law. Concerning general intra-CEE state disputes, in particular those relating to the treatment of national minorities, CEE states frequently seek to articulate their concerns within the framework of international law. As of yet, however, they have not chosen to articulate these legal claims within a formal dispute resolution mechanism. In many instances the attention of CEE states to intra-CEE disputes is diverted by their efforts to complete the transformation from a command to a market economy.

Concerning international disputes with Western states, CEE states appear to be in the early stages of developing a practice of the independent utilization of international law to resolve such disputes. However, since most CEE states currently receive significant amounts of assistance from Western states, which is necessary to sustain their economic transformations, CEE states have chosen in the short term not to emphasize their outstanding international disputes. As CEE states wean themselves from Western assistance programs, however, they will likely begin to more actively employ international law to support their contentions in their international disputes with Western states. Concerning the structuring of mutually beneficial relationships, CEE states are to date responding positively to international agreements containing international legal obligations and are evidencing an understanding of the various nuances of these obligations.

Notably, CEE states are actively intertwining themselves in the network of international environmental treaties. For instance, all CEE states have signed the Long Range Transboundary Air Pollution Convention, which requires parties to endeavor to limit and gradually to reduce and prevent air pollution and to exchange information and engage in consultation and research and monitoring activities for the purpose of combating the discharge of air pollutants.[840] Moreover, all CEE states, except Poland and Romania, have signed and ratified the SO_2 Protocol,[841] and all, except Romania, have signed the NO_x Protocol, which provide for a 30 per cent reduction of SO_2 emission and a reduction of NO_x national annual emissions to 1987 levels.[842] In addition, all CEE states, except Romania, have signed the second SO_2 Protocol, which provides for further reductions of SO_2 emissions,[843] and Hungary and Bulgaria have also signed the Volatile Organic Compounds Protocol, which establishes specific emission reduction targets and sets timetables.[844]

Similarly, all CEE states have ratified the Vienna Convention on Ozone Depletion and the Montreal Protocol, which establish a framework for combating the emission of pollutants likely to modify the ozone layer, and which require the parties to take specified appropriate measures to combat such pollution and set forth specific emission reduction timetables.[845] All CEE states have also signed and/or ratified the Climate Change Convention and the Biodiversity Convention, which commit state parties to stabilize greenhouse gas emissions over the long term and provide for the conservation of biological diversity, the sustainable use of biological resources and the fair and equitable sharing of the benefits of their use.[846]

Specifically with respect to transboundary cooperation, all CEE states have signed the Convention on Environmental Impact Assessment in a Transboundary Context, which requires parties to take all appropriate and effective measures to prevent and reduce significant adverse transboundary environmental impacts from activities occurring on their territory, and further provides a set of procedures and obligations for carrying out a joint EIA of any proposed activity likely to cause a significant adverse transboundary impact.[847] And all CEE states, except for Bulgaria, have ratified the Basel Convention on the Transboundary Movement of Hazardous Waste, which prohibits states from exporting hazardous wastes to states which do not consent to the import of such waste and which sets forth detailed procedural obligations to regulate the transboundary transfer of hazardous waste where such consent exists, and to ensure the transfer takes place in a safe manner and that the hazardous waste can be properly disposed of by the importing state.[848] Only Hungary, however, has ratified the Convention on the Transboundary Effects of Industrial Accidents.[849]

Concerning the regulation of nuclear activity, all CEE states have ratified or approved the Convention on Nuclear Safety and the Convention on Early Notification of a Nuclear Accident. The Convention on Nuclear Safety seeks to ensure the safe operation of nuclear power plants by prescribing a set of construction and operating standards, as well as providing for regular meetings of the contracting parties to discuss additional means for improving safety. The Convention on Early Notification requires parties to notify the IAEA and other states in the event of a nuclear accident and to cooperate with the IAEA and other states in order to minimize the consequences of such an accident.[850]

II. The practice of Central and East European states applying municipal environmental law to resolve domestic resource disputes

Where CEE states have a past practice of applying municipal environmental law to resolve domestic resource disputes, they may have an experience with and recognize the general value of employing environmental law, their domestic stakeholders may be familiar with the legal process for resolving environmental disputes, they may have developed domestic institutions capable of implementing and enforcing domestic compliance with international commitments, and they may therefore be more willing to use and capable of using international environmental law to assist in the resolution of transboundary environmental disputes.

A. Pre-1989 practice

Prior to 1989 the CEE states had developed a distinct view of the role of law as a tool of Marxism designed to further the evolution of society towards communism.[851] Legal rules were thus considered binding upon citizens and could be invoked by them so long as they did not run counter to the interests of the state. Because the state both represented the ultimate legal authority and operated all economic enterprises, CEE states came to lack a core commitment to the resolution of disputes through legal means,[852] and were characterized by a certain degree of legal nihilism, with the domestic legal sphere having no autonomy from the state in which rights could exist.[853]

To create the appearance of a structured legal regime and to provide a legal mechanism for the supervision of activities of state owned enterprises, CEE states adopted a number of domestic environmental regulations.[854] These environmental regulations focused primarily on the setting of fees for the utilization of natural resources and fines for excessive utilization or pollution of those resources. Unfortunately, these regulations were generally symbolic gestures of environmentalism, were seldom enforced, and had almost no impact on the behavior of polluters.[855]

B. Post-1989 practice

1. The evolution from socialist municipal law

Since the change to democratic government and market-based economies, most CEE states have begun the necessary process of abandoning old municipal environmental laws incompatible with a market economy and are developing a new system of environmental protection.[856] As a result, CEE states have adopted more statutes and decrees and have rendered more judicial decisions since the mid 1980s, than in the past 40 years.[857] This activity leads some commentators to contend CEE states are in the active process of institutionalizing law. These commentators further note that because socialist law is discredited, the new CEE constitutional courts look for guidance to international law, which they perceive as representing legitimately agreed upon general principles of justice and fairness.[858] In some instances, however, the plethora of new environmental acts and administrative decisions inhibits the application of environmental law as it is very difficult for those responsible for implementation to determine the precise nature of the law.[859]

Although the rule of law is a generally accepted social norm in CEE, some commentators note that despite the adoption of new environmental legislation, some states retain a legal culture based on state primacy and sovereignty and lack sufficiently trained legal professionals to promote the use of legal norms to resolve disputes.[860] Romania, for instance, until November 1997, had been slow to adopt new environmental legislation and appeared to be experiencing an increasing disregard for the rule of law.⁸⁶¹ While in August 1995, the Deputy Parliamentary Chairman of the governing party of Slovakia, the Movement for a Democratic Slovakia, argued, 'it is a mistake for society to have independent judges [at a time] of economic transformation'.⁸⁶² According to a September 1996 EU assessment, the sluggish approach to strengthening environmental laws and institutions exhibited by some CEE states was the result of budgetary constraints and the limited environmental consciousness of the general public.⁸⁶³

2. The substance of current Central and East European municipal environmental legislation

A general review of the current state of environmental legislation in CEE will aid in ascertaining whether CEE states are now more likely to utilize domestic environmental law to resolve resource disputes, and whether in turn, they may be likely to utilize international law to resolve transboundary environmental disputes. According to a recent World Bank study, a comprehensive environmental legislative package should include a clear definition of the authority of various ministries and local authorities associated with environmental protection, a mix of market oriented instruments and command regulations, the requirement for an EIA for major public or industrial works, provisions for public access to environmental information, and an incorporation of the precautionary and polluter-pays principle.⁸⁶⁴

Since 1989, CEE states have generally been successful at incorporating into their domestic legal regimes a number of general principles and procedures designed to enhance environmental protection.⁸⁶⁵ Notably, a number of CEE states have incorporated a right to a healthy environment in their constitutional provisions, and in 1994, the Hungarian Constitutional Court struck down various amendments to the Law on Agricultural Cooperatives on the basis they violated the constitutional right to a healthy environment by permitting the cultivation of previously protected lands.⁸⁶⁶ Although CEE states have been less successful at implementing specific rules and regulations relating to environmental protection, the municipal environmental protection regimes continue to evolve and are likely to become increasingly proficient, especially as CEE states strive for integration with the EU.⁸⁶⁷ With respect to the specific principles and procedures identified by the World Bank study, CEE states have made some notable progress.

194 *Prospects*

First, many CEE states provide for a relatively clear allocation of regulatory authority between central, regional and local governments. In most of these cases the central government is charged with formulating resource management and environmental protection policies while regional and local governments play a role in implementing these policies, in particular, granting permits to medium and small size industrial enterprises, carrying out enforcement measures and organizing and participating in any EIAs which may be undertaken.[868] Resource management and environmental regulation does, however, suffer in CEE from the lack of clear definition of authority between various central government ministries. For instance, in Hungary the implementation of effective regimes of environmental protection is inhibited by the fact air and water resources are regulated according to an incompatible combination of ambient standards and emission standards, with the public health authorities setting the ambient standards and environmental authorities setting the emission standards.[869]

Second, with respect to a mix of market instruments and command regulations, the implementation of an effective legislative scheme for environmental protection in CEE states is frequently limited by the continuation of elements of the financing system developed under the communist regime, which in the case of both the Czech Republic and Slovakia, undervalues environmental resources, restricts the dissemination of environmentally sound technologies, relies upon a system of punitive fines and appears to treat the environment as a cost-increasing factor.[870] Similarly, current Hungarian environmental administrative law continues to focus on punitive rather than preventative measures. Notably, administrative fines continue to play a key role, as in Poland and Slovakia,[871] but are generally ineffective because the environmental authorities fail to monitor polluting enterprises and pollution levels adequately, there is a general lack of accountability over the expenditure of funds generated from fines, and when enforced, the fines tend to be too low to deter continued violations.[872] To achieve a more effective mix of market oriented instruments and command regulations some CEE states have begun to formulate regulatory programs that rely more heavily on tax exemptions and environmental protection funds.[873]

Third, with respect to EIAs, in Bulgaria the federal, regional and municipal governments are required to conduct EIAs for all activities that may significantly impact the environment. Moreover, in certain circumstances local governments may require private enterprises to conduct EIAs, and in almost all instances the general public is entitled to participate in the 'consideration of the results of the EIA'.[874] The effectiveness of the Bulgarian EIA regime is unfortunately inhibited by the general lack of experience held by the Regional Inspectorates of the Ministry of the Environment. CEE states with similar EIA requirements include Poland, the Czech Republic and Slovakia, with the Czech Republic providing certain administrative bodies are responsible for overseeing the preparation of EIAs and these bodies must hold a specified number of public meetings.[875] The Slovak EIA law further provides members of the public have a right to participate in the environmental decision making process and concerned municipalities are formally charged with ensuring the protection of this right.[876] And although Romania has not adopted a formal EIA regime, it has adopted domestic legislation which provides for a limited degree of environmental assessment in certain circumstances.[877]

Fourth, with respect to public participation, in all CEE states citizens are granted constitutional or legislative rights to obtain information from state bodies and agencies relating to the state of the environment and specific activities that may cause environmental damage.[878] In Bulgaria and Hungary, the actual ability of citizens to obtain such information is hampered by the lack of specific rules providing for access to information. Bulgaria, however, is redrafting laws relating to forestry management, waste treatment and the marine environment, some of which will likely include concrete rules providing for public access to relevant information, and in Hungary the government soon will be required to publish official data relating to pollution and health matters.

Fifth, some CEE states such as Poland, the Czech Republic and Hungary provide that citizens and NGOs may seek judicial review of administrative decisions affecting the environment.[879] Such review is, however, often difficult to carry out as these states have yet to develop a legal standard by which a court may determine whether an administrative agency has acted in accordance with its responsibilities, or the states permit only a narrow review of administrative procedure and specifically preclude a substantive review of the administrative agency's decision, as in the case of Slovakia.[880] Despite these limitations, two important lawsuits have recently been or are about to be filed in CEE domestic courts, one by a consortium of Austrian, German, French and Czech environmental NGOs against the operation of the Temelín nuclear plant, and one by scientists from Germany, Austria and the Black Sea Basin states against Germany and Austria for pollution of the Danube River.[881] In addition, there has been some progress with permitting citizens of affected states to participate in the administrative decision making process. For instance, in 1998 the Czech Government announced Austrian citizens would be allowed to participate in the administrative process relating to the approval of an expansion of the nuclear waste storage facility at the Dukovany nuclear plant.[882]

And finally, many CEE states, and in particular the Czech Republic, Slovakia and Hungary, have incorporated into their environmental legislation key environmental principles such as the precautionary principle and the polluter-pays principle.[883] In some circumstances, however, these principles have been incorporated in such a manner that they are logically detached from or contradict other aspects of the environmental protection regime and may even, in the words of one commentator, represent 'an expansive, vague, and unprioritized list of aspirations' unlikely to be actualized.[884]

Concerning the use of civil litigation to promote environmental protection, a number of barriers continue to exist, including the absence of domestic legislation providing for private property holders to seek compensation for damage caused by pollution,[885] the lack of a legal tradition of using law as a means for resolving disputes, the lack of provisions for class action suits, the lack of precedent, the fact environmental law is almost solely classified as administrative law and the lack of necessary financial resources for NGOs to bring actions against polluters, although they have standing in some CEE states.[886] Hungarian civil law, for example, contains a number of useful provisions for environmental protection related to personal civil rights, intellectual property rights, nuisance, trespass and the utilization of the strict liability standard; however, the civil law does not as of yet play a key role in environmental protection. Hungary has been reluctant to discard the socialist view that environmental protection should be addressed solely through administrative operations with little judicial review of those procedures. According to one Hungarian commentator, civil litigation is further chilled by the requirement the plaintiff pay the costs of the litigation in advance and upon conclusion of the trial, costs are assessed against the losing party.[887]

As the EU includes in its association agreements the requirement CEE states harmonize their legislation with the over 200 EU Directives comprising the EU framework for environmental standards,[888] most observers believe the legislation of many CEE states will soon correspond to Western standards.[889] The Czech Republic, for instance, has amended its 1992 Air Pollution Act and Waste Management Act to comply with EU environmental standards as required by the Czech Republic's 1993 Association Agreement with the EU.[890] The ultimate ability of this new legislation to improve the environment will in large part rest with the efforts of CEE states to comprehend and enforce the legislation, and while some states, such as Poland, are increasingly successful in their enforcement of environmental legislation,[891] states such as Slovakia fail to collect up to 70 per cent of levied fees and fines.[892]

III. The correlation of Central and East European municipal environmental law with international environmental law

Where CEE state municipal environmental law is correlated with international environmental law, CEE states will have a greater understanding of the relevant norms of international environmental law, may be able to give full force and affect to international agreements designed to resolve disputes, may be able to enforce international obligations through their domestic courts, and may therefore be more likely to and capable of using international environmental law to assist in the resolution of transboundary environmental disputes.

A. Pre-1989 correlation

Prior to the 1989 revolution, CEE states subscribed to the view that socialist domestic law took precedence over general international law and general international law was given effect only to the extent it was consistent with socialist domestic law, which, like socialist international law, revolved around the principles of state sovereignty, the ultimate authority of the central government, and mutual cooperation and friendship.[893] Similarly, because of the strictly positivist approach of CEE states to international law, they would consider the implementation of international law within their domestic jurisdiction only if it originated from a multilateral or bilateral treaty or other contractual arrangement.[894] Added to this doctrinal based restriction was the practical restriction that CEE states viewed any international legal obligations with general mistrust,[895] and that due to an emphasis on sovereignty, CEE states preferred to develop additional normative international law rather than to develop municipal mechanisms for monitoring and enforcing existing obligations.[896]

B. Post-1989 correlation

Concerning the consistency of CEE domestic environmental law with international environmental law, most CEE states' municipal law recognizes the principle that the environment should not be contaminated or polluted and resources must be equitably utilized. And as noted above, many CEE states are currently in the process of modifying their domestic environmental laws to include reference to the precautionary principle, the polluter pays principle and the principle of sustainable development.[897] In some instances, CEE states are also beginning to structure their system of fines and other punitive measures so as to promote their consistency with the emphasis of international environmental law on allocative and preventative measures. Moreover, the increasing presence of an EIA requirement is consistent with international environmental law's emphasis on ecosystem management and the interrelation of different sources and types of pollution.

Concerning the implementation of international law, many CEE states have constitutional or municipal law provisions similar to Western states. More specifically, according to Polish and Hungarian municipal law, international agreements are considered a part of municipal law once the government enacts implementing legislation.[898] Romania and Bulgaria, however, have not yet adopted domestic law provisions for the implementation of general international law or international agreements,[899] although there are indications they will soon do so. And, despite the constitutional provision for the implementation of international law in states such as Poland, the Czech Republic and Hungary, the states have been slow to adopt the necessary implementing legislation. Hungary, for instance, is party to numerous international environmental agreements, but has only enacted implementing legislation for two of those agreements, and although the Czech Republic has adopted implementing legislation for most of the international agreements which it has signed, some regional experts note it currently appears to lack the technical and financial capacity to comply with its international obligations.[900]

Concerning the ability of domestic courts to rely upon international law in their deliberations, the Hungarian constitution permits the Constitutional Court to resolve conflicts between administrative regulations and international law.[901] In both the Czech Republic and Slovakia, the Constitutional Court may review a complaint of an individual who alleges a violation of fundamental rights articulated in an international treaty ratified by the Czech Republic or Slovakia, and the Court may suspend legislation if it finds this legislation to be inconsistent with the state's treaty obligations.[902] And in Poland, although the Constitution does not articulate the role of international law in the domestic legal process, Polish courts consider themselves authorized to apply international treaties to the extent they do not conflict with the Polish Constitution.[903]

IV. The accessibility of international judicial dispute resolution mechanisms to Central and East European states

Where CEE states have access to international judicial dispute resolution mechanisms, they may be able to seek an independent clarification of the relevant rules of international environmental law, and they may be able to submit to this mechanism certain aspects of their dispute for resolution, which may then form the basis for resolution of the remainder of the dispute. They may also be able to use the rulings of the mechanism as political cover to implement any unpopular but necessary actions, and they may therefore be more likely to employ international law for the purposes of resolving their transboundary environmental disputes. The willingness of CEE states to utilize dispute resolution mechanisms, however, should be viewed within the general context that states tend to generally avoid judicial and liability-based methods for settling transboundary environmental disputes because of their perceived infringement on state sovereignty.[904]

A. Pre-1989 practice

Prior to 1989, CEE states consistently submitted reservations to all clauses in international agreements relating to the compulsory jurisdiction of the ICJ to resolve disputes arising from the application of the specific treaty.[905] Similarly, prior to 1989, no CEE state accepted the compulsory jurisdiction of the ICJ.[906] The rejection of dispute resolution mechanisms for disputes arising between socialist states and nonsocialist states arose from the socialist states' perception the world was in a perpetual state of conflict between capitalist states and socialist states and since the interests and policies of these states were so opposed, it was not possible to resolve them through impartial adjudication.[907] An important consequence of the rejection of the authority of the ICJ, is that post-1989 CEE states have little practical or analytical experience with international adjudication, and as noted by one socialist international lawyer, 'all books written on [international adjudication] throughout the entire history of the Soviet State can be placed on one little shelf.'[908]

With respect to intra-CEE relations, none of the bilateral agreements concluded between fellow socialist states provided for the compulsory jurisdiction of the ICJ,[909] and except for a few agreements relating to economic relations,[910] there were scant instances of bilateral agreements providing for the submission of a dispute to a third-party dispute resolution mechanism. The doctrinal basis for this absence of dispute resolution mechanisms was the belief that because of the higher nature of their international relationships based upon friendship and cooperation, it was not possible for socialist states to experience disputes.[911] If the socialist states had provided for the resolution of disputes, then they would have admitted to the possibility of the existence of disputes between socialist states and called into question the fundamental basis of socialist integration. A more pragmatic rationale is the Soviet Union was not interested in creating a habit among CEE states of having formal disputes or submitting their disputes to mechanisms that would treat them as equal states, as this would undermine Soviet hegemony in CEE.[912]

B. Post-1989 practice

1. The acceptance of the jurisdiction of dispute resolution mechanisms

In 1987, President Gorbachev of the Soviet Union signaled an impending change in the attitude of socialist states to third party dispute settlement by asserting the Soviet Union should commit to the compulsory jurisdiction of the ICJ.[913] With the 1989 revolutions, CEE states began to reevaluate their attitudes toward dispute resolution, and in particular the ICJ,[914] with Poland, Hungary and Bulgaria submitting to the Court's compulsory jurisdiction.[915] Notably, Poland exempts from the compulsory jurisdiction of the ICJ all 'disputes with regard to pollution of the environment unless the jurisdiction of the ICJ results from the treaty obligations of the Republic of Poland'.[916] This reservation is based on Poland's view that customary international environmental law is not yet 'crystallized', and thus Poland does not wish to expose itself to unpredictable claims of liability.[917]

The Czech Republic, Slovakia and Romania have not, however, as yet submitted to the compulsory jurisdiction of the ICJ. The Czech Republic, though, is actively considering such a submission and is systematically withdrawing reservations to clauses of international treaties providing for the compulsory jurisdiction of the ICJ.[918] CEE states are also beginning to accept the inclusion of third party dispute resolution mechanisms in new international agreements. The Danube River Protection Convention, for example, provides parties to the convention who have a dispute over the application or interpretation of the convention may accept as a means of dispute settlement the jurisdiction of the ICJ or arbitration.[919] Although as of 1998 the Danube River Convention had not yet come into force, it is expected most CEE states who have signed the convention will accept the jurisdiction of either or both the ICJ and arbitration.[920]

With respect to conciliation and arbitration, Poland, Hungary and Romania have signed and ratified the Convention on Conciliation and Arbitration within the Organization for Security and Cooperation in Europe (OSCE), while Bulgaria and Slovakia have signed but not ratified the Convention.[921] The OSCE Convention provides for the establishment of a Court of Conciliation and Arbitration, consisting of Conciliation Commissions and an Arbitral Tribunal.[922] The Convention permits two or more states to consent to submission of a dispute to a Conciliation Commission, or a state party to submit unilaterally to a Conciliation Commission any dispute that has not been settled within a reasonable time through negotiation. Parties may further submit to the compulsory jurisdiction of the Arbitral Tribunal or consent by special agreement to submit a specific dispute to its jurisdiction.[923] As of October 1999, no state party had yet submitted to the compulsory jurisdiction of the Arbitral Tribunal, nor had any of the parties agreed to submit a specific dispute to the Court of Conciliation.[924] In addition to the OSCE conciliation and arbitration mechanisms available to CEE states, certain CEE bilateral and multilateral treaties also provide for submission of disputes to arbitration, as in the Polish–Ukrainian Treaty on Protection of the Environment, and the Danube River Convention.[925]

Concerning good offices and mediation, as members of the OSCE, CEE states may also invoke the Valletta Mechanism in an effort to assist with dispute resolution. The Valletta Mechanism provides that any party to a dispute with another OSCE member state may request the establishment of an OSCE Dispute Settlement Mechanism if they are unable, within a reasonable period of time, to settle their dispute by means of direct consultation or negotiation.[926] The Dispute Resolution Mechanism will then seek to assist the parties in reaching a resolution of their dispute by identifying appropriate procedures for the settlement of the dispute. With the consent of the parties, the Mechanism may also engage in factfinding missions, provide or solicit expert advice, formulate advice and recommend to the parties specific means of resolving their dispute.[927]

2. The operational capacity of dispute resolution mechanisms

Once states submit to the jurisdiction of various international dispute resolution mechanisms, such as conciliation, arbitration or adjudication, the question is whether these mechanisms possess the operational capacity to assist states in resolving their disputes.[928] Although, as noted above, a number of CEE states have adopted the Convention on Conciliation and Arbitration within the OSCE, this mechanism has not yet been invoked by any CEE states to resolve an environmental dispute. This section will therefore briefly examine the operational capacity of the ICJ,[929] which in order to make itself more attractive to parties wishing to resolve environmental disputes, has created an environmental chamber.

As discussed in Chapters 5 and 7, during the resolution phase of a transboundary environmental dispute, parties to the dispute may submit a variety of questions to the ICJ, which may include determining the status of obligations under a preexisting agreement,[930] or identifying principles and rules applicable during the resolution phase.[931] The parties may also request the Court to issue an opinion as to the final resolution of the dispute.[932] In instances when parties refuse to negotiate in the pre-resolution stage or where negotiations break down in the resolution stage, the ICJ may play a role by serving as a catalyst in bringing the parties together in a neutral forum. The parties would then be expected to expand their mutual contact to include negotiations outside the forum of the Court.

Recognizing both that it plays a relatively narrow role as a facilitator of dispute resolution and is not necessarily the entity ultimately responsible for crafting a comprehensive settlement, the Court has held its consideration of the relevant legal aspects of a dispute does not preclude the continuation of negotiations between the parties. The Court has also recognized that a resolution of the legal aspects of a dispute is central to the process of dispute resolution,[933] and has held that the existence of political dispute resolution mechanisms and their concurrent use by the parties does not preclude the Court from considering legal issues raised by the dispute.[934] The Court has similarly held it is not precluded from accepting jurisdiction over a case on the basis that either the legal questions exist within the context of a highly political dispute,[935] or the dispute itself, including its legal dimensions, exists within a wider political context between the parties.[936]

The submission of the Gabčíkovo–Nagymaros Project dispute to the ICJ highlights the potential afforded by the Court for dispute resolution. If Slovakia and Hungary are able to successfully negotiate a settlement within the parameters set by the Court, then other CEE states will likely perceive the Court to possess the capacity to assist in the resolution of their transboundary environmental disputes. This perception of the functionality of the ICJ decision in the Gabčíkovo–Nagymaros Project case is all the more important, since despite the Court's receptive capacity to hear disputes, CEE states appear generally reluctant to submit their transboundary environmental disputes to the Court.

The primary reasons for a perceived reluctance of CEE states to resort to the ICJ to resolve their transboundary environmental disputes relate to the following: the requirement that the need for transboundary co-operation amount to an actual dispute between states,[937] the winner-take-all approach of international adjudication, the as yet unsettled nature of international environmental law, the Court's lack of nonlegal expertise,[938] CEE states' inexperience with assessing the benefits and risks of adjudication,[939] the bilateral nature of the forum,[940] and the time and cost associated with a proceeding before the ICJ.[941]

The Black Triangle dispute highlights some of these concerns. In the case of the Black Triangle dispute, the parties would first have to agree on the subject matter of the 'dispute', which could include local transboundary or regional transboundary air or water pollution and/or specific point sources of pollution, such as the Turów power plant. The parties would also have to determine whether the geographic focus of the dispute should be limited to the Black Triangle, or whether it should include the Silesian Coal Basin, which suffers from similar types of environmental harm, as well as the extent to which Germany should participate in any proceedings. The parties would also be faced with the need to determine whether to allege specific instances of environmental harm or general environmental degradation, the difficulty of proving pollution directly caused the harm, and the pollution causing the harm results from the industrial operations on the opposing state's territory and not industrial operations on its own territory. Finally, the Czech, Polish and German governments would have to determine whether the potential benefit to be derived from a submission of the dispute to the ICJ would outweigh the strain this adversarial process would cause in their bilateral relations.

Concern over the perceived reluctance of CEE states to utilize the ICJ to assist in the resolution of CEE transboundary environmental disputes gives rise to a call by some commentators for CEE states, with EU assistance, to create a European Pollution Control Organization to collect information, monitor and control the levels of transboundary pollution via an International Ecological Police, promote the transfer of pollution control technology, respond to accidents, and resolve disputes, either by apportioning liability or resolving emerging conflicts.[942] CEE states, however, currently exhibit little interest in establishing such a mechanism. Other commentators have suggested the creation of an international environmental court.[943]

Given the uncertain nature of the capacity of the ICJ to assist in CEE transboundary environmental dispute resolution, it is likely CEE states will turn to other types of dispute resolution mechanisms, such as arbitration, as provided in the Danube River Protection Convention, and mediation, as used in the initial stages of the Gabčíkovo–Nagymaros Project dispute. CEE states are also likely to use inquiry increasingly, as with the establishment of the environmental secretariat between Germany, Poland and the Czech Republic in the Black Triangle dispute. And with greater political cooperation among some CEE states, there is likely to be an increase in unassisted bilateral negotiation aimed at resolving transboundary environmental disputes.

V. The incentive for Central and East European states to utilize international law to assist with the resolution of transboundary environmental disputes

Where interested third parties and domestic stakeholders create an incentive for CEE states to resolve their transboundary environmental disputes with the assistance of international law, CEE states may perceive value in employing international law. CEE states may also generally perceive value in employing international law where they determine international law may assist with dispute resolution, as will be discussed in section VI below, or where they perceive they will receive positive idiosyncratic credit from Western states and institutions that indicate they wish CEE states to employ international law to resolve their disputes. Conversely CEE states might suffer a loss of value if they consistently disregard the rule of law. Those states most likely to create value for the employment of international law are those who provide financial assistance to CEE states, or those with whom CEE states seek to associate, such as the EU.

A. Pre-1989 practice

As the primary purpose of socialist international law was to provide legal cover for Soviet political and military hegemony, there was little practical emphasis by the Soviet Union for CEE states to apply socialist international law, or to consider the possible application of general international law. Moreover, prior to 1989, Western states did not place an emphasis on the use of law to resolve CEE transboundary environmental disputes as they were largely unaware of the extent of transboundary pollution within CEE. If Western states had been aware, they nonetheless would have had little interest in their resolution and would have had little opportunity to influence the behavior of CEE states. There was thus almost no idiosyncratic value to be derived by CEE states from employing international law to resolve their transboundary environmental disputes prior to 1989.

B. Post-1989 practice

Since 1989, Western states and organizations have fairly consistently emphasized the need for environmental protection, and have provided significant amounts of assistance to support the efforts of CEE states to remediate environmental degradation.[944] In October 1995, Switzerland, for instance, agreed to write off Bulgarian debts of $17.7 million, in exchange for a commitment by Bulgaria to invest an equivalent amount of local currency in ecological projects.[945] In total, from 1990 to 1997 Western states provided over $3 billion in environmental assistance, which amounted to approximately 10 per cent of the environmental budgets for CEE states.[946] IFIs, although still guided by their mandate to promote development, are attempting to be more conscientious about the environmental impacts of projects proposed by CEE states and, in some instances, make loans specifically for environmental projects. In 1998, for instance, the World Bank loaned Poland $18 million to finance a draft strategy for environmental management.[947]

Unfortunately, Western states and IFIs frequently fail to emphasize transboundary cooperation. Of the many USAID projects in the northern tier of CEE, for instance, only one project specifically focuses on transboundary environmental issues. Other projects are location specific and are frequently designed as pilot or demonstration projects. Similarly, beyond their sponsorship of a limited number of transboundary programs, the EU places primary emphasis on the harmonization of CEE state legislation with EU legislation and does not meaningfully encourage transboundary cooperation between CEE states. The IFIs also fail to promote transboundary cooperation and are tending to lose interest in the northern tier states as these states reorient their environmental investment programs to focus on small scale projects and localized transboundary cooperation. With respect to the southern tier it is difficult for some states to meet the basic requirements for international financial assistance,[948] and thus the IFIs are unlikely to add additional requirements relating to transboundary environmental cooperation.

Since 1989, Western states have emphasized the rule of law in their relations with CEE states, but have placed most of their emphasis on the rule of law as it relates to the promotion of a market economy and democratic reform, the nonproliferation of weapons and respect for human rights, particularly national minority rights. The $16 million EU

Black Triangle program, for instance, contained no provisions for investigating the use of international law to assist with dispute resolution. One notable exception is the pressure brought to bear on both Slovakia and Hungary by the EU to employ international law in the resolution of the Gabčíkovo–Nagymaros Project dispute by submitting it to either arbitration or the ICJ.

Even when the international organizations or Western states do place an emphasis on dispute resolution and the utilization of international environmental law, this influence might still be limited despite large aid programs.[949] CEE states continue to self-finance the vast majority of environmental projects in their respective states, limiting the avenues for constructive pressure from donor states and agencies. In Poland, for instance, 90–95 per cent of expenditures on environmental protection are domestically generated.[950] CEE states are also finding in certain circumstances when international organizations dictate terms perceived as too restrictive, they may be able find alternative sources of private funding, or cut-rate contractors, as in the case of Slovakia's rejection of EBRD conditional funding for the Mochovce nuclear plant and its replacement with funding from the German and French governments and the employment of a Czech contractor.[951]

In addition, CEE states are losing interest in foreign assistance due to a resurgence of national pride and sense of the limited practical value of the assistance. In October 1995, for example, Poland announced as a result of its economic progress, it would no longer seek financial support from the IMF.[952] In some instances, international donors have made promises of direct environmental aid in an effort to influence the policies of a state, but have failed to fulfill their promises. For instance, the government of Ukraine asserts the G7 promised in 1995 to provide $3.1 billion to assist Ukraine in closing the Chernobyl nuclear power plant, but it had received only $250 million as of April 1998, and could therefore not close the plant.[953] The EBRD, however, has provided Ukraine with over $30 million, and received pledges of over $385 million to repair the sarcophagus covering the reactor which exploded in 1986.[954]

The desire of CEE states to attain EU membership, however, will provide a continuing opportunity for the EU to encourage the use of international law to promote a resolution of transboundary environmental disputes.[955] As noted above, although the EU does not yet directly link potential EU membership with the resolution of transboundary environmental disputes, it has indicated to the CEE states its preference to have them resolve their outstanding transboundary environmental disputes. Notably, the EU has conditioned membership on the requirement CEE states cooperate with the EU to bring their levels of nuclear safety up to standards prescribed by international law.[956] Currently, the Czech Republic and Hungary appear to be the most responsive to EU pressure to modify their environmental behavior.[957]

VI. The perceived capacity of international law to assist in the resolution of Central and East European transboundary environmental disputes

To play an effective role in a regime of transboundary environmental protection among CEE states, the set of principles and rules held out as international environmental law must be perceived by CEE states to exist as principles and rules of law, be relevant to their needs and be functional. As noted above, these principles may positively, constructively or indeterminately exist. These principles and rules may also either be subject-matter relevant, temporally relevant or dispositionally relevant and must be both inherently and operationally functional.

Where CEE states perceive relevant and functional principles of international environmental law exist, they may be able to look to these principles to balance their rights of sovereignty with their obligations to neighboring states. They also may be able to use these principles to assist in creating structures that protect their rights within the context of the evolution of authority over natural resource decisions to international organizations and the increasing need to share decision making authority with their neighboring states, and they may therefore be more likely to utilize effectively the substance of international environment law to promote transboundary environmental dispute resolution.[958] This section will not undertake to assess the entire body of international law or international environmental law to determine its capacity for resolving CEE transboundary environmental disputes,[959] but rather it will seek to note where, from the perspective of CEE states, international environmental laws might have the capacity to assist in their efforts to resolve transboundary environmental disputes.

A. Pre-1989 practice

As noted above, prior to 1989 CEE states perceived general international law as serving a narrow purpose and seldom relevant to intra-CEE relations. The field of international environmental law, as a subset of general international law was similarly perceived. Consistent with their perspective that socialist international law should be created through bilateral and multilateral treaties,[960] CEE states adopted a number of bilateral cooperation agreements containing legal obligations to cooperate in the transboundary utilization of natural resources and to share information relevant to that cooperation.[961] Despite their number, the pre-1989 treaties were of little use in promoting actual transboundary environmental protection. For instance, although between 1956 and 1985, Hungary concluded bilateral boundary water agreements with all its neighboring states, these treaties contained only vague and superficial provisions relating to environmental protection.[962]

Similarly, air pollution agreements between CEE states failed to contain target reductions, substantive legal norms, rules of liability, or the identification or creation of dispute resolution mechanisms, but rather relied upon procedural and organizational norms designed to facilitate the reduction of harmful emissions if the parties determined it was in their interest to reduce those emissions.[963] Moreover, as CEE states generally did not implement what substantive provisions did exist in the treaties, it was difficult for them to actually translate these limited norms into pragmatic activities for environmental protection or dispute resolution.[964]

B. Post-1989 practice

Since 1989, CEE states have begun to reassess the capacity of general international law to promote environmental protection and dispute resolution.[965] This reassessment includes a reexamination of existing bilateral and multilateral agreements to determine whether, if given substance, they might prove effective. CEE states have, however, been reluctant to create formal mechanisms to coordinate cooperation on environmental or related matters.[966] In some instances, CEE states have concluded that international law does possess a capacity to assist in the resolution of transboundary environmental disputes, while in other instances they perceive little if any capacity.

CEE states are also attempting to create a basis for the employment of international law by seeking to adopt a number of bilateral[967] and multilateral treaties.[968] In September 1996, for example, Hungary and Romania concluded a bilateral treaty of understanding and cooperation, linking environment and development, and committing themselves to early warning of disasters and extended cooperation in promoting navigation and protection of the Danube and other border waters. Many of these treaties incorporate such requirements as an international EIA where the construction of a new industrial enterprise might negatively affect the environment of a neighboring state. CEE states are currently most active in the area of transboundary water resources, with three notable agreements being the Elbe River Convention, the Oder River Convention and the Danube River Protection Convention.[969] The Danube River Convention adopts not only the basic procedural principles of international law relating to environmental protection, but also the polluter-pays principle and the precautionary principle.[970] The Elbe River Convention and the Oder River Convention create joint commissions for the purpose of monitoring water quality and making proposals for measures to reduce the emission of pollutants into the rivers and for protecting the biodiversity of aquatic species.[971]

1. Positively existing principles

Because of their relatively positive existence in both customary international law and a variety of bilateral and multilateral treaties to which CEE states are party, international environmental legal principles relating to procedural obligations designed to prevent disputes,[972] and certain principles relating to transboundary air pollution,[973] and transboundary water utilization and/or pollution[974] are perceived as possessing the greatest capacity to resolve disputes.[975] Procedural principles relating to the obligation to provide information and consult on operations that might cause transboundary harm are particularly relevant given the historic lack of cooperation and the previous tendency to classify such information as secret. Those principles of international air and water law relating to the duty not to cause harm and the shared utilization of resources are particularly relevant at this time as the Baltic Sea, Black Triangle, Gabčíkovo–Nagymaros Project, Romanian–Bulgarian Danube River Gauntlet and Black Sea disputes all relate to continuing harmful emissions of air and/or water pollutants and the perceived disproportionate use of air and water resources by certain states.

The perceived difficulties with the capacity of environmental procedural obligations and international air and water laws lie with the barriers to their inherent and operational functionality. Inherent barriers to environmental procedural obligations arise from the circumstance that the rules do not specify answers to key questions such as whether a state importing pollution must first request the information, at what stage of the negotiation process this information must be provided, in what language, at what level of technicality and whether the information must contain comparative studies of alternatives and an assessment of the economic benefits of the polluting activity. Concerning operational functionality, although the legal structure for cooperation exists in a number of bilateral cases, the parties often fail to fulfill the basic obligations necessary for such cooperation. In the case of the Romanian–Bulgarian Cooperation agreement, for instance, the parties failed to adopt necessary domestic legislation and initially failed to appoint members to the nuclear and chemical safety teams.[976]

Inherent barriers to the functionality of international air pollution law include the failure of CEE bilateral treaties to require a harmonization of standards for measuring emissions, defining environmental harm and assessing damages.[977] On the operational level, it is difficult to monitor the deposition paths of SO_2, and CEE states frequently accept legal commitments beyond their financial or technical capability to meet.[978] The Czech Republic, for instance, planned to meet its LRTAP commitments by implementing desulphurization mechanisms on a number of large emitters, as well as constructing the Temelín nuclear power plant. The Czech Republic, was unable to achieve the required reductions due to the failed operation of the first desulphurization program at the Tušimice power plant and to substantial construction delays at the Temelín power plant.

The functionality of the law of transboundary watercourses suffers from the same fate. Despite the articulation of a number of factors to be considered in determining an equitable and reasonable utilization, CEE states are generally unable to agree on which particular uses or configuration of uses constitutes an equitable and reasonable utilization of various rivers.[979] With respect to the obligation not to cause significant harm, CEE states have yet to agree whether general environmental degradation constitutes 'harm'.[980] And, as in the case of air pollution, CEE states frequently lack the financial resources to meet their legal obligations to improve water quality.[981] In 1997, for example, the EU estimated it would require between $145–155 billion for the CEE states to comply with EU environmental regulations. This level of investment would constitute 3–5 per cent of the GDP of the CEE states for the next twenty years.

2. Constructively existing principles

Principles possessing a lesser degree of perceived capacity, because of their uncertain status as principles of customary international law, but which nonetheless are considered to exist constructively because of their basis in bilateral and multilateral agreements, include: the procedural right of public participation in decisions affecting the utilization of natural resources,[982] principles of sustainable development,[983] the polluter-pays principle,[984] the precautionary principle,[985] the related obligation to conduct a domestic and/or international EIA,[986] state responsibility/liability for international harm caused by pollution,[987] and the obligation to operate nuclear power plants in a safe manner.[988]

The functionality of the principle of sustainable development is to a certain degree inhibited, because although the phrase frequently appears in CEE legal documents, only a limited number of interested parties understand its meaning, the national government officials generally express little interest in either understanding its meaning or applying its constituent principles, and where there is an attempt to apply the principle it has proven difficult for interested parties to agree on how it should be implemented.[989] With respect to the polluter-pays principle, CEE states, while adopting it in theory, have avoided adopting it in practice. CEE states perceive a decreasing relevance of the precautionary principle and the related requirements of an EIA, as most environmental harm leading to transboundary environmental disputes arises from existing industrial enterprises and not from the construction of new enterprises. Because of the current over-capacity of industrial production in CEE states, they will more likely restructure existing enterprises rather than construct new facilities. In instances where the precautionary principle might be relevant, CEE states have found it difficult to determine its specific substantive definition and the level of potential harm requiring precautionary action.[990]

Current EIA requirements also frequently lack functionality to promote transboundary protection as they tend not to require sufficient notice to those individuals or entities affected by the proposed activity, and there is a lack of institutional capacity for and interest in conducting the necessary technical studies and procedures providing for public participation. When proposals for conducting an EIA were solicited for the construction of a dam on the Elbe River, for example, many bids were received offering to conduct the EIA in a period of two to three weeks for approximately $500. The accepted proposal entailed a slightly longer time frame at a cost of $8,000. Independent assessments from Swedish experts indicated a proper EIA would cost $150,000 and require over one year to complete.

In instances where EIAs are made available for public comment, they often contain excessive amounts of irrelevant technical jargon and statistics, and decision makers are often reluctant to give serious consideration to issues and concerns raised by the public. For instance, although the Slovak government conducted an EIA on the Mochovce nuclear power plant, some environmental NGOs claim the documentation 'fails completely to provide the most important safety-related detail, while providing large quantities of irrelevant information on topics of minor importance'.[991]

Similar problems afflict the capacity of international EIAs, including the absence of bilateral agreements providing procedures for conducting joint EIAs, the lack of some states' technical ability to participate in joint investigations, the uncertainty as to what specifically constitutes an international EIA, and questions concerning which party is responsible for providing for and financing public participation from neighboring states. Slovakia, for example, claims it conducted EIAs for the Gabčíkovo–Nagymaros Project in 1976, 1984 and 1985, and 'international' EIAs were separately conducted by both the Bechtel Corporation and Hydro Quebec. Hungary, however, contended in 1990 it was necessary to conduct a more 'complex and professional' joint EIA, and an EIA conducted by Hungarian scientists in 1985 was in fact itself insufficient as it represented only an environmental impact statement.[992]

Principles of state responsibility/liability are temporally relevant as many states in CEE suffer observable and quantifiable damage as a result of environmental degradation caused by neighboring states, and many of these states refuse to accept responsibility for the consequences arising from their activity. The principles of state responsibility/liability, however, tend to lack dispositional relevance since many CEE states are concerned if they utilize these principles in their efforts to constrain the behavior of or seek compensation from one neighboring state, a different neighboring state might seek to use those principles to constrain the behavior of or seek compensation from that state. The perceived capacity of principles and rules of state responsibility/liability is inhibited by the lack of a generally agreed upon definition by CEE states of what constitutes environmental harm,[993] the difficulty of determining which of a multitude of sources may be responsible for specific environmental harms,[994] and the difficulty of precisely calculating the lost monetary value associated with environmental damage.[995]

The obligations to operate nuclear power facilities in a safe manner, although highly relevant to the circumstances of all CEE states except Poland, are difficult to monitor unless the nuclear facilities receive IAEA assistance, and beyond an obligation to consult, they do not provide neighboring states with a right to request an equitable balancing of interests.[996] Unlike Western European states, CEE states have not yet entered into bilateral treaty relationships specifically requiring states constructing nuclear facilities to share relevant information and consult with neighboring states.

3. Indeterminate principles

Principles not binding upon CEE states, because they are neither reflected in customary international law nor international agreements, but which might be relied upon by CEE states to structure a peaceful resolution of transboundary environmental disputes include: human rights principles providing for a right to a healthy environment,[997] an ethnonational minority's right to participate in an equitable and fair utilization of a state's natural resources, certain transnational rights that could be of use to individuals or groups seeking to protect the environment,[998] and principles and rules relating to the use of regional economic instruments, such as regional tradable emissions permit schemes to promote environmental protection.[999]

Human rights principles relating to special protection for national minorities are becoming more relevant as CEE states continue both to engage in activity causing environmental degradation in border regions where national minorities reside, and to enact legislation restricting the rights and privileges of national minorities. The general disinclination of CEE citizens to use domestic legal mechanisms to resolve domestic disputes and the varying degrees of independence of the judiciary in CEE states decrease the relevance of environmental legislation providing for access of foreign nationals to participate in the administrative decision making process or to bring claims against industrial enterprises responsible for transboundary pollution affecting those foreign nationals.[1000]

Although the Czech Republic acknowledges the existence of the concept of tradable permits,[1001] the use of such mechanisms is illegal in the Czech Republic, Slovakia and Poland.[1002] The relevance of regional economic instruments may, however, increase, as for the next twenty years CEE states possess the greatest opportunity for energy-bargaining due to their large and inefficient consumption patterns.[1003] This increased capacity for energy-bargaining coupled with their lack of financial resources creates a condition ripe for the use of market-based instruments, which in many instances will be more relevant than in West European states.[1004] Regional economic instruments also have an increasing degree of dispositional relevance, as CEE states, particularly in the northern tier, are relying to a greater extent on domestic economic instruments and are coming under pressure by Germany to consider their use on a regional level.[1005] Barriers to the use of regional economic instruments include the operation of polluting enterprises such as subsidized monopolies and their directors' limited interest in participating in a regional environmental scheme entailing restrictions on production in exchange for economic incentives, as well as the reluctance of state governments to share technical and economic information required for effective operation of a regional tradable permit scheme.

Conclusion to Part IV

Prior to 1989, the ability of CEE states to resolve transboundary environmental disputes was inhibited by the reluctance of socialist doctrine to admit to the possibility of environmental degradation and to the possibility of environmental disputes between fellow socialist states. The incidence of environmental pollution was, however, exacerbated by the centrally planned nature of economic development, the neglect of cost-accounting for economic externalities, and the heavy industrialization of socialist states.

Since the initiation of the transformation in CEE, the CEE states more readily identify existing transboundary environmental degradation and seek to engage relevant states in the process of resolving the disputes arising from this degradation. CEE states are also significantly more willing to identify cases of potential environmental degradation, or threats to environmental security, as in the case of the Temelín, Mochovce, and Kozloduy nuclear power plants, and to seek to establish some process for resolving these disputes. As CEE states continue with their economic transitions they will experience a general improvement in the condition of their environment and the increased possibility of using market-based mechanisms to enhance transboundary cooperation and to further reduce pollution. They, however, will also find increasing economic growth will lead to new pressures on the environment, and possibly to increasingly competitive bilateral relations.

As the CEE states continue their political transformation, the enhanced levels of public participation will increase the pressure to take action to improve the environment and to resolve transboundary environmental disputes. Increased public participation may also promote the ability of CEE states to devise and to implement mechanisms necessary to bring about a reduction of the pollution and a resolution of the dispute.

Moreover, the increasing development of relations among CEE states and between CEE states and West European states will likely enhance the funding and institutional support necessary to improve environmental protection and engage in a meaningful process of dispute resolution. The enhanced ties with Western Europe may, however, also provide for an increasing opportunity for some Western sub-state actors to export their environmental problems to CEE. The continued tension relating to the treatment of national minorities will influence the inclination of certain CEE states to cooperate with other states to resolve transboundary environmental disputes, and may in some instances provide a common ground from which to build integrative cooperative mechanisms.

Prior to 1989, the role of international law in the resolution of CEE transboundary environmental disputes was inhibited by the fact CEE states operated under a system of socialist domestic and international law that in many instances explicitly sought to minimize the role of law in the regime of transboundary environmental protection. The minimal role of socialist international law was largely based upon its primary goal of regulating the relationship between socialist and capitalist states, which was irrelevant for the purposes of resolving intra-CEE transboundary environmental disputes and of providing a bulwark against West European and American hegemony.

Since 1989, CEE states have begun to take steps that promote the role of international law in the regime of transboundary environmental protection. Particularly, CEE states are establishing a practice of using general international law to regulate their international relations and are redrafting and employing municipal environmental law to resolve domestic resource conflicts. Moreover, as their legal systems develop in a parallel manner, particularly in the case of Poland, the Czech Republic, Slovakia and Hungary, there is likely to be an increased opportunity for transboundary cooperation at both the systematic and sub-systematic levels.

CEE states are also beginning to correlate their municipal environmental law with international environmental law and recognize certain interested third parties are creating an incentive for them to resolve their transboundary environmental disputes consistent with the rule of law. Similarly, CEE states increasingly perceive third party dispute resolution mechanisms as having some capacity for assisting in the application of international law to their disputes. The presumably successful resolution of the Gabčíkovo–Nagymaros Project dispute will play a role in strengthening this perception. Finally, CEE states perceive that international environmental law might have some capacity to assist in the resolution of their disputes, with the principles appearing to have the greatest capacity at present being those relating to transboundary air and water pollution and the obligation to cooperate in the reduction of pollution.

Barriers to the functionality of international law, and thus its increased role, include the relative inexperience of CEE states in the field of transboundary environmental dispute resolution, the continued lack of serious enforcement of municipal environmental law and the slow maturation of municipal environmental institutions. CEE states are also hesitant to give full force and effect to international agreements, and as they wean themselves from international aid programs, they may be less influenced by pressure from donor states to employ international law to resolve their transboundary environmental disputes.

As CEE states, in particular Poland, the Czech Republic and Hungary, approach potential EU membership, the EU will likely be able to counteract the antithetic influence of lax domestic enforcement and decreasing aid programs by exerting increased pressure on CEE states to use international law to resolve their disputes. Similarly, as international environmental laws relating to the precautionary principle and the polluter-pays principle become more distinct and generally accepted in international practice, and as CEE states become parties to the increasing number of international environmental agreements, CEE states will likely perceive a greater capacity of these principles to assist in reducing pollution and resolving transboundary environmental disputes.

Conclusion

The forced industrialization of CEE states and the wholesale neglect of environmental protection under the former communist regimes from 1945 to 1989, have left CEE states suffering from substantial environmental degradation. The consequences of this environmental neglect include serious threats to human health and safety and the substantial destruction of local and regional ecosystems. As much of this environmental degradation occurs in border regions, it has given rise to a number of transboundary environmental disputes between CEE states, encompassing disputes over transboundary air and water pollution, the transport and storage of radioactive and hazardous waste, the protection of marine resources, and the operation of nuclear power plants.

To ensure adequate environmental protection in CEE transboundary regions, CEE states must cooperate to move these disputes from the pre-resolution phase through the resolution phase and into the implementation phase, as is necessary in the Black Triangle and Black Sea disputes. And in cases where a formal dispute has been articulated, to move the dispute out of the resolution phase and into the implementation phase, as is necessary in the Romanian–Bulgarian Danube River Gauntlet dispute. Since international law may play a central role in assisting CEE states in resolving these disputes, it is important to understand the functions of international law throughout the process, the situational circumstances which influence the resolution of disputes and the utilization of international law, the role of sub-state actors and interested third parties, the current degree of reliance upon international law, and the future prospects for the increased use of international law to assist in the resolution of CEE transboundary environmental disputes.

The functions of international law

In the dispute formation phase, international law serves a social function by providing a means for states to indicate they wish to resolve the dispute in a peaceful means consistent with established international or regional norms. International law further serves procedural and substantive functions by providing a set of rules governing when and how a state must share information about its actions that may pose a threat to another state and a set of principles against which the parties can ascertain whether a certain activity may violate the rights of one of the parties, such as in the Romanian–Bulgarian Danube River Gauntlet dispute. In the dispute formation phase, dispute resolution mechanisms, such as inquiry and good offices or mediation, often assist the parties in ascertaining the exact nature of the activity giving rise to the dispute, whether this activity constitutes a real threat to either of the parties and which means may be most appropriate for resolving the dispute. In some instances, a third party facilitates the commencement of negotiations by providing technical capacity or by providing financial or political incentives for resolving the dispute by peaceful means. Third parties, such as the EU in the case of the Gabčíkovo–Nagymaros Project dispute, may even use the norms of international law to persuade the parties to submit the dispute to a more formal means of dispute resolution, such as arbitration or adjudication.

In the pre-resolution phase, international law helps the parties ascertain the parameters of the dispute and identify physical solutions to the dispute by indicating their rights and responsibilities, obligating them to share relevant information and engage in consultations, and in some instances obligating them to conduct joint investigations in order to determine the most practicable and equitable means of resolving the dispute, as in the Romanian–Bulgarian Nuclear Corridor dispute. International law also assists the parties in articulating their negotiating positions and signaling sub-state actors as to the negotiating priorities of the national government. During the pre-resolution phase, inquiry and mediation continue to play a role, and in some instances, conciliation mechanisms may also be employed as they provide a means to enlist more formally the participation of third parties and identify possible physical or political solutions to the dispute, as with the proposed involvement of EU in the Silesian Coal Basin dispute.

In the resolution phase, international law assists the parties in reaching a formal agreement as to the specific course of action necessary to resolve the dispute by providing the substantive norms against which to measure the positions of the respective parties and by assuring the parties they are obliged to fulfill the obligations contained in the agreement, as with the Baltic Sea Conventions and the Convention on the Protection of the Black Sea. International law also provides a means for creating additional norms that may be necessary to resolve a unique or unforeseen dispute. During the resolution stage, arbitration and adjudication frequently play a role by deciding for the parties which substantive legal rules govern the dispute and how those rules constrain the behavior of the parties, as evidenced by the desire of Austria to submit the Temelín dispute to arbitration or the ICJ. Although arbitrators and adjudicators seldom articulate the exact terms and modalities of a settlement, they do identify the principles governing the settlement and significantly narrow the options available to the parties. They also provide the parties with the political cover necessary to persuade sub-state actors to accept the settlement, as was possibly the initial intention of the Hungarian government in the case of the Gabčíkovo–Nagymaros Project dispute.

In the implementation phase, international law aids the parties in formally adopting and implementing the settlement by obligating the states to fulfill their agreements, including the obligation to take all necessary domestic measures to give domestic effect to the settlement. States are moreover obligated to cooperate in the implementation of the agreement and to share information regarding the effect of the agreement on the circumstances initially giving rise to the dispute, such as in the case of the operation of HELCOM in the Baltic Sea dispute. Where the parties disagree as to their obligations under the settlement, they may again involve a third party mediator or invoke arbitration or adjudication mechanisms.

Circumstances influencing the utilization of international law

Notably, a number of the political, economic and ecological circumstances present in CEE tend to favor the positive utilization of international law in the region, and the likelihood CEE states will begin to move their disputes through the dispute resolution process. For instance, since the 1989 revolutions, CEE states appear to be reconceptualizing their view of sovereignty from a Hobbesian view, defined in instrumental terms, where sovereignty is an institution that exists to protect the citizens of a state from internal and external violence, to a Rousseauian view of sovereignty, defined in expressive terms as the embodiment of social and political order. Although CEE states are reconceptualizing sovereignty in terms of social and political order, they also desire to exercise their sovereign rights actively after forty years of subjugating those rights to the former Soviet Union, and they have a strong heritage of considering the utilization of natural resources to be a high order sovereign right. CEE states may thus be attracted to international regimes as a means of facilitating their reintegration into the international social process, yet they may be hesitant to embrace these regimes fully given their historical inertia.

Similarly, since 1989 CEE states have undergone a transformation with respect to their definition of welfare gains, shifting from a desire to preserve the unity of the socialist bloc and assert socialist hegemony, to a desire to pursue economic development, integration with the West and, to some degree, greater environmental protection. This transformation, which is still occurring, has brought CEE states into closer social alignment with Western states and to a varying extent enabled CEE states to share norms and social institutions with Western states.

The prospects for the operation of international law in the resolution of CEE transboundary environmental disputes are enhanced by a high degree of collective interdependence among CEE states as the causes and effects of environmental degradation in CEE and the benefits from a more reasonable utilization of natural resources are interlinked across national boundaries and experienced by the local community of states. Environmental degradation also affects a range of other interrelated transboundary interests important to CEE states, such as recreation, economic development and public health. This interdependence of CEE states on the North–South axis is mirrored by the growing interdependence of CEE states and West European states along an East–West axis. The EU may play a particular role here in promoting the operation of regimes. Given that all CEE states seek EU membership, the EU can exercise power in terms of how it may require CEE states to behave in order to deserve further association with the EU. Specifically, the EU may promote an inclusive CEE regime by emphasizing its interaction with CEE states as a community.

The role of sub-state actors and interested third parties

Turning to an assessment of the influence of sub-state actors and interested third parties, it is useful to recall that although states are the primary actors in the international arena and are both subject to and capable of using international law, whether a CEE state chooses to and is capable of using international law is to a significant extent affected by the interests and capabilities of a collection of sub-state actors including government officials, economic actors, environmental actors and representatives of the general public. Moreover, the interest and capability of CEE states to use international law is affected by a range of interested third parties, including the EU and its member states, the United States and IFIs.

As the CEE national governments guide the transformation of their states to democracy they increasingly reflect public opinion as well as their own interest in retaining power. In some states, such as the Czech Republic and Poland, where the democratic transformation has made significant progress the national government is inclined to use international law to resolve its disputes and to participate more fully in existing legal regimes, such as the Baltic Sea conventions. It should be noted, however, these governments reflect the view of economic actors as well as environmental actors and may equally seek to use international law to protect their economic interests. In instances where the attention of the national government has not focused on transboundary disputes in regions far from the center of government, such as the Black Triangle dispute, regional or local governments have stepped in to provide leadership and establish mechanisms for resolving the dispute.

Since 1989, CEE environmental actors have continued to increase their ability to advocate for their interests effectively, although they have generally been unable to sustain formal parliamentary representation as the economic costs of the transition to a market economy become more apparent. As such, environmental actors have been most influential in putting environmental concerns on the public agenda, as in the Romanian–Bulgarian Danube River Gauntlet dispute and the environmental legacy of Soviet occupation. Environmental actors have also played a role in the informal monitoring of the implementation of dispute settlements, as will likely be the case with the Gabčíkovo–Nagymaros Project dispute.

Economic actors have worked effectively to preserve their interests in a number of disputes, particularly those relating to nuclear power, such as the Temelín Nuclear Power Plant dispute and the Romanian–Bulgarian Nuclear Corridor dispute. In these instances, the economic actors have sought to commit their national governments to working within the established IAEA mechanisms and treaty regimes so as to avoid opportunities for those opposed to the operation of the nuclear power plants to build a consensus to submit the disputes to formal arbitration or adjudication. In other cases economic actors have become pitted against each other, as in the case of the Baltic Sea Basin dispute in which fishing interests have sought to increase restrictions on the emission of pollutants, while the hydrocarbon industry has sought to prevent further restrictions. Notably, both sets of actors have worked through the legal process established by the Baltic Sea conventions.

Immediately after the velvet revolutions of 1989, the general public, which was quite politically active, supported measures designed to reduce pollution and incidentally to resolve transboundary environmental disputes. As the economic costs of the transition have set in, the enthusiasm of the general public for environmental protection has waned. As the CEE states go through the transition toward a rule of law based society, however, they expect their national governments to act according to the law, including international law. And, as the CEE states strive for membership in the EU, the general public increasingly perceives the value of resolving transboundary disputes and acknowledging certain compromises may have to be reached in order to resolve a dispute and promote regional political development.

As for third parties, the EU and its member states have been and will be deeply involved in promoting the use of international law to resolve CEE transboundary environmental disputes. For instance, the EU mediation of the Gabčíkovo–Nagymaros Project dispute played a key role in the agreement of the parties to submit the dispute to the ICJ. Similarly, the role of the EU in funding and directing an inquiry into the causes and consequences of transboundary air and water pollution in the Black Triangle will likely provide a basis for a subsequent legal regime designed to reduce pollution and ease transboundary tension. With the invitation to Poland, the Czech Republic and Hungary to enter negotiations to join the EU, it will play an even greater role as disputes affecting these parties will automatically involve the EU. Moreover, these new members may use the threat of veto over future membership of their neighbors as a means of inducing them to make concessions on specific transboundary disputes. The new members, however, will also be inclined to act with caution and reason in dealing with their non-EU neighbors, as they will wish to be perceived as responsible new members.

The United States has been and will be involved in CEE transboundary environmental dispute resolution to the extent it funds environmental remediation efforts and to the extent it wishes to promote regional stability. On the whole, the United States prefers the EU to take the lead in promoting the resolution of CEE transboundary environmental disputes, but will provide funding of measures designed to resolve these disputes and will apply political pressure to encourage the parties to invoke formal dispute resolution mechanisms rather than take unilateral action.

The IFIs are also interested in promoting transboundary environmental dispute resolution, but generally do not design their projects with this aim in mind. Rather, the IFIs support certain projects designed to promote environmental protection, which may incidentally promote

224 *International Law and Transboundary Disputes*

the resolution of a transboundary dispute. In other instances, the IFIs provide funding for activities that may generate pollution or cause environmental degradation and thus may aggravate or create transboundary environmental disputes, although the IFIs are less and less inclined to support such projects as they have come under pressure from environmental actors.

The contemporary use of international law to resolve CEE transboundary environmental disputes

Turning to the specific transboundary environmental disputes involving CEE states, it is clear in some disputes, such as the Baltic Sea dispute and the Gabčíkovo–Nagymaros Project dispute, the parties have relied extensively on international law and the legal process to promote dispute resolution. In the Baltic Sea dispute the parties have developed a series of treaties and oversight mechanisms incorporating traditional public international law principles, such as the duty to consult and share information, as well as environmental law principles, such as the precautionary principle and sustainable development. The Baltic Sea dispute could, however, benefit from increased regulation of non-point sources of pollution and more tailored articulation of environmental law principles such as equitable utilization of shared natural resources. In the Gabčíkovo–Nagymaros Project the parties successfully invoked various international legal processes, including inquiry, limited mediation and adjudication to narrow the options on the negotiating table. In narrowing the options, the ICJ relied upon traditional principles of public international law as well as more recent principles of international environmental law.

In other disputes, such as the Black Sea and the Danube River Gauntlet, the involved parties have to date made unsuccessful attempts to construct effective regimes for transboundary environmental protection. The current insufficiency of these regimes appears to result more from the states' inability to use international law effectively and less from the insufficient capacity of international law itself. In these circumstances, the strengthening of one or two additional situational circumstances might be sufficient to bring the regime into effective operation and bring about a resolution of the dispute and improvement in the environmental condition of the area. These disputes may also benefit from an invocation of a legal process, such as arbitration or adjudication, to clarify the application of various legal principles to the dispute.

In the Black Triangle, Silesian Coal Basin and Temelín Nuclear Plant disputes, greater use of international law and the legal process may assist in moving the disputes through the dispute resolution phases and may more readily promote environmental protection. In the Black Triangle and Silesian Coal Basin disputes, an international legal regime could structure a regional energy program and multinational linkages between regional power plants, and possibly a regional tradable emission permit scheme. These disputes could also benefit from a more prevalent application of international law principles such as sustainable development, precautionary action and the polluter pays, as well as domestic principles such as the devolution of environmental decision making authority to municipal authorities and market rate pricing for energy resources.

The Temelín Nuclear Power Plant dispute, in particular, may benefit from submission to mediation, arbitration or adjudication. If the dispute is not resolved via a formal legal process, it is likely Austria will use its position as a member of the EU to force the Czech Republic to make significant concessions on operation of the Temelín plant in order for the Czech Republic to attain EU membership. Although potentially effective at limiting the operation of the Temelín plant, such a purely political process would likely generate significant animosity between two neighboring states and is not likely to result in an outcome that balances the economic and environmental needs of the Czech Republic with its international obligations not to cause or threaten to cause harm to a neighboring state.

Future prospects for the increased utilization of international law

Turning finally to the factors promoting the functionality of international law within the regime of transboundary environmental protection, it is important to note prior to 1989, CEE states participated in a system of socialist international law which, although providing a framework for interaction with nonsocialist states, did not constitute a functional regime for the regulation of intra-CEE environmental relations. With the dissolution of the Soviet Union and the ebbing of influence of socialist international law, CEE states have found themselves in transition toward adopting the framework of Western legal principles. Notably, CEE states have also begun to structure their relationships with Western states according to the principles of international law and are articulating more general intra-CEE concerns in legal terms. To a certain extent, however, their ability to comprehend and assimilate the principles, norms and rules of general international law is inhibited by their

226 *International Law and Transboundary Disputes*

primary focus on completing the transformation of their states to liberal democracies with market economies.

To further the employment of international environmental law, action may be taken by CEE states, their sub-state actors and interested third parties. As CEE states continue to adopt and use municipal environmental law designed to promote the resolution of domestic resources and correlate their municipal environmental law with international law, they will strengthen their understanding of the value of international environmental law and develop a proficiency with its use, thereby enhancing the prospects for its future utilization. Domestic actors may promote this process by taking full advantage of their opportunities for participation in the domestic political process, particularly as it relates to the integration of economic and environmental decision making, and by exercising their rights within the domestic judicial process to ensure the enforcement of municipal environmental legislation.

Interested third parties may promote the operation of international environmental law by encouraging CEE state participation in international environmental agreements, emphasizing the rule of law in their relations with CEE states, particularly where they fund economic or environmental projects, and encouraging CEE states to utilize international judicial dispute resolution mechanisms to resolve the legal aspects of environmental disputes. Interested third parties, however, must be careful not to encourage CEE states to seek the judicial resolution of political and economic issues properly lying with nonjudicial or hybrid dispute resolution mechanisms, since the ultimate failure of the judicial mechanisms to resolve such disputes adequately will lead CEE states to conclude that these institutions lack the capacity to assist in the resolution of their disputes. The EU in particular, will play a key role in the near future in affecting the extent to which CEE states employ international law, since EU membership is the top foreign relations priority of most CEE states. Conversely, the influence of the United States will wane as CEE states attain EU membership and their need for bilateral assistance diminishes.

As international environmental law continues to evolve and take on a more positive existence, and as states and institutions responsible for the evolution of international environmental law concern themselves with its relevance and functionality, CEE states will perceive a greater capacity of international law to assist in dispute resolution. Similarly, CEE states themselves may adopt regional multilateral treaties designed to overcome the perceived deficiencies in the existing structure of international environmental law and may consider the creation of a regional dispute resolution mechanism to assist in the application of that law.

As noted in the introduction, there has been an attempt in this book to indicate to CEE states, their sub-state actors and interested third parties that international law and the legal process may play a wide variety of roles in promoting transboundary environmental dispute resolution. The discussion contained herein aims to assist CEE states in recognizing how international law and the legal process can constructively contribute to transboundary environmental dispute resolution. Moreover, it is hoped CEE states will take from this analysis an understanding of how better to use international law and the legal process, and thus reduce transboundary tensions arising from environmental disputes and advance the spirit of regional cooperation.

Notes

1. European Commission, Regional Environmental Program Evaluation, at 2.
2. It is estimated that if the level of air pollution in only 18 cities classified as hot spots was lowered to EU standards, at least 18,000 lives and 65 million working days could be saved annually. ODD, No. 208 (25 October 1995).
3. Fidas & Gershwin, at 5.
4. Fidas & Gershwin, at 24–5.
5. Although the scope of this book is limited to the geographic region of CEE, the analysis presented is broadly applicable to the general operation of the norms and structures of international law in other regions.
6. To ensure the official 'nonexistence' of environmental degradation and any associated transboundary environmental disputes, CEE states often classified state sponsored ecological studies as secret and suppressed attempts by local entities or individuals to release their own studies. For a detailed explanation of the relationship between communist ideology and environmental protection, *see* Cole(f), at 159–64.
7. Tunkin(a), at 427–8.
8. For a traditional identification of the forms of international systems, *see* Kaplan, at 21.
9. *See generally* Sands(h).
10. For a detailed explanation of the social science methodology of descriptive inference, *see* King *et al.*, at 35–74.
11. *See* King *et al.*, at 63–74.
12. Kinnear & Rhode, at 6; REC, Strategic Environmental Issues in CEE, at 14; Environmental Action Programme at I-7. *Cf.* Vidlakova.
13. Richard Bilder defines an international environmental dispute as 'any agreement or conflict of views or interests between States relating to the alteration, through human intervention, of natural environmental systems'. Bilder(a), at 153. The definition adopted by this book is consistent with Bilder's definition as it excludes disputes between sub-state entities, but is narrower than Bilder's definition in that it also excludes global disputes. For a criticism of the exclusion of sub-state entity disputes, *see* Sand(c), at 126; Bhagwati, at 443.
14. John Carroll defines transboundary environmental disputes as arising when benefits from a pollution-emitting activity disproportionately accrue to one state with the costs disproportionately accruing to a neighboring state. Carroll, at 1. The PCIJ defines a dispute as constituting a 'disagreement on a point of law or fact, a conflict of legal views or interests between two persons', Mavrommatis Palestine Concessions Case, at 11, and notes the dispute need not manifest itself in a formal manner, rather the parties need only have opposing views. Chorzów Factory Case, at 10–11.
15. *See also* Fidas & Gershwin, at 24–5.
16. M. Fitzmaurice(b), at 24; It is estimated the long term cleanup of the Baltic Sea will require $23 billion over the next twenty years. Reuters Textline (2 June 1995).

17. Kinnear & Rhode, at 12; M. Fitzmaurice(b), at 24–5; Regional Environmental Program, at 6; Baltic News Service (25 May 1998); Russian Press Digest (5 March 1998).
18. Regional Environmental Program, at 6. For more information on the ecological state of the Baltic Sea, *see* Brodecki; International Legal Problems of the Environmental Protection of the Baltic Sea; Lindwall; Ten Years of Environmental Co-operation in the Baltic Sea: An Evaluation and Look Ahead.
19. Lloyds List (19 May 1997); The European, *Pollution Tide Turns in the Baltic Sea* (20 March 1997). In addition, the German government has established an Institute for Baltic Sea Research in the former East German city of Warnemunde for the purpose of increasing the effectiveness of various regimes for monitoring and protecting the Basin. The activities of the Institute include advancing knowledge about the Baltic's hydrodynamics, and monitoring Germany's compliance with the Baltic Sea Conventions and conducting research concerning the impact of natural and human influences on the future development of the Baltic Sea. FBIS-WEU (16 August 1994).
20. RFE/RLN, No. 37, Part 1 (24 February 1998); *see also*, Speech by Ritt Bjerregaard, Member of the Commission responsible for Environment Baltic Agenda 21 Foreign Ministers' meeting – Council of the Baltic Sea States in Nyborg, Denmark (22 June 1998).
21. *International Market Insight Reports* (11 September 1997).
22. FBIS-EEU (1 December 1994). For a review of the efforts of the EU to assist Baltic Sea Basin states in meeting their obligations under the Baltic Sea Conventions, *see* East European Markets (26 April 1996).
23. Fidas & Gershwin, at 21.
24. For a review of the international law governing the extraction of natural resources in the marine environment, *see* DeMestral; Evans; Kindt.
25. FBIS-EEU (22 July 1995).
26. *See* Boczek(a); M. Fitzmaurice(a); Johnson.
27. For a review of the regional efforts to reduce marine pollution, *see* Yturriaga; Hayward; Pallemaerts(c). For information concerning the Mediterranean, *see* P. Haas(a), and for the North Sea, *see* Pallemaerts(c).
28. For a review of the general principles of international environmental law, *see* Pallemaerts(b); R. Dupuy(b); Birnie & Boyle; Hughes; Lang *et al*.; Magraw; Munro; Sand(a, f); Sands(g, h); Schachter(b); Soto; Stephen. For a review of principles of international law relevant to marine pollution, *see* Ardia; Boczek(b); Soni; Timagenis; Tharpes.
29. International law facilitated the movement of the dispute to this phase in part by requiring the parties to negotiate in good faith to reduce emissions and effluents which caused harm to other Baltic Sea Basin states. For more information on the Baltic Sea Conventions, *see* M. Fitzmaurice(a); Johnson.
30. For a general review of the legal principles relating to pollution of the marine environment by vessels, see Bodansky(b); Sasamura, and by dumping, *see* Duncan; International Maritime Organization, The London Dumping Convention: the First Decade and Beyond; Susman (discussing EU regulations); Winter; Zeppetello.
31. 1974 Baltic Sea Convention, at arts. 3, 7, 11. The Convention does not, however, apply to the international waters of the Baltic Sea. M. Fitzmaurice(a), at 25–6.

32. For a review of the use of institutions to protect the marine environment, *see* Kimball.
33. Environmental Action Programme, at II-12.
34. Regional Environmental Program, at 7.
35. Environmental Action Programme, at V-29; *Royal Society of Chemistry Newsbrief* (22 April 1997). For a review of international legal principles governing the disposal of radioactive waste in the marine environment, *see* Curtis.
36. For more information on the polluter-pays principle, *see* Gaines; Sands(h), at 213–9; OECD, The Role of Economic Instruments, at 42.
37. For more information on the precautionary principle, *see* Sands(h), at 208–13; Bodansky(a); Cameron & Abouchar; Gündling(b); Freesonte; Hey(a, b).
38. For more information on EIAs, *see* Sands(h), at 579–94; Convention on Environmental Impact Assessment Colombo; Futrell; Kennedy; OECD, Application of Information and Consultation Practices; Pallemaerts(a); Pineschi; Sheate; Stewart; Wathern.
39. 1992 Baltic Sea Convention, at arts. 3, 6, 7, 13, 17.
40. 1992 Baltic Sea Convention, at arts. 5, 16.
41. For a general review of international efforts to reduce land based pollution of the marine environment, *see* Bliss-Guest; Kwiatowska; M'Gonigle.
42. Brunnée(a), at 5. Land-based sources contribute nitrogen, phosphorus, heavy metals and organic substances to the Baltic Sea.
43. *Europe Environment* (7 April 1998); Brunnée(b), at 12. For a review of the operability of these conventions, *see* Hey(a).
44. *See* Brunnée(b), at 12.
45. For more information on the obligation not to cause serious injury to neighboring states and the principle of equitable use of natural resources, *see* Sands(h), at 347–54; Handl(c).
46. For a review of international law applicable to transboundary pollution, *see* Barros & Johnston: P.-M. Dupuy(c); Magraw; Springer(a, b); Teclaff & Teclaff(b).
47. Environmental Action Programme, at II-7.
48. Schöpflin, at 250; Mloch. In addition to indigenous pollution, the Black Triangle is at times affected by long range NO_x emissions from Germany, France and the United Kingdom.
49. Mloch, at 43; Environment of The Czech Republic, 41, 192–3.
50. Fidas & Gershwin, at 11.
51. Jancarova, at 363. The ten largest sources of pollution in the Czech Republic (0.04 per cent of the large source polluters) generate 35 per cent of the SO_2 emissions. The Czech Republic even boasts the second largest point source emission in all Europe and the former USSR: the sister power plants of Pruněřov and Tušimice. Environment of the Czech Republic, at 37, 39, 107, 238.
52. The varied matrix of industrial enterprises at the local level in the Black Triangle often results in the emission of a multi-layered cocktail of dangerous pollutants. In the Liberec region, for instance, the environment suffers from the large source Turów power plant, a substantial number of local industrial sources consuming low quality lignite, the industrial pollution of the Nisa and Jablonec rivers and threats from uranium mining runoff into the Ploučnice River, as well as radioactive runoff in the Ralsko, Trutnuv and Vesecko municipal districts. 1992 EYBCR, at 260.

53. Kramer(e) at 1–3; EYBCR 1992, at 89–90.
54. Mloch, at 43; EYBCR 1992, at 273–4. For a review of the consequences of the pollution in the Black Triangle, *see* Hinrichsen.
55. EYBCR 1992, at 195. The environmental degradation also negatively affects local recreation and international tourism. The acute die-back of forest in the Jizera Mountains near German and Polish thermal power plants, for instance, causes a substantial economic loss to the local community. Environment of The Czech Republic, at 192–3. *See also* FBIS-EEU (16 November 1994).
56. FBIS-EEU (14 December 1994).
57. Strategic Environmental Issues in CEE, at 52.
58. At least 31 per cent of the sewage discharged from Czech towns with a population over 50,000 enters the Elbe River in Northern Bohemia untreated, and all of the sewage discharge from towns with populations less than 50,000 enters the river basin untreated. Environment of the Czech Republic, at 53. By comparison, in Germany only 6.5 per cent of the towns with populations over 50,000 discharge untreated sewage into the Elbe River. EYBCR 1992, at 166–7; Water Research Institute, Elbe Project, at 2.
59. Environment of the Czech Republic, at 194. World Bank, Czech and Slovak Federal Republic Joint Environmental Study, V. II, at 48. For instance, in late September 1998 a significant quantity of cyanide spilled into the Oder River from a Czech steel works plant. Polish Press Agency (1 October 1998).
60. ZumBrunnen, at 399. *See also* Environmental Action Programme, at II-15.
61. Environment of the Czech Republic, at 43, 245.
62. The Turów plant is responsible for the direct and indirect employment of over 70,000 Polish workers. Like most power plants in CEE, the Turów power plant currently does not utilize desulphurization mechanisms, costing approximately $2 million per 200 MW unit, but rather relies on electrostatic precipitators, which generally do not reduce SO_2 emissions. Over the course of the next ten years, with $20 million in assistance from the Polish EcoFund and the United States, Turów plans to replace its existing boilers with more efficient boilers and to install FGD equipment on a number of blocks. With these retrofits, the Polish government intends to continue to operate the Turów power plant until 2040 for political and economic reasons. Both the Czech and German governments have pressured Poland to close the Turów power plant.
63. FBIS-EEU (14 November 1994). For similar statements *see also* FBIS-EEU (16 September 1994).
64. For a review of recent investments in the Turów power station, *see* King (5 April 1996); Reuters Textline (21 February 1995).
65. CTK News Wire (20 April 1998); *see also East European Energy Report* (19 December 1995).
66. By the end of 1998 the Czech electric utility had invested close to $1.2 billion in 500 desulphurization and conservation projects throughout the Czech Republic in an effort to meet strict pollution limits scheduled to come into effect on 1 January 1998. CTK Czech News Wire (13 November 1998).
67. For a review of legal principles applicable to air pollution, *see* Flinterman *et al.*; Gündling(c); Haigh(b); Reitze. For a review of legal principles applicable

to transboundary water pollution, *see* Caponera; Gleick; Florio; Handl(a); Kiss(a); Korbut & Baskin; Lammers(a, b); McCaffrey(a, b, d–g); Sette-Camara; Szilgyi; Teclaff; Teclaff & Teclaff(a); Zavadsky.
68. Generally speaking, the mechanism of inquiry involves a formal attempt by the parties to discover the facts and circumstances relevant to the formation and resolution of a dispute. The parties may engage in unilateral inquiries, coming together to then share their findings, they may establish a bilateral commission to conduct the inquiry, or they may establish a trilateral commission involving the participation of a designated third party. In some instances the parties may contract out the inquiry to a third party state or organization, agreeing to accept as authoritative the findings of that party or organization. *See generally* United Nations, *Handbook on the Peaceful Settlement of Disputes between States*, at 74–100.
69. Regional Environmental Program, at 7. FBIS-EEU (9 March 1995).
70. FBIS-EEU (21 October 1994).
71. FBIS-EEU (14 December 1994).
72. FBIS-EEU (30 November 1996); CTK News Wire (2 September 1997).
73. Elbe River Convention, art. 1.
74. Elbe River Convention, art. 2.
75. Oder River Convention.
76. There is a basic precedent for such a policy as the Czech Republic annually imports a small amount of Polish coal. For example, the Czech Republic imports 600,000 tons annually for the Dedmarovice power plant in Ostrava.
77. Stypka, at 31.
78. FBIS-EEU (1 December 1994).
79. Generally speaking, the mechanisms of good offices and mediation involve a third party, which in the case of good offices generally provides the disputing parties with a neutral forum for meeting and serves as a conduit of information and positions if the parties are initially unwilling to engage in direct contact. In the case of mediation, the third party frequently makes informal nonbinding suggestions regarding how best to avoid or resolve the dispute and may, in some circumstances, present additional information to the parties of which they were unaware. *See generally Handbook on the Peaceful Settlement of Disputes*, at 101–22, 123–39.
80. Agreement Between the Government of the United States of America and the Government of Canada on Air Quality. For a review of the operation of this agreement, *see* Glode & Glode.
81. In order to construct the confidence necessary for a regional tradable permit scheme Black Triangle states might first engage in more modest schemes, such as a local regional tradable permit scheme with a transparent verification process. Once this scheme is functional, the parties might develop increased confidence in the behavior of other parties, become familiar with the operation of the regime and assess the benefits of a broader scheme. Notably, the more frequently states engage in a certain type of rule-governed behavior yielding benefits, the more inclined they will be to expand the scope of that behavior.
82. Framework Convention on Climate Change. From 1996 to 1998 three American utility companies funded the installation of a gas burning power plant in the Northern Bohemian town of Déčín in exchange for offsets under

the American Clean Air Act and possibly under a future protocol to the Climate Change Convention. *Los Angeles Times* (1 November 1998). For more information on the GEF, *see* Helland-Hansen.
83. For more information on environmental liability, *see* ILC, *Report on International Liability for Injurious Consequences Arising Out of Acts Not Prohibited by International Law* (1990) Allott; d'Arge & Kneese; Boyle(b); Brownlie(a); de Sola; Doeker & Gehring; P.-M. Dupuy(a, b); European Commission, Communication on Environmental Liability; Francioni & Scovazzi; Gilbert; Handl(a, e); Hoffman; OECD, Responsibilities and Liabilities of States; OECD, Report by the Environment Committee; OECD, Pollution Insurance and Compensation Funds; Kiss(c); McCaffrey(c); Rest(a, b, c); Rosas; Szasz(b); Tomuschat; Vicuna; Zemanek.
84. Juergensmeyer *et al.*, at 835.
85. Mloch; Environmental Action Programme, at II-7. For a review of the extent of coal mining carried out in the Silesian Coal Basin, *see* J. Chadwick, Mining Magazine (March 1996).
86. The primary watercourses affected include the: Odra, Olse, Ostravice, Becva, Opava and Morava Rivers. Environment of the Czech Republic, at 196–7.
87. FBIS-EEU (13 May 1995). FBIS-EEU (31 March 1995).
88. Fidas & Gershwin, at 14.
89. EYBCR 1992, at 261; Environment of the Czech Republic, at 196–7, 308. The region's indigenous environmental problems are exacerbated at times by imported environmental pollution. In August 1988, for instance, the civil authorities of the Bogatynia region of Upper Silesia signed an agreement to import 900,000 tons of highly polluted Rhine River sludge into the region. ZumBrunnen, at 400.
90. FBIS-EEU (13 May 1995).
91. Environmental Needs Assessment, at 52; ODD, No. 31 (13 February 1996); *Polish News Bulletin* (12 November 1998).
92. Poborski, at 2.
93. FBIS-EEU (2 June 1995).
94. *Polish News Bulletin*, Weekly Supplement: Pollution (12 November 1998).
95. Juergensmeyer *et al.*, at 833–5; Poborski, at 2. *See also* Air Pollution Study in Northern Bohemia and Silesia, at 1–2.
96. FBIS-EEU (13 May 1995).
97. ODD, No. 31 (13 February 1996); RFE/RLN, Polish Parliament Adopts Mining Restructuring Plan (30 November 1998).
98. Juergensmeyer *et al.*, at 836.
99. Juergensmeyer *et al.*, at 836.
100. RFE/RLN (7 January 1998). By the year 2000, it is projected the Temelín and Dukovany plants will provide 50 per cent of the Czech Republic's energy needs. In June 1998, the Czech Republic decided to upgrade the Dukovany plant's instrumentation and control systems to meet Western standards. *Nucleonics Week* (10 September 1998).
101. Austrian concerns are heightened because it believes the Czech Republic is covering up construction difficulties and flaws. *See* World Bank, *Czech and Slovak Federal Republic Joint Environmental Study*, V. II, at 69. For a review of the risks associated with the operation of nuclear power plants in CEE, *see* Andrew Steele, *Jane's Intelligence Review* (1 December 1997).

102. CTK News Wire (21 June 1995).
103. FBIS-EEU (2 November 1994); CTK News Wire (5 September 1997).
104. FBIS-WEU (19 October 1994); FBIS-WEU (17 March 1995); CTK News Wire (25 November 1997).
105. Kramer(e), at 14. *See also* Environmental Needs Assessment, at 30.
106. RFE/RLN, No. 64, Part II (2 April 1998).
107. FBIS-EEU (17 April 1995); CTK News Wire (7 July 1997).
108. RFE/RLN, No. 84, Part II (4 May 1998).
109. Kramer(e), at 14–15. *See also* FBIS-EEU (31 May 1996); FBIS-EEU (17 October 1996); FBIS-EEU (2 April 1997). For a review of the efforts of CEE states to upgrade the safety of their reactors, and of Western companies to secure the contracts for this work, *see* Colin Woodard, *Bulletin of Atomic Scientists* (15 May 1996).
110. RFE/RLN (7 January 1998).
111. Pehe, at 17; *See also* Austrian View of Temelín Power Station Explained, JPRS-TEN-95-003) (2 November 1994). FBIS-EEU (3 January 1995).
112. Kramer(d), at 62; Kramer(f).
113. For a review of international legal principles relating to the operation of nuclear power plants, *see* Adede; A. Boyle(a, c); Handl(f); Handl & Lutz; Sands(a); Szasz(a).
114. Austria signed the convention in late 1994 and ratified it on 26 August 1997, while the Czech Republic approved the convention on 18 September 1995.
115. Convention on Nuclear Safety, *see* arts. 4, 5, 8, 16, 18, 19. *See also* FBIS-WEU (20 September 1994); Convention on the Early Notification of a Nuclear Accident.
116. For more information on the obligation to share information, *see* Sands(h), at 596–612; Pallemaerts(a); OECD, Application of Information and Consultation Practices; Partan; Smets; Weber.
117. FBIS-WEU (7 June 1995).
118. For a legal basis for Austria's arguments, *see* Handl(d).
119. CTK News Wire (9 October 1998); CTK News Wire (2 April 1998).
120. CTK News Wire (6 January 1998).
121. CTK News Wire (2 April 1998).
122. CTK News Wire (3 April 1998); CTK News Wire (16 April 1998).
123. A. MacLachlan, *Nucleonics* (30 April 1998).
124. CTK News Wire (9 October 1998); CTK News Wire (2 July 1998); CTK News Wire (14 April 1998); CTK News Wire (1 April 1998).
125. CTK News Wire (3 April 1998).
126. CTK News Wire (10 September 1997). For more information on environmental litigation by NGOs, *see* Führ & Roller; Geddes.
127. CTK News Wire (14 November 1997).
128. For a review of the status of all the nuclear reactors operation in CEE, *see* *Nuclear Engineering International* (30 June 1998).
129. FBIS-EEU (30 September 1994).
130. Environment of the Czech Republic, at 287. The Dukovany plant receives its fuel from a French, German and Russian joint venture. FBIS-EEU (30 September 1994).
131. FBIS-EEU (20 April 1992); FBIS-EEU (31 May 1993).

132. FBIS-EEU (12 April 1997).
133. FBIS-EEU (4 January 1995); FBIS-EEU (5 February 1995); Ann MacLachlan, *Nuclear Fuel* (19 May 1997); CTK News Wire (24 March 1998).
134. M. Ledford (1 September 1995). The Austrian government has offered to supply energy to Slovakia if Slovakia were to close the Bohunice nuclear power plant. D. Fisher, at 74–5. The Bohunice power plant is identified by a 1995 United States intelligence report as one of the ten most dangerous Soviet designed reactors still in operation. W. Broad, IHT, at 1 (24 July 1995).
135. Frydman *et al.*, at 318. Slovak NGOs substantially criticized the decision to construct the Mochovce nuclear power plant, arguing Slovakia is failing to consider alternative energy options adequately. Statement of the Slovak Non-Governmental Organizations Concerning the Mochovce Nuclear Power Plant Project (17 February 1995). On 31 October 1995, Slovakia and Russia signed an agreement providing for the Russian firm of Atomenergoexport to complete the first two of four nuclear reactors at the Mochovce power plant, with Russia providing $150 million in financing. The Slovak government also signed agreements with German and French companies relating to the construction of the Mochovce power plant. ODD, No. 241 (13 December 1995); No. 213 (1 November 1995).
136. In the winter of 1995, over 1.2 million Austrians and Slovaks signed a petition against the completion of the Mochovce nuclear power plant. M. Ledford, *Multinational Monitor* (1 September 1995). Slovakia's decision to construct the Mochovce nuclear power plant has been substantially criticized by Slovak NGOs for failing to consider least cost options, such as combined heat and power production, modernization of coal burning power plants, conservation, and increased efficiency of production, and for pandering to a plan which would prohibit Slovakia from realizing sustainable energy policies based on conservation, increased efficiency and the use of alternative energy sources in the foreseeable future. Statement of the Slovak Non-Governmental Organizations Concerning the Mochovce Nuclear Power Plant Project Documentation Submitted for the Public Discussion (17 February 1995).
137. *Financial Times* (10 February 1997); CTK News Wire (2 May 1998).
138. RFE/RLN, No. 93, Part II, (18 May 1998); *Europe Energy* (5 June 1998); CTK News Wire (8 June 1998).
139. FBIS-EEU (14 July 1993).
140. ODD, No. 219 (9 November 1995); ZumBrunnen, at 402.
141. R. Frydman *et al.*, at 317.
142. *See* Czech and Slovak Technical Study, at 66. The Slovak Republic reports former Soviet military forces polluted over 80 different localities and it has already expended over $20 million to remediate some of these sites. RFE/RLN, No. 77 Part II (22 April 1998). Interestingly, Slovakia has decided to reuse many of the Soviet bases for military purposes despite their contamination. Fidas & Gershwin, at 9.
143. CEE states do not have the financial resources to meet the cost of clean up. In the early 1990s, for instance, the former Czechoslovakia budgeted only $8.2 million for clean up operations on 92 sites.
144. Fidas & Gershwin, at 9.

145. EYBCR 1992, at 261; Czech and Slovak Technical Study, at 66.
146. King (5 April 1996).
147. *See* CTK News Wire (6 February 1997). CEE states are also attempting to pool their information with respect to how best to remediate the environmental degradation. In August 1995, for instance, Poland and Estonia signed a defense cooperation agreement, in part providing for cooperation in remediating environmental damage at former Soviet military bases. ODD, No. 171 (1 September 1995).
148. In 1996, the government of Poland allocated $11.67 million for the restoration of former Soviet military facilities, with the program targeting restoration of airports, housing stock and hospitals. ODD, No. 47 (6 March 1996). The Czech Republic has closed 28 sites and is attempting to privatize 35 other sites.
149. For a review of post transition developments concerning environmental liability in CEE states, *see* Goldenman *et al.*, at 179.
150. For a review of legal principles concerning liability for damage caused by hazardous substances, *see* Sands(h), at 673–8; Bethlem & Faure; Wagner & Popovic (discussing liability for US contamination of its Panama bases).
151. FBIS-EEU (20 February 1992). *See also* FBIS-WEU (31 March 1995), where Polish scientist Feliks Pieczka claims 'In settling accounts with Russia, the Polish government agreed to the so-called zero option. We thus have to rely on our own resources in counteracting the effects' of Soviet bases, a cost which Pieczka estimates at more than $2.5 billion dollars.
152. For a review of the relationship between privatization and the assessment of environmental liability, *see* Thomas.
153. For a comprehensive review of the industrial utilization and environmental quality of the Danube River, *see* Equipe Cousteau(b). For a more specific review of the factors contributing to environmental degradation in the stretch of the Danube River forming the border between Romania and Bulgaria, *see* Romania Environment Strategy Paper, at 39.
154. Environmental Needs Assessment, at 16–17.
155. Romania Environment Strategy Paper, at iii.
156. Bulgarian Environment Strategy Study, at 11.
157. FBIS-EEU (15 July 1993). In Silistra, acute respiratory diseases increased over the last ten years by 120 per cent, hydrogen sulphide levels exceed health safety standards by 600–1000 per cent, and phenol levels exceed those standards by 200–400 per cent. Sofia BTA (3 April 1993). In an attempt to focus the attention of third parties on the dispute, in August 1994, the Confederation of Independent Trade Unions of Silistra solicited the involvement of the European Trade Union Confederation. Although the Confederation declined to become involved in the dispute, the two municipalities began pre-resolution consultations on 6 October 1994.
158. FBIS-EEU (12 July 1993).
159. As a result of their role in the 1989 revolution, Bulgarian environmentalists obtained key positions in the new government and consequently were able to persuade the government to host a major CSCE conference on the environment in 1990. D. Fisher, at 39.
160. D. Fisher, at 40. The Romanian government has, however, taken steps to monitor and improve water quality in the Danube River and to carry out

provisions of the Bucharest Declaration signed in 1985 designed to improve environmental protection in the Danube River Basin. 2 YIEL, at 279.
161. Grabowska, at 138.
162. D. Fisher, at 40.
163. For a review of recent legal and institutional developments concerning the Danube River basin, *see* Murcott. For a general review of efforts to construct regimes to manage freshwater resources, *see* Brunnée & S. Toope.
164. D. Fisher, at 39. Western agencies have also provided $400,000 worth of monitoring equipment to both Romania and Bulgaria. Romania Environment Strategy Paper, at 84.
165. Bulgarian–Romanian Convention on Environmental Protection (9 December 1991), Sofia BTA (7 February 1992).
166. Romania Environment Strategy Paper, at 84.
167. Romania Environment Strategy Paper, at 40. In 1985, the Danube River Basin states signed a Declaration on the Cooperation of the Danube River Countries on the Problem of the Danube River Water Management. Závadský, at 36.
168. Parties to the convention include Austria, Bulgaria, Croatia, Germany, Hungary, Moldova, Romania, Slovakia, Ukraine, Slovenia, the Czech Republic and the European Union.
169. *See also* Závadský, at 37.
170. Sofia BTA (5 May 1993).
171. FBIS-EEU (12 July 1993).
172. FBIS-EEU (24 April 1995).
173. FBIS-EEU (27 October 1995).
174. FBIS-EEU (25 November 1995).
175. FBIS-EEU (28 May 1997).
176. *See* Fidas & Gershwin, at 24. For more information on the principle of sustainable development, *see* Sands(h), at 198–208; Clark & Munn; Conable; Di Leva; Munro & Holdgate; Munro & Lammers.
177. For more information on the right to public participation in environmental decision making, *see* Anderson; OECD, Application of Information and Consultation Practices; Pallemaerts(a); Partan; Raustiala; Sheate; Weber.
178. In this regard, on 19 September 1995, Romania adopted a new law on environmental protection providing for substantial criminal penalties for individuals or organizations causing 'massive pollution', importing toxic waste or releasing such waste into natural waters. ODD, No. 183 (20 September 1995).
179. For a review of the irradiation effects of these nuclear power plants on the environmental quality of the Danube River, *see* Equipe Cousteau(a), at 127–42.
180. *See* FBIS-EEU (22 April 1992).
181. D. Fisher, at 15. Because of the relatively limited storage facilities at the Kozloduy site, there is significant concern by United States experts nuclear waste materials might contaminate the Danube River. Bulgaria Environment Strategy Study, at 147–8.
182. D. Fisher, at 76. For a critique of Canadian policies relating to the export of Canadian built nuclear reactors, *see NYT* (3 December 1997), which

cites IAEA officials critical of Canadian designed reactors which, although simpler to build than American and other reactors, are 'far more complex to keep running safely'. *See* Reuters Textline (30 April 1995).
183. Sofia BTA (2 February 1992).
184. 5 YIEL, at 400.
185. FBIS-EEU (5 August 1997); AFP (23 May 1998).
186. RFE/RLN, No. 36, Part II (23 February 1998).
187. ODD, No. 185 (22 September 1995); the *Christian Science Monitor* (6 December 1996).
188. IAEA is particularly concerned as the Kozloduy plant lacks common safety features such as 'redundant systems for high-pressure coolant injection, back-up feedwater circuits and a full scale containment structure'. Bulgaria Environment Strategy Study, at 148. In addition, Romanian environment officials assert the Kozloduy plant is built in an earthquake zone. D. Fisher, at 77.
189. In total, 60 Soviet designed reactors continue to operate in the former Soviet Union and CEE. W. Broad (24 July 1995). For a review of international law governing liability for nuclear accidents, *see* de La Fayette; Malone; OECD, Nuclear Third Party Liability; Politi; Sands(c); Von Busekist.
190. In addition to the additional funds necessary for modernization, the power plant is owed over $70 million by its industrial customers. The cost for an alternative coal fired power plant would be $450 million. M. Milner, the *Guardian* 35 (3 September 1994).
191. FBIS-EEU (20 September 1995).
192. D. Fisher, at 77; BBC World Broadcasts (2 June 1996); BBC World Broadcasts (28 February 1997).
193. ODD, No. 194 (5 October 1995).
194. ODD, No. 190 (29 September 1995); No. 188 (27 September 1995); No. 185 (22 September 1995).
195. ODD, No. 208 (25 October 1995); No. 196 (9 October 1995); No. 193 (4 October 1995). In August 1997, operators at Kozloduy experienced difficulties with Unit 1, drawing attention back to IAEA criticism about the operation of the outdated plant. RFE/RLN, No. 98, Part II (19 August 1997).
196. RFE/RLN, No. 64, Part II (2 April 1998).
197. RFE/RLN, No. 87, Part II (7 May 1998).
198. BBC World Broadcasts (4 December 1997).
199. FBIS-EEU (5 April 1995); RFE/RLN, No. 73, Part II (16 April 1998).
200. FBIS-EEU (26 June 1995).
201. PR Newswire (19 February 1998).
202. On 30 November 1995, for instance, Romania seized a barge and tug carrying 106 containers of nuclear fuel for the Kozloduy power plant. Romania subsequently released the fuel shipment after protests by Bulgaria. ODD, No. 233 (1 December 1995); No. 235 (5 December 1995).
203. ODD, No. 234 (4 December 1995). For a review of the international law principles applicable to the transport of nuclear materials, *see* Sands(a), at 472–4.
204. RFE/RLN (18 November 1997); BBC World Broadcasts (17 August 1998).
205. D. Fisher, at 41; BBC World Broadcast (23 July 1998).

206. *See* Reuters Limited (18 December 1995). If the G7 continue to pursue their opposition to the project, they can likely threaten to diminish the amount of financial aid currently provided to Bulgaria. The IMF, for instance, loaned Bulgaria $400 million, and on 21 March 1994, the EU pledged balance of payments support of $171 million, which was matched by loans from the Group of 24, with total Western financial assistance to Bulgaria reaching $1 billion in 1994. 4 BNA Eastern Europe Reporter 304 (11 April 1994).
207. Kinnear & Rhode, at 12.
208. Regional Environmental Program, at 5; The Danube Task Force, Environmental Programme for the Danube River Basin, chs 2, 6, 7; ODD, No. 44 (1 March 1996). *See also Offshore* (July 1998).
209. RFE/RL Caucasus Report, No. 7 (14 April 1998); BBC World Broadcasts (13 November 1997). *See also* BBC World Broadcasts (1 April 1997); *Pipeline and Gas Journal* (1 September 1998).
210. RFE/RLN, No. 8, Part II (23 April 1998).
211. Environmental Action Programme, at II-12; Regional Environmental Program, at 5; Financial Times Asia Intelligence Wire (15 December 1997). By some estimates the ecological collapse of the Black Sea causes damages of over $1 billion to the littoral states. *Turkish Daily News* (6 November 1998). For a review of international legal principles relating to the protection of biodiversity, *see* Hubbard.
212. For a review of the treatment of the Bulgarian Turkish national minority, *see* Bowers, at 45.
213. Black Sea Convention; 4 YIEL, at 395. The GEF Fund allocated $8.5 million to support the implementation of the convention. Regional Environmental Program, at 6. The EU and GEF also initiated the Black Sea Environmental Management Program, seeking to identify the principal sources of pollution of the Black Sea. This $794,000 program, however, relates only to the collection of scientific evidence and does not create any mechanisms of cooperation or establish any legal obligations. Regional Environmental Program, at 11.
214. For a review of international law relating to the disposal of radioactive and hazardous waste, *see* Mawson; Moisé; Cassidy; Forster.
215. For a review of international law relating to the transport of hazardous waste, *see* Hackett; Kurkela; Kiss(b); Kummer & Rummel-Bulska; Kummer; Long.
216. Black Sea Convention, arts. 4–16, 25, at 112–20.
217. Ministerial Declaration on the Protection of the Black Sea (7 April 1993).
218. The *Guardian* (26 September 1997).
219. Oral Pleadings of Slovakia, Peter Tomka, at 1 (24 March 1997).
220. For a comparison of the Gabčíkovo–Nagymaros Project dispute with other transboundary water disputes, *see* Graffy.
221. The preamble of the 1977 Agreement cites the mutual interest of Czechoslovakia and Hungary in the development of 'water resources, energy, transport, agriculture and other sectors of the national economy' that will result from the Gabčíkovo–Nagymaros Project.
222. Republic of Hungary, Gabčíkovo–Nagymaros File, at 1. In 1975, Hungary was promised a loan of 100 million convertible rubles from the Soviet

Union to assist in the funding of the Project. Although Hungary and the Soviet Union eventually signed a loan agreement on 30 November 1977, the Soviet Union failed to fulfill its promise of assistance.
223. For a more detailed review of the legal nature of the Joint Contractual Plan, *see* Oral Pleadings of Hungary, Boldizsar Nagy, at paras. 23–7 (4 March 1997); Dissenting Opinion of Judge Oda, in Case Concerning the Gabčíkovo–Nagymaros Project, at 2–4.
224. According to the Agreement, the Hrusov/Dunakiliti dam, located on the Czechoslovakian–Hungary boundary at the mouth of the canal, would divert the Danube into the canal; the Gabčíkovo dam, located within Czechoslovakian territory in the mid region of the 25 km canal, would regulate the water level and generate hydroelectric power; and the Nagymaros works, located within Hungarian territory 100 km downstream of the canal, would return the water levels to run-of-the-river and generate additional hydroelectric power. 1977 Agreement, at 237, 248.
225. 1977 Agreement, art. 5, at 239–40; Galambos, at 74; Vodohospodarska Vystavba, The Gabčíkovo–Nagymaros Project: Part Gabčíkovo, Solution According to the 1977 Treaty. According to the cost calculations used by Czechoslovakia and Hungary in 1977, 60 per cent of the cost of the Project was attributable to energy production, 16 per cent to navigation, 11 per cent to water management and 14 per cent to infrastructure. Fitzmaurice, at 77. During the course of oral arguments, the Agent of Slovakia asserted the cost of the Project amounted to $2.5 billion. Pleadings of Slovakia, Peter Tomka, at 1 (24 March 1997).
226. *See* Slovak Republic Government Office, Press and Information Department, The Gabčíkovo Water Works Project: Conserving the Danube's Inland Delta; National Council of the Slovak Republic, Resolution on the Gabčíkovo–Nagymaros Hydroelectric Project. The primary beneficiary of the improved navigation would be the Austrian iron and steel industry. Equipe Cousteau(a), at 44.
227. Solution According to the 1977 Treaty, at 2. The power generated from the Gabčíkovo plant was planned to meet approximately 7 per cent of Slovakia's electricity needs, *NYT* (8 November 1992).
228. Enyedi & Zentai, at 215. Ministry of Environment of the Slovak Republic, Background for Environmental Policy: An Overview of the Environmental Situation, at 5.
229. Vodohospodarska Vystavba, Gabčíkovo–Nagymaros Project – Basic Information About its Actual State and Perspectives, at 13. Republic of Hungary, Information About the Scope of Problems Connected With the System of Gabčíkovo–Nagymaros Waterworks, at 2.
230. Fitzmaurice, at 75.
231. ZumBrunnen, at 402.
232. Galambos, at 81; ZumBrunnen, at 402.
233. *See* Fitzmaurice, at 76–7.
234. World Wildlife Fund, Repercussions of the Power Station, at 2. *See also* Republic of Hungary, The Danube Story: Educational Summary, at 31–5; The Danube Defense Action Committee, The Danube Blues: Questions and Answers About the Bos (Gabčíkovo)–Nagymaros Hydroelectric Station System, at 5.

235. *See* Fitzmaurice, at 78; Galambos, at 75; *see also* The Danube Blues, at 6. *See also* Oral Pleadings of Hungary, Laszlo Valki, at para. 4 (3 March 1997).
236. Galambos, at 79; Republic of Hungary, Information About the Scope of Problems Connected With the System of Gabčíkovo–Nagymaros Waterworks, at 32; Basic Information, at 14, and Resolution of the National Council of the Slovak Republic on the Gabčíkovo–Nagymaros Hydroelectric Project adopted 24 March 1993; Galambos, at 78.
237. Basic Information, at 13; Information About the Scope of Problems of Gabčíkovo–Nagymaros, at 1.
238. Galambos, at 79; Repercussions of the Power Station, at 2.
239. Repercussions of the Power Station, at 2; The Danube Blues, at 4; Danube Story, at 31; Galambos, at 79.
240. Basic Information, at 13 and 26; Vodohospodarska Vystavba, Gabčíkovo–Nagymaros Project: Standpoint of the Czecho–Slovak Side and Answers to Questions, at 14.
241. Preamble to the 1977 Agreement, at 236.
242. Oral Pleadings of Hungary, Laszlo Valki, at para. 8 (3 March 1997).
243. For a review of Hungary's environmental concerns as subsequently articulated to the ICJ, *see* Memorial of the Republic of Hungary in the Case Concerning the Gabčíkovo–Nagymaros Project (Hungary/Slovakia), at 137–78, Appendices 1–3 (2 May 1994); Counter-Memorial of the Republic of Hungary in the Case Concerning the Gabčíkovo–Nagymaros Project (Hungary/Slovakia), at 32–82, 153–72, Volume 4 Scientific Reports (5 December 1994); Reply of the Republic of Hungary in the Case Concerning the Gabčíkovo–Nagymaros Project (Hungary/Slovakia), at 38–44, 90–8, Volume 3 Scientific Reports (20 June 1995).
244. For a review of the environmental position of Slovakia as subsequently articulated to the ICJ, *see* Memorial of the Slovak Republic in the Case Concerning the Gabčíkovo–Nagymaros Project (Hungary/Slovakia), at 38–43, 50–3, 87–101, 191, Volume 3 Select Reports (2 May 1994); Counter-Memorial of the Slovak Republic in the Case Concerning the Gabčíkovo–Nagymaros Project (Hungary/Slovakia), at 186–95, 216–29, Volume 2 Select Reports (5 December 1994); Reply of the Slovak Republic in the Case Concerning the Gabčíkovo–Nagymaros Project (Hungary/Slovakia), at 295–311, 323–31, Volume 3 Data and Monitoring Reports (20 June 1995).
245. Danube Story, at 25; Gabčíkovo–Nagymaros File, at 10; *NYT* (8 November 1992).
246. Danube Story, at 22.
247. Gabčíkovo–Nagymaros File, at 12; Danube Story, at 23; and The Danube Blues, at 4.
248. Danube Story, at 23; and The Danube Blues, at 4.
249. Repercussions of the Power Station, at 2; Gabčíkovo–Nagymaros File, at 11. *See also* Slovak Union of Nature and Landscape Protectors & Slovak Rivers Network, Damming the Danube: What Dam Builders Don't Want You To Know; Cousteau(b).
250. The Danube Blues, at 4.
251. Danube Story, at 24; *see also* The Danube Blues, at 4; and Gabčíkovo–Nagymaros File, at 11.
252. Repercussions of the Power Station, at 2.

253. The Danube Blues, at 4.
254. Enyedi & Zentai, at 227; Environmental Needs Assessment, at 43.
255. Significance and Impacts of Gabčíkovo–Nagymaros Project, at 7; Solution According to the 1977 Treaty, at 2; Standpoint of Czecho–Slovak Side, at 13.
256. According to Czechoslovakia, 'international experts from Bechtel Environmental, Inc., Hydro Quebec International and an ECC Working Group have confirmed the Gabčíkovo Dam does not threaten the region's environment'. Slovak National Agency for Foreign Investment and Development, Press Release: The Gabčíkovo Dam Q&A Fact Sheet, at 2. *See also* Resolution of the National Council, para. 5, at 16.
257. Danube Story, at 19; *see also* and The Danube Blues, at 4; and Gabčíkovo–Nagymaros File, at 11.
258. Gabčíkovo–Nagymaros File, at 11.
259. Kinnear & Rhode, at 12.
260. Basic Information, at 25; Solution According to the 1977 Treaty, at 2.
261. Gabčíkovo–Nagymaros File, at 13.
262. Gabčíkovo–Nagymaros File, at 13; *see also* The Danube Blues, at 4; Danube Story, at 18.
263. Basic Information, at 24; Solution According to the 1977 Treaty, at 3.
264. The Gabčíkovo Dam Q&A Fact Sheet, at 2.
265. For a review of the Hungarian national minority situation in Slovakia, *see* Height; Schwabach, at 303–4.
266. The Danube Blues, at 5; Republic of Hungary, Declaration of the Government of the Republic of Hungary on the Termination of the 1977 Treaty, at 4.
267. Declaration, at 5.
268. Declaration, at 5.
269. Galambos, at 80–1; *see also* The Danube Story, at 3.
270. Galambos, at 81.
271. Protocol Amending the Treaty Between the Czechoslovak Socialist Republic and the Hungarian People's Republic Concerning the Construction and Operation of the Gabčíkovo–Nagymaros System of Locks.
272. Information About the Scope of Problems of Gabčíkovo–Nagymaros, at 4; Declaration, at 6.
273. A report conducted by an American environmental organization presented the Hungarian government with similar conclusions. Ecologia, The Bös–Nagymaros Barrage Study: Program Operations and Impacts.
274. Protocol of 6 February 1989 amending the Agreement between the Government of the Hungarian People's Republic and the Government of the Czechoslovak Socialist Republic for Mutual Assistance in the Construction of the Gabčíkovo–Nagymaros System of Locks.
275. Application of the Republic of Hungary v. The Czech and Slovak Federal Republic on the Diversion of the Danube River, at 3. The Hungarian Parliament subsequently approved this suspension on 2 June 1989. Declaration, at 7.
276. ZumBrunnen, at 401.
277. Application of the Republic of Hungary, at 3; Standpoint of the Czecho–Slovak Side, at 1.
278. Application of the Republic of Hungary, at 3; Galambos, at 82–3. The Provisional Solution provided for construction of a reservoir and diversion

dam similar to the original plan, but upstream of Dunakiliti at a point (river kilometer 1851.75) where both banks of the Danube were within the territory of Czechoslovakia. Vodohospodarska Vystavba, Gabčíkovo–Nagymaros Project Counter-Proposal of Operation Gabčíkovo, at 3–4; Standpoint of Czecho–Slovak Side, at 17. The Provisional Solution required the construction of an 11 km earthen dam from the point of the new diversion dam, within Czechoslovakia, through the midsection of the Danube River, parallel to the Hungarian border, to the mouth of the power canal. Counter Proposal, at 3–4, and Standpoint of Czecho–Slovak Side, at 17. The effect of the Provisional Solution was to divert the flow of the Danube River from its natural course and into the power canal before it reached Hungarian territory.

279. Declaration, at 8.
280. For a review of the Gabčíkovo–Nagymaros Project dispute in terms of conflict between various political, economic and social forces, *see* Fleischer.
281. Fitzmaurice, at 93–6.
282. Fitzmaurice, at 102.
283. In February 1989, the Hungarian Parliament was presented with over 140,000 signatures on a petition calling for the abandonment of the Gabčíkovo–Nagymaros Project. Fitzmaurice, at 94. For a review of the various positions taken by the Hungarian political parties with respect to the Gabčíkovo–Nagymaros Project, *see* Fitzmaurice, at 100–2.
284. Standpoint of Czecho–Slovak Side, at 2; *see also* Solution According to the 1977 Treaty, at 4. The action of the Parliament was in large part attributed to the economic and environmental analysis conducted during the initial suspension of the Project. *See* Hardi *et al.* eds.
285. Standpoint of Czecho–Slovak Side, at 2; Galambos, at 84.
286. Application of the Republic of Hungary, at 3; Galambos, at 84–5; Fitzmaurice, at 96.
287. D. Fisher, at 32–3.
288. Fitzmaurice, at 96–7. *See* Czech and Slovak Federal Republic Lustration Law, in *Transitional Justice: How Emerging Democracies Reckon With Former Regimes*, at 312 (Kritz ed., 1995).
289. For a review of Slovakia's slow pace of economic and political transformation, *see* R. Frydman *et al.*, at 55–66.
290. Galambos, at 84–5; Fitzmaurice, at 68–9, 96–7.
291. Environmental Needs Assessment, at 66.
292. *See* Fitzmaurice, at 99–100.
293. D. Fisher, at 19. The World Wildlife Fund also contended the Project operated for 6 months without the necessary permits. World Wildlife Fund, A New Solution for the Danube, at 7. *But see* Ground Water Consulting Ltd., Gabčíkovo–WWF: the Pros and Cons, claiming most of these conditions were met.
294. *See* Fitzmaurice, at 87; Binder, at 77.
295. Galambos, at 81; *see also* ZumBrunnen, at 401. For an examination of the political aspects surrounding the decision of Hungary to contract with Austrian contractors, *see* Galambos, at 89.
296. ZumBrunnen, at 402.
297. ZumBrunnen, at 402; Declaration, at 9.

298. Galambos, at 86.
299. Fitzmaurice, at 81. *See also* Galambos, at 81, 86; ZumBrunnen, at 402. Interestingly, Czechoslovakia contended a primary reason for constructing the project was prior to Austria's construction of nine hydroelectric dams, the Danube River would transport up to 5 million cubic meters per year of gravel into Slovak territory, which would then either settle on the floor of the Danube, providing protection of the regional aquifer, or would be mined by the Slovak construction industry. From 1993, Slovakia received only 50,000 cubic meters of gravel per year.
300. Fitzmaurice, at 82.
301. Declaration, at 13.
302. Galambos, at 84.
303. Application of the Republic of Hungary, at 3.
304. Standpoint of Czecho–Slovak Side, at 2; *see also* Liška(b), at 4. As recognized by Judge Oda in his Dissenting Opinion, at 6, 'If no campaign had been launched by environmentalist groups, then it is my firm conviction that the Project would have gone ahead as planned.'
305. Declaration, at 11.
306. Fitzmaurice, at 103.
307. Declaration, at 11.
308. Declaration, at 12.
309. Declaration, at 13.
310. *See* Fitzmaurice, at 104.
311. Declaration, at 2, 23–30.
312. *See* Fitzmaurice, at 104.
313. Basic Information. *See also* Gabčíkovo–WWF: the Pros and Cons; The Gabčíkovo Dam Q&A Fact Sheet, at 2; Standpoint of the Czecho–Slovak Side.
314. As noted above, both Czechoslovakia and Hungary sought the intervention of the European Union, its members states and the United States. At various times during the dispute, Hungary also appealed to the Danube River Commission, the OSCE, the UN Secretary-General and the international community more broadly. Fitzmaurice, at 104.
315. *See* Fitzmaurice, at 105. The Commission did, however, express some reticence to becoming involved in the dispute as it feared too much attention to CEE disputes might distract it from intra-EU integration and it might expose itself to hostile political criticisms if its mediation efforts failed. *Id.* at 110.
316. *See* Fitzmaurice, at 111.
317. *See* Report of the European Commission Fact Finding Mission. A New Solution for the Danube, at 13.
318. *Where the Danube Runs Dry*, at A1.
319. ZumBrunnen, at 402.
320. Notably, Hungary engaged in a similar activity, when in July 1993 it began to restore the area around the Nagymaros works. FBIS-EEU (14 July 1993).
321. London Agreement on the Gabčíkovo–Nagymaros Project between the Czech and Slovak Federation, the European Commission and Hungary, arts. 1–3; Diversion, at 2.
322. London Agreement, art. 4, at 1291. It appears the European Commission had hoped the parties would pursue arbitration by an independent inter-

national water engineering institute or an ecological institute, rather than the ICJ. A New Solution for the Danube, at 19.
323. Constitutional Law on the Termination of the Czech and Slovak Federal Republic, art. 1 (25 November 1992).
324. Brussels Agreement on the Gabčíkovo–Nagymaros Project between the Czech and Slovak Federation, the European Commission and Hungary.
325. *See* Report of the Working Group of Independent Experts.
326. Agreement Concerning Certain Temporary Technical Measures and Discharges in the Danube and Mosoni Branch of the Danube.
327. Resolution of the European Parliament of the European Communities on the Gabčíkovo–Nagymaros Barrage System. The January 1993 report recommended a 50/50 division of the water resources.
328. Special Agreement: For Submission to the International Court of Justice of the Differences Between The Republic of Hungary and the Slovak Republic Concerning the Gabčíkovo–Nagymaros Project, art. 2 (7 April 1993). Although Hungary had submitted to the compulsory jurisdiction of the Court, in 7 October 1992, neither Czechoslovakia (pre-January 1993), nor Slovakia (post-January 1993) had done so and thus it was necessary for Hungary and Slovakia, pursuant to Article 36, paragraph 1 of the Statute of the ICJ, to enter into a special agreement providing for submission of the dispute to the Court. Yearbook of the ICJ, at 93.
329. Sommer(c), at 214; Grieves, at 327–8. For a review of other CMEA structures created to address the question of environmental pollution, *see* Jakubowski.
330. *See* Szilagyi.
331. *See* Enyedi & V. Zentai, at 222; Cole(b); Kozhevnikov; Cassese, at 109; Schöpflin, at 265; Fletcher, at 152–3.
332. *See* Bándi.
333. *See* Bándi *et al.*, at 2–47; Hardi & Galambos, at 18.
334. For Hungary 50 per cent of its international assistance is received from the EU, 20–25 per cent from OECD bilateral assistance, 15–20 per cent from Germany and 5–10 per cent from USAID. Environmental Needs Assessment, at 42.
335. *See* Report on Temporary Water Management Regime from the Working Group of Monitoring and Water Management Experts for Gabčíkovo System of Locks; Report of the Working Group of Water Management Experts for Gabčíkovo System of Locks.
336. FBIS-EEU (14 July 1993). *See also* A New Solution for the Danube, at 5. *See also* Oral Pleadings of Slovakia, Peter Tomka, at 59–60 (15 April 1997).
337. Special Agreement, art. 2.
338. Special Agreement, art. 5.
339. Although relevant to the outcome of the case, this section does not discuss the issue of treaty continuity as the findings of the Court are not significantly relevant to other CEE transboundary environmental disputes.
340. For more information on the decision of the ICJ, *see* Bekker; Bostian; Eckstein; Schwabach.
341. Hungarian Memorial, at 266–80; Hungarian Counter-Memorial, at 208–9. *See also* Oral Pleadings of Hungary, Pierre-Marie Dupuy (4 March 1997).
342. Hungarian Memorial, at 271.

343. Hungarian Counter-Memorial, at 201–7. Slovakia and Hungary also discussed in their pleadings the extent to which various articles of the Vienna Convention reflected customary international law. *See* for example, Slovak Memorial, at 313–15, Hungarian Memorial, at 295–6.
344. Hungarian Counter-Memorial, at 189–96.
345. Draft Articles on State Responsibility, ILC Yearbook Vol. II (1977); ILC Yearbook (1994); 'Report of the ILC on the work of its forty-eight session, 6 May to 26 July 1996', Official Records of the General Assembly, Fifty-first Session, Supplement No. 10 (A/51/10).
346. Hungarian Memorial at 269, 287. For a review of the essential interests of Hungary affected by the Project, *see* Oral Pleadings of Hungary, Gabor Vida (3 March 1997); for a review of the threats to these essential interests, *see* Oral Pleadings of Hungary, Klaus Kern, at paras. 22–84 (4 March 1997).
347. Hungarian Memorial, at 288–9. *See also* Hungarian Counter-Memorial, at 208–9; Hungarian Reply, at 121–7. The Hungarian government further noted in a 1995 PHARE report compiled by Slovakia a number of essential interests were identified as being under threat from the Project. *See* Oral Pleadings of Hungary, James Crawford, at 34–5 (5 March 1997).
348. Letter from Czechoslovak Prime Minister, Mr. Marian Calfa, to Hungarian Prime Minister, Mr. József Antall (23 April 1992).
349. Hungarian Memorial, at 148–67; *see also* Hungarian Counter Memorial, at 34–5, 44–82. Hungary cited as evidence of the imminent threat the fact it did in fact suffer significant environmental harm immediately after Czechoslovakia diverted the Danube River on 24 October 1992. Hungarian Memorial, at 289–90.
350. Hungarian Memorial, at 90–1; Hungarian Reply, at 158.
351. Hungarian Memorial, at 291–3.
352. *See* Slovak Memorial, at 313–20; Slovak Reply, 83–93; Oral Pleadings of Slovakia, Alain Pellet, at 41–6 (25 March 1997); Oral Pleadings of Slovakia, Peter Tomka, at 53–4 (15 April 1997). Hungary asserted in its oral argument Slovakia had recanted this perspective in its own oral arguments. *See* Oral Pleadings of Hungary, James Crawford, at para. 6 (10 April 1997); Oral Pleadings of Hungary, Philippe Sands, at para. 25 (10 April 1997).
353. *See* Oral Pleadings of Slovakia, Stephen McCaffrey, at 61–6 (25 March 1997) and 10–17 (26 March 1997).
354. Slovak Memorial, at 320.
355. Slovak Reply at 106–10; Oral Pleadings of Slovakia, Samuel Wordsworth, at 27–33 (14 April 1997).
356. Slovak Memorial, at 320–4. *See also* Oral Pleadings of Slovakia, Stephen McCaffrey, at 19–32 (25 March 1997); Oral Pleadings of Slovakia, Igor Mucha, at 34–9 (25 March 1997).
357. Slovak Memorial, at 324–32; Slovak Reply, at 104–6.
358. Slovak Counter-Memorial, at 298–9.
359. Slovak Reply, at 110.
360. The findings of the Court were reached by a vote of fourteen to one, with Judge Herczegh dissenting. Decision, at 77. *See* Dissenting Opinion of Judge Herczegh for more detailed examination of the argument of the state of ecological necessity.

361. For support of this conclusion, the Court cited Legal Consequences for States of the Continued Presence of South Africa in Namibia Opinion, at 47; Fisheries Jurisdiction Case (Merits), at 18; Interpretation of the Agreement of 25 March 1951 between the WHO and Egypt Opinion, at 95–6.
362. Decision, at 32, *citing* article 73 of the Vienna Convention; Interpretation of Peace Treaties with Bulgaria, Hungary and Romania, Second Phase Opinion, at 228; Article 17 Draft Articles on State Responsibility, at 32.
363. Decision, at 33.
364. Decision, at 33.
365. Decision, at 33–4.
366. In concurring with the conclusions of the Court, Judge Oda added Hungary was aware of the environmental deterioration which would be caused by the implementation of the Gabčíkovo–Nagymaros Project at the time it signed the 1977 Agreement and an environmental assessment conducted in the 1980s would not have yielded any significant additional information. Dissenting Opinion of Judge Oda, at 6.
367. Decision, at 34–9.
368. Decision, at 35, *citing* YILC, 1980, Vol. II, Part 2, at 49, para. 32; at 35, para. 3; at 39, para. 14.
369. Legality of the Threat or Use of Nuclear Weapons, 1996 ICJ 226, at 241–2, para. 29 (Advisory Opinion, July 8).
370. Decision, at 35.
371. Decision, at 35, *citing* the ILC's commentary which indicated the imminent peril must have been 'a threat to the interest at the actual time'. YILC, 1980, Vol. II, Part 2, at 49, para. 33.
372. Decision, at 36.
373. Decision, at 37.
374. Decision, at 37, *citing* the Report of the *ad hoc* Committee of the Hungary Academy of Sciences. See Hungarian Academy of Sciences, Report on Environmental, Ecological, Water Quality and Seismic Aspects of the Nagymaros Barrage Construction or its Cancellation (Budapest, 23 June 1989) in Hungarian Memorial Volume 5 (Part I) Scientific Reports.
375. Decision, at 38.
376. Hungary acknowledged the doctrine of necessity addressed the question of whether a state could avoid responsibility for an unlawful act, but chose to discuss the doctrine as a ground for termination alongside the various grounds provided in the Vienna Convention. Hungarian Reply, at 116–18, 157. Slovakia too acknowledged the doctrine of necessity related to the responsibility for an unlawful act, but concluded from this fact Hungary was therefore prohibited from invoking this doctrine as a basis for terminating the 1977 Agreement and could only rely on the grounds provided in the Vienna Convention. Slovak Memorial, at 313–16; Slovak Reply, at 84–100.
377. The Court reached its conclusions regarding the validity of Hungary's invocation of the state of necessity and certain treaty grounds for the termination of 1977 Agreement by a vote of eleven to four, with Judges Schwebel, Herczegh, Fleischhauer and Rezek dissenting. Where Judges dissented to a specific factual finding or legal holding of the Court, the specific nature of their dissent will be noted below where appropriate.

248 *Notes*

378. Hungary also argued its termination of the treaty was based on the doctrines of impossibility of performance and joint repudiation. Hungarian Memorial, at 293–7, 320–1; Slovak Counter-memorial, at 311–13, 282–9. Given the Court's determinations of Hungary's inability to meet the criteria for invoking impossibility of performance or joint repudiation are not particularly relevant to other CEE transboundary environmental disputes, it is not discussed here. Decision, at 54, 58. For information on the claim of impossibility of performance as it relates to the dispute, *see* Williams.
379. Elias(a), at 119, *see also* Vamvoukos, at 70–117.
380. Sinclair, at 193.
381. Sinclair, at 199–200.
382. Vienna Convention, art. 62, at 212.
383. Sinclair, at 193. For a detailed examination of the Court's rulings on change of circumstances *see* Vamvoukos, at 153–74.
384. Fisheries Jurisdiction Case, at 18.
385. Fisheries Jurisdiction Case, at 21.
386. Hungarian Reply, at 141; Slovak Memorial, at 337–8.
387. Hungarian Reply, at 142.
388. Hungarian Memorial, at 299–308; Hungarian Counter-Memorial, at 212–14; Slovak Memorial, at 337–9; Slovak Reply, at 110–13; Oral Pleadings of Slovakia, Alain Pellet, at 30–6 (26 March 1997).
389. Hungarian Memorial, at 309; Hungarian Reply at 144–52.
390. *See* Oral Pleadings of Hungary, Philippe Sands, paras. 40–3 (6–7 March 1997).
391. For a review of this transformation *see* Fitzmaurice, at 44–69.
392. Hungarian Memorial, at 309–13; Hungarian Reply, at 143.
393. *See* Gadourek. By 1989, 'many Hungarians had come to regard [the Gabčíkovo–Nagymaros Project] as a symbol of high-handed Communist rule and Soviet domination'. *NYT* (3 November 1992).
394. Mayors Conference at 1, contained in Gabčíkovo–Nagymaros File, at 17, *see also NYT* (4 February 1990) (100,000 people formed a 100-mile human chain to protest the Project); *NYT* (29 August 1991) (Police intervention to break up citizen demonstration near pumping station), and *NYT* (5 December 1990) (Environmental groups in Czechoslovakia call for halt to construction of Gabčíkovo–Nagymaros Project on the expectation it will contaminate ground water sources).
395. *See* Delaney. *See also* Gadourek. Although the environmental dangers of the Gabčíkovo–Nagymaros Project were known to a small group of scientists and hydrologists when the agreement was negotiated, their views were suppressed by the ruling communist party, leaving the populace unapprised of these effects. Danube Blues, at 3 and 5; Fitzmaurice, at 93.
396. Hungarian Memorial, at 309–13; Hungarian Reply, at 143–4.
397. Repercussions of the Power Station, at 3; ISTER; Future of the Danube 8 (1991).
398. Hungarian Memorial, at 309–13; Hungarian Reply, at 144. An interesting question raised by this discussion is: 'What party should bear the cost of unpredictable changes in societal values?' The unpredictability of the way in which societal values will change suggests the parties to the agreement

would, under a Rawlsian veil of ignorance, have distributed the risk evenly between the two. On the other hand, the society whose values have changed may rightly pay the full price for indulging its changed values.
399. *See* Oral Pleadings of Slovakia, Vaclav Mikulka, at 26–35 (24 March 1997).
400. Slovak Reply, at 110–13; Oral Pleadings of Slovakia, Alain Pellet, at 30–6 (26 March 1997).
401. Oral Pleadings of Slovakia, Peter Tomka, at 1 (15 April 1997).
402. Decision, at 55.
403. Decision, at 55.
404. Decision, at 55. Article 15 of the Agreements provides, 'The Contracting Parties shall ensure, by the means specified in the joint contractual plan, that the quality of the water in the Danube is not impaired as a result of the construction and operation of the System of Locks.' 1977 Agreement, at 244. Article 19 provides, 'The Contracting Parties shall, through the means specified in the joint contractual plan, ensure compliance with the obligations for the protection of nature arising in connection with the construction and operation of the System of Locks.' *Id.*, at 245. And Article 20 provides, 'The Contracting Parties, within the framework of national investment, shall take appropriate measures for the protection of fishing interests in conformity with the Danube Fisheries Agreement . . .' *Id.*, at 245.
405. Vienna Convention, art. 60, at 211–12.
406. Hungarian Memorial, at 316; Slovak Memorial, at 343.
407. *See* Gabčíkovo–Nagymaros File, at 14.
408. These related conventions and general rules of international law included the 1976 Boundary Waters Convention, the precautionary principle and the obligation to cooperate. Hungarian Memorial, at 196–7, 198–203, 203–9.
409. Hungarian Memorial, at 316–17; Hungarian Reply, at 140. For a review of the specific nature of Czechoslovakia's breaches of Articles 15, 19 and 20, as alleged by Hungary, *see* Hungarian Memorial, at 183–8. For a review of the specific nature of Czechoslovakia's breaches of related international conventions and rules of general international law, *see* Hungarian Memorial, at 196–209. *See also* Oral Pleadings of Hungary, Alexandre Kiss (6 March 1997); Oral Pleadings of Hungary, Philippe Sands, paras. 62–9 (6–7 March 1997).
410. Hungarian Reply, at 140–1.
411. Hungarian Reply, at 140. *See also* Hungarian Memorial, at 213–19; Hungarian Counter-Memorial, at 214, 239–47; Hungarian Reply, at 139–41.
412. Oral Pleadings of Slovakia, Alain Pellet, at 25–7 (26 March 1997); Oral Pleadings of Slovakia, Vaclav Mikulka, at 41–7 (14 April 1997).
413. Slovak Memorial, at 340–2; Slovak Counter-Memorial, at 315–18.
414. Slovak Memorial, at 342–3; Slovak Reply, at 114–15; Oral Pleadings of Slovakia, Alain Pellet, at 28–30 (26 March 1997); *see also* Standpoint of Czecho–Slovak Side, at 18. *See* Vodohospodárska Výstavba, Gabčíkovo, at 20 (1992), and Letter by Ivan Gasparovic, Chairman of the National Council of the Slovak Republic, to Egon Klepsh, Chairman of European Parliament, published in FBIS-EEU, at 21 (17 March 1993).
415. Decision, at 56.
416. Decision, at 56.

417. Decision, at 56. Judge Fleischhauer disagreed with this finding of the Court on the bases that in his view, the act of initiating the construction of the Provisional Solution frustrated the object and purpose of the 1977 Agreement in contravention of Articles 18 and 26 of the Vienna Convention and the implementation of the Provisional Solution was a continuation of this wrongful breach, giving rise to a right of Hungary to terminate the Agreement. Dissenting Opinion of Judge Fleischhauer, at 3. Judge Fleischhauer further noted in the event Czechoslovakia did not breach the 1977 Agreement until it implemented the Provisional Solution in October 1992, the premature act of terminating the Agreement would have become mature and effective at that time. Dissenting Opinion of Judge Fleischhauer, at 6.
418. *See* Decision, at 46.
419. Decision, at 57, *citing* Chorzów Factory, at 31. Judge Schwebel disagreed with this particular conclusion, declaring 'I am not persuaded Hungary's position as the party initially in breach deprived it of a right to terminate the Treaty in response to Czechoslovakia's material breach, a breach which in my view ... was in train when Hungary gave notice of termination.' Schwebel Declaration, at 1.
420. Decision, at 57. Judge Fleischhauer took exception to this finding of the Court, noting the construction and implementation of the Provisional Solution was not the only possible response to Hungary's breach of the Agreement and the Provisional Solution was a disproportionate response to that breach, as noted in the findings of the Court relating to Slovakia's justification of countermeasure. Dissenting Opinion of Judge Fleischhauer, at 7.
421. Hungarian Memorial, at 317–18. For a review of Hungary's arguments relating to newly developed norms of international law as a basis for terminating a treaty and its consistency with the Vienna Convention, *see* Hungarian Reply, at 160–1; Oral Pleadings of Hungary, Philippe Sands, paras. 70–6 (6–7 March 1997).
422. Hungarian Counter-Memorial, at 195–6. For a detailed review of Hungary's argument these various principles reflected new norms of international law, *see* Hungarian Memorial, at 198–209, 219–33; Hungarian Counter-Memorial, at 220–6; Hungarian Reply, at 130–1.
423. Hungarian Memorial, at 318; Hungarian Counter-Memorial, at 194–5.
424. Hungarian Reply, at 161–2.
425. More specifically, Slovakia argued the principle of *lex specialis derogat legi generali* served as a principle of interpretation, not as a basis for treaty termination. Slovak Memorial, at 350.
426. Slovak Memorial, at 347–8; Slovak Counter-Memorial, at 313; Slovak Reply, at 77–9; Oral Pleadings of Slovakia, Stephen McCaffrey, at 21–37 (25 March 1997); Oral Pleadings of Slovakia, Alain Pellet, at 18–21 (14 April 1997). For the Hungarian rebuttal to these arguments, *see* Oral Pleadings of Hungary, Philippe Sands, at paras. 2–19 (10 April 1997).
427. *See* Oral Pleadings of Slovakia, Stephen McCaffrey, at 49, 55–8 (24 March 1997).
428. Decision, at 57.
429. Decision, at 57.

430. Decision, at 58. In a Separate Opinion, Judge Bedjaoui emphasized the doctrine of 'evolutionary interpretation' should be viewed through the requirements of Article 31 of the Vienna Convention on Treaties and should only be applied in vary narrow circumstances. Separate Opinion of Judge Bedjaoui, at 1–2.
431. Decision, at 58.
432. Declaration, at 58, *citing* ICJ Reports 1996, para. 29. Judge Oda, agreeing with the finding of the Court, took the opportunity to note 'any construction work relating to economic development would be bound to affect the existing environment to some extent but modern technology would, I am sure, be able to provide some acceptable ways of balancing the two conflicting interests'. Dissenting Opinion of Judge Oda, at 7.
433. Decision, at 58.
434. Decision, at 58.
435. *See* Hungarian Memorial at 90–1, 148–67, 287, 269, 288–90; Hungarian Counter-Memorial, at 34–5, 44–82, 208–9; Hungarian Reply, at 158, 121–7. *See also* Oral Pleadings of Hungary, James Crawford (5 March 1997).
436. *See* Slovak Memorial, at 313–20.
437. Slovak Memorial, at 320; Slovak Reply, at 106–10.
438. Slovak Memorial, at 320–4.
439. Slovak Counter-Memorial, at 298–9; Oral Pleadings of Slovakia, Stephen McCaffrey, at 67–71 (14 April 1997).
440. Decision, at 54.
441. The arguments relating to 'approximate application' of a terminated/breached treaty are not discussed in this chapter as the ruling of the Court was such that it was not required to address the merits of this argument, the confirmation or rejection of the doctrine would have little effect on existing CEE transboundary environmental disputes, and there is in fact scant precedent to support the existence of such a doctrine.
442. The Court's findings and holdings with respect to whether Czechoslovakia was entitled to put into operation, from October 1992, the Provisional Solution, were reached by a vote of ten to five, with Judges Oda, Koroma, Vereshchetin, Parra-Aranguren and Skubiszewski dissenting. Judge Koroma disagreed with the conclusions of the Court, finding the Provisional Solution represented a genuine attempt to achieve the object and purpose of the 1977 Agreement as nearly as possible given Hungary's wrongful acts of breach and the decision of the Court did not adequately consider the financial damage and environmental harm suffered by Czechoslovakia nor the proper nature of the doctrine of equitable utilization. Separate Opinion of Judge Koroma, at 4–6. Judge Oda similarly based his dissent on the view Czechoslovakia was forced to construct the Provisional Solution as the only means possible of accomplishing the purpose of the 1977 Agreement. *See* Dissenting Opinion of Judge Oda, at 9–11. *See also* Dissenting Opinion of Judge Parra-Aranguren, at 1, which emphasized the environmental consequence of leaving the project unfinished as a justification for the Provisional Solution. *See further* Dissenting Opinion of Judge Skubiszewski, at 2–5.
443. Decision, at 44.
444. Decision, at 46. Judge Parra-Aranguren drew particular attention to the fact the 2–3 November 1992 Report of the Working Group of Independent

252 Notes

Experts on Variant C of the Gabčíkovo–Nagymaros Project found the 'on-going activities with [the Provisional Solution] could be reversed'. Dissenting Opinion of Judge Parra-Aranguren, at 2.

445. Decision, at 46, *citing* the ILC's commentary on Article 41 of the Draft Articles on State Responsibility, in the 'Report of the ICL on the work of its forty-eight session, 6 May to 26 July 1996', at 141, and YILC, 1993, Vol. II, Part 2, at 57, para. 14. The Court reached this conclusion by a vote of nine to six, with Judges Schwebel, Bedjaoui, Ranjeva, Herczegh, Fleischhauer and Rezek dissenting.

446. Declaration of Judge Schwebel, at 1. *See also* Separate Opinion of Judge Bedjaoui, at 5; Dissenting Opinion of Judge Ranjeva.

447. Hungarian Memorial, at 211–13; Hungarian Counter-Memorial, at 235–9, *citing* Articles 1 and 10 of the 1977 Agreement; Articles 3, 7 and 11 of the 1976 Boundary Waters Convention; Article 4 of the 1976 Agreement Concerning Cooperation and Mutual Assistance along the Czechoslovakian–Hungarian Border; and Article 3 of the Danube River Convention, and Article 5 of the 1958 Danube Fisheries Agreement. *See also* Application of the Republic of Hungary, at 9–10, and Gabčíkovo–Nagymaros File, at 16.

448. Hungarian Memorial, at 214.

449. Hungarian Memorial, at 213–14, *citing* Articles 15, 19, and 20 of the 1977 Agreement. Hungary also cited Article 11 of the Boundary Waters Convention which required the parties to undertake efforts to maintain the purity of the boundary waters. *Id. See also* Hungarian Reply, at 131; and Oral Pleadings of Hungary, Roland Carbiener and Klaus Kern (5 March 1997).

450. Hungarian Memorial, at 216–17, *citing* Article 22 of the 1977 Agreement and Article 4 of the Boundary Waters Convention.

451. Slovak Counter-Memorial, at 207–32.

452. Slovak Memorial, at 340–2; Slovak Counter-Memorial, at 76–89, 333–4.

453. Slovak Memorial, at 295–9, *citing* Article 22 of the 1977 Agreement; Slovak Counter-Memorial, at 334–6.

454. Hungarian Memorial, at 219–23, *citing* Principle 21 of the Declaration of the United Nations Conference on the Human Environment; the Trail Smelter Arbitration; the Corfu Channel Case; the Lac Lanoux Arbitration; the Convention on Long Range Transboundary Air Pollution; the Helsinki Convention on the Protection and Use of Transboundary Watercourses and International Lakes; the Helsinki Rules on the Uses of the Waters of International Rivers; and the Convention on the Law of Non-Navigational Uses of International Watercourses. *See also* Hungarian Counter-Memorial, at 220–2; Hungarian Reply, at 133–4.

455. Hungarian Memorial, at 223–8, *citing inter alia* Principle 6 of the 1978 UNEP Principles of Conduct; Declaration of the United Nations Conference on Environment and Development; Convention on Environmental Impact Assessment. *See also* Hungarian Counter-Memorial, at 220–2, 228–34.

456. Hungarian Memorial, at 228, *citing* Principle 13 of the UNEP Principles of Conduct. *See also* Hungarian Counter-Memorial, at 220–2.

457. Hungarian Memorial, at 228–32, *citing* the Articles on the Law of Non-Navigational Uses of International Watercourses; Article 2 of the Helsinki

Convention; Article 5 of the Helsinki Rules. *See also* Hungarian Counter-Memorial, at 223–6; Hungarian Reply, at 135–7.

458. Hungarian Memorial, at 232–4, *citing* Article 1 of the International Covenant on Economic and Social and Cultural Rights; Article 1 of the International Covenant on Civil and Political Rights; Fisheries Jurisdiction Cases. *See also* Application of the Republic of Hungary, at 8, and Gabčíkovo–Nagymaros File, at 15.
459. Slovak Memorial, at 211–17, 301–6; Slovak Counter-Memorial, at 236–40, 255–77, 318–19; Oral Pleadings of Slovakia, Igor Mucha, at 30–9 (27 March 1997).
460. Decision, at para. 45, *citing* Articles 1, 8, and 10 of the Agreement. Judge Ranjeva emphasized the unlawfulness of the Provisional Solution was directly related to replacing an international project by a national project. Dissenting Opinion of Judge Ranjeva, at 4.
461. Decision, at 45.
462. Decision, at 45.
463. Slovak Counter-Memorial, at 348–9.
464. Slovak Counter-Memorial, at 352.
465. Oral Pleadings of Slovakia, Sir Arthur Watts, at 17–18 (27 March 1997); Slovak Counter-Memorial, at 352–3, *citing* Nauillaa Arbitration; Military and Paramilitary Activities Case (Merits).
466. Hungarian Memorial, at 234–6.
467. Hungarian Memorial, at 237–43; Hungarian Reply, at 138.
468. Decision, at 47. In support of these criteria the Court cited, Military and Paramilitary Activities, at 127, para. 249. The Court also referenced the Air Service Agreement Arbitral Award; Articles 47 to 50 of the Draft Articles on State Responsibility adopted by the ILC on first reading, 'Report of the ILC on the work of its forty-eight session, 6 May to 26 July 1996', at 144–5.
469. Decision, at 47.
470. In support of its recognition of the right to equitably use and share riparian natural resources, the Court cited International Commission of the River Oder Case, at 27, and the Convention on the Law of Non-Navigational Uses of International Watercourses.
471. Decision, at 47.
472. Hungarian Memorial, at 329–30, *citing* Article 70 of the Vienna Convention on Treaties.
473. Hungarian Memorial, at 229–30. *See also* Hungarian Counter-Memorial, at 259–62.
474. For a more detailed explanation of this argument *see* Hungarian Reply, at 174–6.
475. Hungarian Memorial, at 331–5.
476. Hungarian Memorial, at 336. *See also* Hungarian Reply, at 179–80; Oral Pleadings of Hungary, Gyorgy Szenasi, at paras. 7–11 (11 April 1997).
477. Hungarian Reply, at 176–7.
478. Hungarian Reply, at 177.
479. *Citing* Articles 2, 3, 17 and 18 of the Rio Declaration, Article 2 of the Danube River Protection Convention, and Operational Directives of the European Bank for Reconstruction and Development. Hungarian Counter-Memorial, at 266–9.

480. Slovak Memorial, at 353–4, *citing* the International Status of South West Africa Case, at 132; United States Diplomatic and Consular Staff Case, at 7; and Free Zones of Upper Savoy and the District of Gex Case, at 96 to support the argument the Court could find a treaty continued in force despite the claim of one of the parties that it had lapsed or was otherwise no longer valid.
481. Slovak Memorial, at 356, *citing* Chorzów Factory, at 46–7.
482. Slovak Memorial, at 355–60; Slovak Counter-Memorial, at 363–4; Oral Pleadings of Slovakia, Alain Pellet, at 40–54 (27 March 1997).
483. Slovak Memorial, at 364–70.
484. Slovak Counter-Memorial, at 358–63.
485. Slovak Reply, at 353.
486. Judge Rezek disagreed with this conclusion, finding the 1977 Agreement was no longer in existence as it had been abrogated by the actions of both Hungary and Czechoslovakia. Judge Rezek concluded, however, both parties were obligated under the duty of good faith to seek to fulfill the reciprocal duties not yet performed under the Agreement and to effectively utilize those structures which had been constructed. *See* Declaration of Judge Rezek. Judge Koroma on the other hand, took the opportunity of a separate opinion to reinforce the position under the principle of *pacta sunt servanda* the 1977 Agreement must be considered to continue in force despite Hungary's actions, and any finding based on Hungary's arguments and evidence relating to a state of necessity or fundamental change of circumstances the Agreement did not continue in force would have created a precedent whereby states would seek to extricate themselves from obligations they found to be inconvenient. Separate Opinion of Judge Koroma, at 1–2.
487. Decision, at 65. The Court emphasized, however, the facts did not determine the law, but rather the principle of *ex injuria us non oritur* (from a wrong or injury a right does not arise) controlled and as such the 1977 Agreement could not be voided by the unlawful acts of the parties. Decision, at 65. *See also* Decision, at 66. For a more detailed analysis of a legal basis for incorporating the structures of the Provisional Solution into the parameters of the 1977 Agreement, *see* Separate Opinion of Judge Bedjaoui.
488. Decision, at 68.
489. Decision, at 68.
490. Decision, at 68, *citing* the Convention on the Law of the Non-Navigational Uses of International Watercourses, art. 4, para. 2. The Court reached these conclusions by a vote of thirteen to two, with Judges Herczegh and Fleischhauer dissenting. Judge Fleischhauer asserted as the 1977 Agreement was properly terminated by Hungary, it no longer held a share of the works of the Provisional Solution and was not required to operate jointly nor contribute to their maintenance, and Slovakia was entitled to operate solely the Provisional Solution so long as it adopted a water management regime that adequately provided for Hungary's ecological needs and rights under the doctrine of equitable utilization. Dissenting Opinion of Judge Fleischhauer, at 9–10.
491. Decision, at 66.

492. Decision, at 67, *citing* the North Sea Continental Shelf Case, at 47, para. 85.
493. Judge Bedjaoui, in a Separate Opinion, emphasized the obligation to renegotiate in good faith must be seen as a strict obligation as required by the 1977 Agreement and by the evolution of international law relating to the environment and to transboundary watercourses. Separate Opinion of Judge Bedjaoui, at 8–9.
494. Decision, at 67. In the view of Judge Oda, during the course of subsequent negotiations, the parties should pay particular attention to the technological solution available for remedying any environmental damage caused by the operation of the Provisional Solution. Dissenting Opinion of Judge Oda, at 13.
495. Decision, at 69, *citing* Chorzów Factory Case, at 47.
496. Decision, at 69.
497. Decision, at 69. The Court reached this conclusion by a vote of twelve to three, with Judges Oda, Koroma and Vereshchetin dissenting. Judge Oda dissented in part on the basis Hungary's abandonment of the Nagymaros works did not cause any practical damage to Czechoslovakia. Dissenting Opinion of Judge Oda, at 13. Judge Vereshchetin dissented on the basis as the Provisional Solution was a valid countermeasure, Slovakia was not obligated to pay damages to Hungary. Dissenting Opinion of Judge Vereshchetin, at 1. Some of the concurring Judges indicated they preferred to have a separate vote on the issues of reparations to be made by Hungary and reparations to be made by Slovakia, but were unable to convince the majority to permit such a vote. *See* Dissenting Opinion of Judge Parra-Aranguren, at 4. Presumably, if the Court had separated the question, it might have rendered a determination Hungary was liable for damages while Slovakia was not, and this would have inhibited the ability of the parties to agree upon a zero-sum assessment of damages, which the Court clearly favored.
498. Decision, at 70.
499. Decision, at 70.
500. *But see* Separate Opinion of Judge Verramantry, which seeks to elaborate on the meaning of the term sustainable development.
501. *See* RFE/RLN, No. 127 Part II (29 September 1997).
502. RFE/RLN, No. 134, Part II (8 October 1997).
503. RFE/RLN, No. 134, Part II (8 October 1997).
504. RFE/RLN, No. 180, Part II (16 December 1997).
505. RFE/RLN, No. 194, Part II (13 January 1998).
506. RFE/RLN, No. 9, Part II (15 January 1998).
507. RFE/RLN, No. 14, Part II (22 January 1998).
508. RFE/RLN, No. 17, Part II (27 January 1998).
509. AFP (6 February 1998); RFE/RLN, No. 41, Part II (2 March 1998).
510. RFE/RLN, No. 27, Part II (10 February 1998).
511. RFE/RLN, No. 45, Part II (6 March 1998).
512. CTK National News Wire (25 March 1998).
513. RFE/RLN, No. 64, Part II (2 April 1998). Separately, the Prime Minister of Slovakia threatened to take the dispute back to the Court if Hungary had not signed the protocol by 15 May 1998. BBC World Broadcasts (23 March 1998).

514. BBC World Broadcasts (3 April 1998).
515. RFE/RLN, No. 60, Part II (27 March 1998).
516. Hungarian News Agency (24 July 1998).
517. Special Agreement, art. 5(3).
518. Submission of Slovakia to the International Court of Justice in the matter of Gabčíkovo–Nagymaros Project (Hungary/Slovakia) (3 September 1998).
519. Hungarian News Agency (17 November 1998).
520. CTK National News Wire (18 November 1998).
521. Notably, in the view of the Slovak Prime Minister, the protocol permits both parties to save face with the international community and with domestic political constituencies. CTK National News Wire (25 March 1998).
522. ODD, No. 176 (11 September 1995). The Slovak–Hungarian Basic Agreement was signed on 19 March 1995. A subsequent protocol on implementation was signed in November 1998. BBC World Broadcasts (20 November 1998).
523. Fitzmaurice, at 68.
524. *See* RFE/RLN, No 191, Part II (8 January 1998).
525. Environmental Needs Assessment, at 72.
526. Fidas & Gershwin, at 5. Environmental Needs Assessment, at 69.
527. *See* CTK National News Wire (25 March 1998).
528. *See* Decision, at 69.
529. RFE/RLN, No. 41, Part II (2 March 1998); AFP (6 February 1998).
530. RFE/RLN, No. 51, Part II (16 March 1998).
531. RFE/RLN, No. 44, Part II (5 March 1998).
532. RFE/RLN, No. 27, Part II (10 February 1998); RFE/RLN, No. 25, Part II (6 February 1998).
533. RFE/RLN, No. 39, Part II (26 February 1998).
534. RFE/RLN, No. 127, Part II (29 September 1997).
535. RFE/RLN, No. 60, Part II (27 March 1998).
536. RFE/RLN, No. 69, Part II (9 April 1998).
537. RFE/RLN, No. 65, Part II (3 April 1998); CTK National News Wire (2 April 1998).
538. *See* M. Jordon, *Christian Science Monitor* 6 (29 September 1997), for a similar assessment about the influence of potential NATO and EU membership on the dynamics of the post ICJ decision Slovak–Hungarian relationship.
539. Fitzmaurice, at 122.
540. BBC World Broadcasts (23 March 1998).
541. For a review of the recent literature examining the relationship between international relations theory and the operation of international law, *see* Aceves; Alvarez; Beck; Beck *et al.*; Fassbender; Franck(a); Herbert; Horohoe; Koh; Keohane(a, b); Kratochwil; A. Slaughter *et al.*; Ehrlich, Setear(a, b, c); Brown Weiss(d). For a review of ecological theory and international law, *see* Pirages. For a review of recent innovative developments in international relations theory, *see* Doyle & Ikenberry.
542. Krasner(b), at 2.
543. For a concise review of the development of regime theory from the lawyer's perspective, *see* Slaughter(a), at 217–18, who refers to the early regime theory as modified structural realism. For a more detailed examination of the development of regime theory and its limitations, *see* Kratochwill &

Ruggie. For a review of the history of states relying upon norms and institutions to regulate their international behavior, *see* Luard, at 282–3, 312–14.
544. Krasner(b), at 2; *See also*, Keohane(c); Young(a).
545. P. Haas(b), at 174. For more information on the formation of rules, *see* Siegal.
546. Regimes operate with respect to specific 'issue areas' that are part of 'theoretically determined policy areas'. List & Rittberger, at 90. For a more detailed examination of what constitutes an issue area, *see* Keohane(a).
547. For a review of the means by which regimes reduce these transaction costs, *see* Keohane(a), at 244–5.
548. Keohane & Nye(b); Keohane & Nye(c).
549. Ausubel & Victor. *See also* Hurrell.
550. *See* Keohane & Nye(c).
551. List & Rittberger, at 89; Rittberger(a), at 3.
552. For a review of the specific role of transnational sub-state actors networks in the operation of international regimes, *see* Keck & Sikkink. *See also* Garner.
553. Slaughter(a), at 218–20.
554. *See* List & Rittberger, at 90; Rittberger(a), at 3.
555. For a perspective of international law as a language of international relations, *see* Kritsiotis. For an evaluation of the feminist theory of international law and its relationship to international relations, *see* Joyner & Little.
556. *See* Nadelman; Kratochwill.
557. Hurrell & Kingsbury, at 12.
558. Byers, at 126.
559. Slaughter investigates the extent to which there is significant overlap between the functions and benefits of international regimes as identified by international relations scholars, and the functions and benefits of international law as identified by international law scholars. Slaughter(a), at 218–20. *See also* Byers, who attempts to construct a link between regime theory/institutionalism and the operation of international law.
560. *See* Byers, at 137, which, based upon an examination of various international relations theories, theories of international law and a number of specific rules, 'explains the process of customary international law as a social institution in a political system made up of socially unequal, self-interested states.'
561. Schwarzenberger, at 41.
562. Kelsen, at 17.
563. Henkin, at 5, 24, 29, 41–7.
564. Brierly, at 56.
565. Rubin & Brown, at 2.
566. Spector & Korula. *See also* Korula; McDonald & Bendahmane; Stein; Ross; Saunders. For an application of the tri-sected negotiation process to a modeling of preferential orders for European integration, *see* Schneider, at 133–40.
567. The deviation from standard social science terminology expressed in this book is attributed to a shifting of emphasis away from negotiated dispute resolution toward the broader phenomena of dispute formation and

258 *Notes*

resolution in the context of which negotiation is but one means of dispute definition and resolution.
568. *See* Spector & Korula, at 3.
569. *See* Irvin, at 65–6.
570. These mechanisms are not mutually exclusive. For instance inquiry may frequently form a key component of mediation, conciliation or arbitration. Merrills(b), at 43.
571. For a general review of the merits of inquiry, *see* Merrills(b), at 55–8.
572. *See* Oral Pleadings of Slovakia, Christian Refsgaard, at 41–51 (26 March 1997).
573. *See* Merrills(b), at 35. The mechanisms of mediation and good offices may be invoked while the parties are pursuing the process of inquiry, and these mechanisms frequently facilitate that inquiry by providing a stable forum in which to share information produced as a result of the inquiry.
574. *See* Merrills(b), at 39–42.
575. *See* Schneider, at 133.
576. *See* Schroeer, at 188–90.
577. Henkin, at 1.
578. *See* Henkin, at 15.
579. Although rare in practice, the ICJ has held it may accept jurisdiction of a case prior to attempts by the parties to achieve a negotiated solution. *See* Continental Shelf Case, at 218. *See also* Chorzów Factory Case, at 10–11.
580. For a review of the circumstances in which conciliation has proved most useful, *see* Merrills(b), at 76–9.
581. Spector & Korula, at 3.
582. *See* Irvin, at 64–5.
583. Henkin, at 18–19.
584. *See* Irvin, at 68.
585. *See* Aegean Sea Continental Shelf Case, at 12.
586. *See* Military and Paramilitary Activities Case, at 439.
587. Haya de la Torre Case, at 79.
588. Ambatielos Case, at 10, where the Court held the United Kingdom was required to resolve its dispute with Greece through a previously designated arbitration mechanism.
589. North Sea Continental Shelf Case, at 6.
590. In some limited instances, these determinations may lead to a direct settlement of the dispute, as where the Court is requested to draw a land or maritime boundary. Gulf of Maine Case. For a review of the role of the Court as perceived by the Hungarian party to the Gabčíkovo–Nagymaros Project dispute, *see generally* Oral Pleadings of Hungary, James Crawford, at para. 5 (3 March 1997).
591. McNair(a), at 303–4.
592. *See* Oral Pleadings of Hungary, James Crawford, at paras. 9–12 (10 April 1997); Oral Pleadings of Hungary, Alexandre Kiss (10 April 1997); Oral Pleadings of Slovakia, Peter Tomka, at 18 (24 March 1997).
593. Military and Paramilitary Activities Case, at 436–7, 439.
594. United States Diplomatic and Consular Staff in Tehran Case, at 19–20; Aegean Sea Continental Shelf Case, at 13.
595. McNair(a), at 307. *See* Haya de la Torre Case, at 79.

596. *See* Handl(g), at 71; Chinkin & Sadurska, at 70.
597. *See generally Handbook on the Peaceful Settlement of Disputes*, at 168–95.
598. For a general review of possible unilateral state action, *see* Bilder(b).
599. For a review of the literature concerning compliance with and enforcement of international legal obligations, *see* Chayes & Chayes; Downs; Kingsbury; Jacobson; Rottem.
600. For a review of the process for creating international environmental norms, *see* Caron; Scott *et al.*(a).
601. Henkin, at 18–19. *See also* Birnie(a), at 105.
602. *See* Applicability of the Obligations to Arbitrate Under Section 21 of the United Nations Headquarters Agreement Opinion; Interpretation of the Agreement of 25 March 1951 Between the WHO and Egypt Opinion; Interpretation of Peace Treaties with Bulgaria, Hungary and Romania (Second Phase) Opinion; Interpretation of Peace Treaties with Bulgaria, Hungary and Romania (First Phase) Opinion.
603. *See generally* Ordeshook; Radinsky; Schmidt. For a review of the application of game theory to environmental problems relating to the global commons, *see* Ward(a) and (b).
604. Cederman, at 51–3. *See also* Schneider, at 128; Putnam; Iida; Aggarwal & Allan, at 10. For a detailed examination of the use of game theory in modeling and predicting international cooperation, *see* Wagner; Snyder & Diesing.
605. Schroeer, at 190–1, 202–3; Prittwitz, at 7–10. *See* Kremenyuk & Lang, at 3 further asserting if environmental agreements are to be effectively implemented, they must meet the needs of all parties.
606. *See generally* P. Haas(b); Haufler; Dahlberg *et al.*
607. Some game theorists contend although states serve to aggregate the national interest, governments do not necessarily act solely as the agents of their constituents, but rather have the additional agenda of political self-preservation. Schneider, at 127. *See also* Schelling.
608. Galambos, at 86–90.
609. Aggarwal & Allan, at 11. For a detailed examination of the concept of 'nesting', *see* Tsebelis. For a brief application of the nesting of issues in the case of the Danube River Basin, *see* Linnerooth.
610. *See* Wilson; Halperin; Allison.
611. Galambos, at 90.
612. For an examination of the role of energy consortiums in the transition to a market economy and the conduct of national politics which may affect transboundary environmental disputes, *see* Frydman *et al.*, at 299–316.
613. Political science studies indicate agreements dealing with the prevention or control of pollution and industrial regulation invariably require more time than other treaties for ratification and implementation. Spector & Korula, at 16. *See also* P. Haas(b), at 172.
614. For a review of the role of multinational corporations in promoting environmental security, and in particular the creation of transboundary institutions designed to advance ecological security and promote sustainable development, *see* Rogers.
615. *See Wall Street Journal* (23 October 1995). For a more detailed review of the activities of Western corporation, as well as international organizations

260 *Notes*

such as the EBRD, IAEA and EU in the modernization of nuclear plants in CEE, *see* Reuters Textline, Euromoney (25 April 1995).
616. Pleadings of Slovakia, Peter Tomka, at 14 (24 March 1997).
617. *See generally* Lowe & Goyder; Charnovitz; Raustiala. For an examination of the role of NGOs in international trade disputes, *see* Esty; Shell.
618. Ausubel & Victor, at 18. In 1997, for instance, a coalition of environmental NGO's attempted to pressure the EU into requiring all new CEE meet EU environmental standards soon after admission to the EU. Greenwire (19 August 1997).
619. Spector(a), at 4–5.
620. Hurrell & Kingsbury, at 20.
621. Hurrell & Kingsbury, at 28.
622. CTK News Wire (18 March 1997). For a discussion of the role of NGOs in environmental litigation, *see* Führ & Roller; Geddes.
623. *See generally* Slaughter(a, c). For a review of the various means by which the general public can affect environmental decisionmaking, *see* Casey-Lefkowitz.
624. Spector(a), at 5. For more information on the value and effects of enhanced public participation, *see* Anderson; Charnovitz; D. Kennedy; Partan; Popovic; Raustiala; Weber. For a general review of the utility of public participation at the grass roots level, *see* Foster.
625. *NYT* (4 February 1990).
626. Zartman. *See also* Keck & Sikkink.
627. Hurrell & Kingsbury, at 29.
628. The bilateral trade volume in 1995 between Poland and Denmark, for instance, exceeded $1.3 billion, with Danish investment in Poland reaching $150 million. ODD, No. 49 (10 March 1996).
629. The Scandinavian and Nordic states also are involved in more general environmental protections efforts. *See, e.g.*, Danish–Polish and Danish–Romanian Agreements on Environmental Conservation, 5 YIEL, at 373, 393.
630. ZumBrunnen, at 396. Approximately 5–10 per cent of the pollution deposited on Finnish territory originates from Poland.
631. Environmental Action Programme, at IV-22.
632. For a review of the efforts of the IFIs to promote environmental protection, *see* Di Leva, at 507–12.
633. Hungary, for instance received a $387 million standby loan from the IMF in March 1996, and became eligible for a $200 million loan from the World Bank. It is expected Hungary will eventually qualify for between $400–$500 million in World Bank loans. ODD, No. 55 (18 March 1996).
634. For a general description of the IFIs' efforts to promote environmental protection, *see* Piddington; Rice.
635. For a brief review of the functions of the Committee on the Challenges on Modern Society as they relate to environmental protection, *see* Birnie & Boyle, at 72.
636. ODD, No. 60 (25 March 1996). For a brief review of the environmental activities of the OECD, *see* Sands(h), at 87–8.
637. ODD, No. 42 (28 February 1996); No. 241 (13 December 1995).
638. For example, the German consortium RWE-Energie-Versorgung Schwaben, Bayernwerk and Isar Amperwerk, the French Electricité de France and

Gaz de France, and the Belgian concern Powerfin S.A own minority stakes in the Hungarian electricity companies and majority stakes in Hungary's gas companies totaling over $1.5 billion. ODD, No. 239 (11 December 1995).
639. Galambos, at 81, 86; ZumBrunnen, at 402.
640. For a review of the application of international law within the EU, see Lasok & Bridge; Wyatt & Dashwood.
641. Environmental Needs Assessment, at 32–3, 41. By the end of 1995, all CEE states had applied for EU membership. Although the EU intends to impose the same conditions for membership on all of the applicants, some CEE states are projected to achieve membership by the year 2002, with France and Germany having declared they are in favor of early admission for Poland. ODD, No. 244 (18 December 1995); No. 210 (27 October 1995).
642. ODD, No. 248 (22 December 1995).
643. The Danube Task Force, Strategic Action Plan at 1–3. On 18 September 1995, the Environment Ministers from the Baltic states, Poland, the Czech Republic, Slovakia, Romania and Bulgaria met with EU representatives and agreed to establish a regular process to assess their progress in meeting EU environmental standards. ODD, No. 183 (20 September 1995). For a case specific review of the effect of potential EU membership on Polish environmental legislation, see Cole(c), at 285–6.
644. The EU recently provided CEE states $30 million to carry out 22 energy-saving and environmental projects. The EU's Poland–Hungary Assistance for the Reconstruction of the Economy (PHARE) program for economic assistance also allocates significant sums for environmental protection, with the Czech Republic, for instance, receiving $17 million in 1994. Zum-Brunnen, at 403.
645. D. Fisher, at x. Environmental Needs Assessment, at 18.
646. Mloch, at 44; The EU also provides PHARE assistance for projects relating to the Elbe, Oder and Morava Rivers. Elbe Project Bulletin, at 1.
647. Deutsche Presse-Agentur (17 May 1995). For critical reviews of the role of the IFIs in promoting environmental protection, see Goldberg & Hunter; Rich; Sands(e); Shihata; Stein & Johnson; World Bank, The World Bank and the Environment.
648. The United States also established a $20 million program for reducing pollution in the Silesian Coal Basin, and supports efforts by United States utilities to install SO_2 abatement mechanisms on CEE state utilities in exchange for credit under the United States Clean Air Act and the Global Warming Convention. See also Read.
649. Environmental Needs Assessment, at 62.
650. In a similar project, the World Bank funded a Polish forestry program where there is significant concern harvest levels required to meet the debt are not sustainable, and the equipment purchased upon recommendation of the World Bank is not suitable for Polish forests. Environmental Needs Assessment, at 57, 71.
651. The IFIs are most likely to apply pressure to employ international environmental law to resolve disputes in transboundary regions where the institutions are providing assistance to remedy a particular environmental problem, as frequently the environmental remediation program will not be

effective without transboundary cooperation. *See, e.g.*, Bulgarian–Rumanian Danube Delta World Bank Program.
652. Goldberg & Hunter, at 1.
653. Environmental Needs Assessment, at 57. In a recent example, $50 million worth of illegal pesticides were provided to Poland under an EU assistance program. Juergensmeyer *et al.*, at 835. *See also* Park, at 670, discussing the shipment of hospital waste to Poland under the guise of charity.
654. 3 YIEL, at 401.
655. ODD, No. 24 (2 February 1996); Environmental Needs Assessment, at 20. Austria too has been criticized by some environmental experts for engaging in environmental neo-colonialism by shifting its energy production east, and by actively supporting its industrial enterprises in their efforts to export less than desirable technology to CEE. The Austrian government, for example, supported the efforts of an Austrian industrial enterprise seeking to construct a hazardous waste incinerator in Bratislava, which would then be funded through fees charged for the incineration of Austrian waste. Environmental Needs Assessment, at 70; Galambos, at 93.
656. In 1993, Germany provided $46 million to upgrade the efficiency of Polish and Czech coal-fired power plants in Tušimice, Počerady and Turów, which are principle polluters in the Black Triangle. Mloch, at 44.
657. Improving water quality in the Elbe River is particularly important to Germany since although it is subject to intensive treatment prior to use, its quality remains low and the average unit of water is used by Germany two and one half times before it reaches the sea. ZumBrunnen, at 399.
658. *See* Strategic Environmental Issues, at 15.
659. For a general recognition of the need to examine the social and political context of an international dispute for the purpose of understanding the operation of international law, *see* Chinkin & Sadurska, at 48, who identify fourteen 'situations' affecting the operation of international law. For a review of situational circumstances proposed by social scientists, *see* Ausubel & Victor, at 2; McDonald; Korula; Wettestad & Andresen; Schroeer; Kay & Jacobson; Shaw.
660. *See* Goldberg *et al.*, at 404.
661. Spector(a), at 2; *Improving Environmental Negotiations*.
662. *See* Goldberg, at 404.
663. Spector & Korula, at 10. *See also* Frydman *et al.*, at 319, which notes at a per capita GDP of $5,600 concern for the environment becomes a priority of the general public.
664. *See* Goldberg, at 404. According to the World Bank it will cost Poland $1.6 billion annually over the next 25 years to meet EU air pollution standards, and according to United States Department of Commerce, it will cost Poland $200 billion to meet all of its environmental priorities. Fidas & Gershwin, at 20.
665. Spector(b), at 11.
666. Spector & Korula, at 10. Notably, CEE states currently spend between 0.5 to 1.3 per cent of their annual GDP on measures to protect the environment, whereas spending by EU members states on average accounts for 2 per cent of annual GDP. Fidas & Gershwin, at 18.
667. List & Rittberger, at 105.

668. *See* Madeo.
669. For an examination of the relationship between political and economic power and international relationships, *see* Morgenthau; Carr; Kennan. *See also* Irvin, at 58, 69, wherein an international legal scholar argues states are more inclined to use international law to assist in the resolution of their disputes when they possess equal bargaining strength and when their domestic economic structures are similar.
670. Potier, at 69–78.
671. Fidas & Gershwin, at 24.
672. Sprinz & Vaahtoranta, at 80.
673. For a further explanation of the realist perspective on international environmental relations, *see* Hurrell & Kingsbury, at 37, 43.
674. Environmental costs include lost human capital and productivity, diminished physical and natural capital and reduced environmental quality (amenity costs). Environmental Action Programme, at II-2 to II-4.
675. *See* Oral Pleadings of Hungary, Katherine Gorove (4 March 1997) for the implicit argument states have a right to conduct a cost–benefit analysis of joint projects and for an example of the substance of such an analysis.
676. Fidas & Gershwin, at 20. To meet EU environmental standards it is estimated it will cost Hungary approximately $7 billion, the Czech Republic $16 billion, and Poland up to $52 billion. Hungarian News Agency (10 August 1998); CTK News Wire (16 April 1998); Greenwire (25 March 1998).
677. Fidas & Gershwin, at 7, 16.
678. Fidas & Gershwin, at 24.
679. Sprinz & Vaahtoranta, at 100.
680. *See* Darmstadter & Fri.
681. For a more detailed exploration of the interconnection between environmental degradation and economic costs to society, *see* Goldmark Jr. & Larocco, at 84.
682. Reisman, at 119–23 (1992). *See also* Lasswell & McDougal; Chen, at 14–15; McDougal & Associates; McDougal. For a review of the origins and subsequent development of the policy-oriented approach, *see* Slaughter(a), at 209–12.
683. *See* Goldberg, at 404. For a review of the manner in which domestic politics affect regime formation, *see* Zürn(b).
684. G. Schneider, at 126.
685. Slaughter(a), at 228.
686. Potier, at 80–1.
687. Spector & Korula, at 6.
688. *See* Dupont(b), at 156–7.
689. Spector & Korula, at 23.
690. Scholars subscribing to the liberal approach to legal theory argue states with homogeneous domestic structures will be more likely to adopt and use international law in their reciprocal relations. Slaughter(a), at 233–4, 237.
691. *See* Spaulding for a discussion of the value of transparency and public participation for promoting a resolution of transboundary environmental disputes.
692. Prittwitz, at 19, 24.

693. Zürn(b), at 282.
694. For a review of the role of law in promoting political transformation, *see* Teitel.
695. Sprinz & Vaahtoranta, at 77. For a general review of the influence of international politics on the application of international law, *see* Arend.
696. For a review of the implementation of regional norms and agreements in Latin America, *see* Holley.
697. 6 YIEL, at 450–1.
698. *Europe Energy* (25 July 1997). For a review of the source and purpose of EU structural funds, *see* Baldock & Wenning.
699. For a review of the role of environmental matters in international relations, *see* Barnes; Busterud; Kay & Jacobson; Lang; MacNeill *et al.*; Marvin; Nalven.
700. *See* Schwarzenberger, at 13–15, 31. For an assessment of the occurrence of sub-regional coalescence in CEE, *see* Hartnell.
701. *See* the International Covenant on Civil and Political Rights; Sieghart, at 359–61.
702. *See* the International Covenant on Civil and Political Rights, art. 27; Sieghart, at 376–7.
703. *See* Montville.
704. For a general review of preconditions which support the use of law, *see* Cassese; Henkin; Deutsch & Hoffmann; Falk; Schwarzenberger.
705. List & Rittberger, at 103; Rittberger & Zürn, at 42–3.
706. Hurrell.
707. *See* Ausubel & Victor, at 28–9.
708. List & Rittberger, at 104–5. *See also* Kiss & Shelton(a).
709. Irvin, at 69–70.
710. Ausubel & Victor, at 2.
711. *See* Timoshenko, at 667.
712. *See* Nye; E. Haas.
713. *See* Sommer(a), at 227.
714. *See* Rittberger & Zürn, at 44.
715. In limited circumstances, the structural-systematic approach could also be included in this category.
716. Vukas. Vukas does not comment on the question of the possible existence of regional international law.
717. Cassese, at 20; Henkin, at 60–2; R. Fisher, at 1135–9.
718. Henkin, at 63.
719. Hurrell & Kingsbury, at 30; Ausubel & Victor, at 3.
720. For a review of the circumstances that enhance the ability of mechanisms of international adjudication, *see* Helfer & Slaughter.
721. Cassese, at 13–14; Henkin, at 55; Irvin, at 69.
722. Cassese, at 13–14.
723. Hurrell & Kingsbury, at 27.
724. For a review of the various means and forums of international dispute resolution, *see Handbook on the Peaceful Settlement of Disputes*; Janis; Merrills(b). For a review of dispute resolution mechanisms particularly relevant for environmental disputes, *see* Craik; Koskenniemi(a); Stein. *See also* E. Lauterpacht, exploring the growing use of international tribunals and

some of the general issues faced by these tribunals. For a general review of judicial remedies in international law, *see* Gray.
725. *See* Jennings(a), at 309; R. Fisher, at 1138.
726. *See* Henkin, at 49, 68–74; Cassese, at 17. *But see* R. Fisher, at 1138, who argues a state may use international law to promote fair and wise dispute settlement in preference to an advancement of their short term financial or institutional interests.
727. A state pursuing a maximum strategy attempts to maximize the minimum pay-off by acting so as to achieve the best possible outcome in light of the worst possible pay-off that could result from the action taken by the other state. A nash equilibrium is achieved when no state can defect from the current pattern of behavior without causing self-injury. A nash equilibrium thus occurs when all states are maximizing their net benefit. Rittberger & Zürn, at 39. *See* Oye(b); Snidal; Snyder & Diesing.
728. A pareto-optimal result exists when no state can become better off without another state being made worse off. A sub-pareto-optimal result exists when it is possible to make one state better off without making another state worse off.
729. This type of regime is supported by procedural rules of international law relating to notification, consultation, international EIAs and joint management.
730. *See* Rittberger & Zürn, at 40.
731. *See* Fidas & Gershwin, at 5.
732. *See* Henkin, at 52–4; Spector & Korula, at 26.
733. *See* Liamco v. Libya Arbitration, at 58–9, finding Libya was entitled to nationalize certain industrial assets because of its sovereignty over natural resources; and UNGA Resolution 626 (21 December 1952), declaring sovereignty encompasses the right of peoples to use their natural resources.
734. *See* Corfu Channel Case, at 4; Trail Smelter Arbitration, at 716; Principle 21 of Declaration of the United Nations Conference on the Human Environment. *See also* Perrez.
735. For a review of the socialist perspective of international environmental law, *see* Timoshenko.
736. Handl(g), at 85–6; Caldwell(c).
737. *See* Irvin, at 68. As one dispute resolution expert notes, environmental dispute resolution is an area in 'which existing applicable rules of international law – for example, principles of good neighborliness and equitable utilization of shared resources – do not stipulate specific obligations or measures on the part of states, but at best provide an obligation and a framework for states to negotiate agreement on concrete and effective actions'. Albin(b), at 7. In a comprehensive social science analysis of the US–Canadian acid rain negotiations, international law was deemed worthy of one sentence, 'international law, such as the Stockholm Declaration, is reasonably toothless'. Schroeer, at 203.
738. For an explanation of the concept and authority of soft law, *see* Burhenne; Chinkin; P.-M. Dupuy(d); Gruchalla-Wesierski; Steven.
739. Birnie(c), at 83.
740. *See generally* Chinkin.
741. Kinnear & Rhode, at 7.

742. Kinnear & Rhode, at 7, 9–10.
743. Kinnear & Rhode, at 9. For a detailed description of the nature and effects of acid rain, *see* Schroeer, at 183; Environmental Action Programme, at II-19; Impacts of Long-Range Transboundary Air Pollution.
744. Schroeer, at 184.
745. Environment of the Czech Republic, at 12.
746. Environmental Action Programme, at II-20–3. For a narrative of the general environmental degradation in CEE, *see* Carter & Turnock; Manser; Thompson, *National Geographic* (June 1991); Weinstein; International Environment Report (BNA); French; IUCN East European Program, Environmental Status Reports; Schultz & Crockett; DeBardeleben; Zvosec; Kramer(a); Kirkpatrick; Volgyes.
747. An Overview of the Slovak Environmental Situation, at 5.
748. For a specific calculation of the projected reduction in emissions, *see* Environmental Action Programme, at III-8 to III-14.
749. Environmental Action Programme, at III-15.
750. Haigh *et al.*, at 25.
751. Haigh *et al.*, at 26.
752. Juelke, at 1378–9; Enyedi & Zentai, at 223–4.
753. Haigh *et al.*, at 26; Cole(c), at 316.
754. Environmental Action Programme, at III-22.
755. Haigh *et al.*, at 28.
756. Bándi *et al.*, at 3–4; Cole(c), at 305.
757. Schöpflin, at 190. As a consequence of the use of inefficient mining technology in Poland, approximately 60 per cent of recoverable coal was wasted during the extraction process. In geologically similar states, approximately 30 per cent of recoverable coal is wasted. Cole(c), at 307–8.
758. Environmental Action Programme, at II-10-11.
759. Schöpflin, at 165–6; Cole(c), at 512.
760. Haigh *et al.*, at 28; Enyedi & Zentai, at 223–4; Schöpflin, at 179.
761. Until recently, for example, Bulgaria's environmental legislation restricting discharge of pollution into watercourses was not applicable to international watercourses such as the Danube River and Black Sea. Djolov & Dimitrova, at 56.
762. *See* Galambos, at 75–6. *See* McGee & Block for an evaluation of the benefits of tradable emission permits.
763. For a review of the role of law in post-1989 CEE economic systems, *see* Rubin.
764. Schöpflin, at 249.
765. *See* Juelke, at 1379.
766. Mloch, at 44.
767. Environment of the Czech Republic, at 98–9.
768. Cole(c), at 297–8; Environmental Needs Assessment, at 15, 59, 66.
769. Higher energy prices generally lead to an increase in industrial efficiency and a consequential reduction in pollution. Environmental Action Programme, at II-23, III-5. In the long term, however, it is estimated energy consumption in CEE will grow by more than 60 per cent by the year 2010. Goldmark & LaRocco, at 87–8 (1992).
770. Hard budget constraints facilitate the ability of governments to more effectively induce enterprises to reduce pollution emissions by imposing meaningful environmental fees and fines. Cole(b), at 352.

771. Environmental Action Programme, at I-7, II-3, III-1.
772. Enyedi & Zentai, at 223–4; Vidlakova, at v. For a review of the utilization of market-based incentives in Poland, *see* Smith.
773. Enyedi & Zentai, at 223–4. In order to avert a purely economic calculation, CEE states should undertake a dualist policy of fines and strict emission standards.
774. Cole(c), at 319.
775. D. Fisher, at 15–16.
776. EYBCR 1992, at 293, 298.
777. *See Use of Economic Instruments in Environmental Policy in Central and Eastern Europe: Case Studies of Bulgaria, the Czech Republic, Hungary, Poland, Romania, the Slovak Republic, and Slovenia* (published by the Regional Environmental Center 1995).
778. Danube Strategic Action Plan, at chs. 7, 10.
779. Poland, for instance, requires capital investment of between $3 and $9.7 billion to meet its 1997 emission standards. Environmental Action Programme, at I-3. In Bulgaria, capital investment for the environmental sector is currently $6 million, with Bulgaria requiring $24 million simply to undertake the most urgent pollution abatement measures. D. Fisher, at 11. In the Czech Republic and Slovakia only 3–4 per cent of industrial capital costs are allocated for environmental protection compared with on average 12–15 per cent for OECD states. Czech and Slovak Joint Environmental Study, at 41.
780. Goldenman, at 2.
781. In 1995, CEE states received the following amounts of foreign direct investment: Hungary – $2.5 billion, the Czech Republic – $2.2 billion, Poland – $850 million, Romania – $450 million, Slovakia – $200 million and Bulgaria $100 million. *Emerging-Market Indicators, The Economist* 120 (2 March 1996).
782. Wicke, at 34.
783. Environment of the Czech Republic, at 14.
784. *See* Stewart.
785. Schöpflin, at 225. For a more philosophical review of the relationship between Marxist ideology and lack of environmental protection in CEE, *see* Cole(b). For a review of the economic and environmental decision making process in CEE, *see* Sidman *et al.*
786. G. Bándi *et al.*, at 23. For a more detailed analysis of the extent to which the former communist states suppressed information and data on environmental quality, *see* Juergensmeyer *et al.*, at 831–2.
787. The Danube Blues, at 3, 5; Galambos, at 75–6. Environmental information in CEE states was frequently classified as 'secret' for alleged national security reasons. *See* Cole(b), at 53–4.
788. Environment of the Czech Republic, at 30–5.
789. Sommer(c), at 227.
790. Haigh *et al.*, at 28–9; Enyedi & Zentai, at 224–5.
791. *See* Delaney; Gadourek; Djolov & Dimitrova, at 66–7.
792. Schöpflin, at 169, 174–5, 204.
793. Schöpflin, at 196; Environmental Action Programme, at IV-22.
794. Schöpflin, at 256–8, 275, 286; D. Fisher, at 2, 8.

795. Environmental Action Programme, at I-2. In Slovakia, for instance, where environmental protection ranked as a first priority in February of 1990, in July 1991 it ranked 7th, with some social groups branding environmentalists as anti-market and anti-nation. D. Fisher, at 14.
796. Environmental Action Programme, at III-22; Czech and Slovak Joint Environmental Study, at 26.
797. Sand(b), at 31.
798. Prittwitz, at 4–5.
799. EYBCR 1992, at 327. *See also* Polish–Ukrainian Treaty on Protection of the Environment, which expressly encourages cooperation between state, regional and local governments, and NGOs. 5 YIEL, at 372–3, art. 6.
800. An Overview of the Slovak Environmental Situation, at 10. *See also* Inotai.
801. Environmental Action Programme, at IV-8.
802. Galambos, at 75.
803. In mid September 1995, for instance, Russia explored the possibility of forming a new economic–political bloc in CEE similar to the communist-era CMEA. The CEE states universally rejected this overture. ODD, No. 181, Part I (18 September 1995); No. 180 (15 September 1995).
804. Strategic Environmental Issues, at 14.
805. *See* Pehe, at 14–16.
806. Strategic Environmental Issues, at 14.
807. For a review of the various levels of environmental cooperation between CEE states, *see* Strategic Environmental Issues, A 25–6.
808. The EC Commission, GEF Fund, EBRD, USAID, Austria, Netherlands and the World Bank provide financial support for this program. Environmental Action Programme, at II-12.
809. Environmental Action Programme, at II-12.
810. Sand(b), at 22. *See also* Brunnée & Toope, at 35.
811. Schultz & Crockett, at 59.
812. Euroregion Neisse-Nisa-Nysa 4 (1992); presentation of Ms. Kordela-Borczyk & Suchanek.
813. It should be noted national minorities may either be associated with a state, e.g., Hungarians, or they may be stateless minorities, for example, gypsies and Ruthenians. Presentation of Dr Pilinski (Cambridge, 3 March 1995).
814. Presentation of Dr Adamus-Matuszynska (Cambridge, 2 March 1995); Presentation of Dr Ira (Cambridge, 2 March 1995).
815. Grabowska, at 139.
816. Kosary & Vardy, at 222. The new state of Czechoslovakia, for instance, hosted populations of Bohemians, Moravians, Silesians and Slovaks, as well as Hungarians, Germans and Poles. Wallace, at 128–35, 148–50.
817. Peace Treaty of Trianon, art. 27, para. 4, at 3558; Wojatsek.
818. *See* Mocsy; Heinrich.
819. Magocsi, at 160, 164. Poland, for example, hosts 921,300 national minorities; the Czech Republic and Slovakia host 833,400 national minorities, with the 600,000 strong Hungarian minority in Slovakia accounting for 12 per cent of the population; Hungary hosts 351,300 national minorities; Romania hosts 2,525,500 national minorities, with the 1.7 million ethnic Hungarian minority accounting for 8 per cent of the Romania population;

and Bulgaria hosts 1,303,800 national minorities, with the Turkish national minority comprising 11 per cent of the Bulgarian population.
820. Czechoslovakia and Hungary, for example, entered into a bilateral treaty on 27 February 1946, providing for a one-to-one repatriation of ethnic Slovaks and Hungarians. Janics, at 133–4. Between 1946 and 1947, Czechoslovakia also expelled 3 million Germans from the Sudeten land. Magocsi, at 167.
821. *See* Bowers, at 30–1, 36. Human rights law is designed to alleviate much of this tension, yet in the pre-1989 CEE communist regimes law relating to human rights were seldom if ever applied. Hartwig, at 450.
822. Schöpflin, at 143–4.
823. Schöpflin, at 291. *See also* Müllerson(b), at 100–3.
824. Judt, at 826–7.
825. World Wildlife Fund, Repercussions of the Power Station 3 (1991).
826. *See* Bowers, at 31–4.
827. Müllerson(a), at 102–3.
828. Presentation of Dr Grodzinski (Cambridge, 2 March 1995).
829. Bowers, at 40–3.
830. *See* ODD, No. 239 (11 December 1995); No. 234 (4 December 1995); No. 187 (26 September 1995). The treaty was subsequently signed in October 1996.
831. ODD, No. 187 (26 September 1995).
832. ODD, No. 32 (14 February 1996).
833. As a result of this heightened ethnic tension in transboundary environmental hot spots, some CEE commentators propose the designation of Ethnic Environmental Areas for the purpose of linking environmental planning with special protection for minority rights and concerns. Presentation by Dr Grodzinski (Cambridge, 2 March 1995).
834. EYBCR 1992, at 10.
835. Pehe, at 17.
836. For an explanation of the principles of proletarian internationalism and socialist internationalism, *see* Tunkin(a), at 4–7, 433–4.
837. Tunkin(a), at 435–6. For a candid view of the Soviet use of socialist international law, *see* Góralczyk, at 478–81.
838. *See* Tunkin(a), at 282.
839. *See* Tunkin(a), at 281, 294–7, 438. For a review of the historical development of socialist international law, *see* Müllerson.
840. Convention on Long Range Transboundary Air Pollution, arts. 2, 3, 4, 7, 8.
841. First Sulphur Dioxide Protocol to the Convention on Long Range Transboundary Air Pollution.
842. Nitrogen Oxide Protocol to the Convention on Long Range Transboundary Air Pollution, art. 2.
843. Second Sulphur Dioxide Protocol to the Convention on Long Range Transboundary Air Pollution.
844. Volatile Organic Compounds Protocol to the Convention on Long Range Transboundary Air Pollution.
845. Ozone Convention, arts. 2, 4; Protocol to the Ozone Convention, arts. 2, 3.
846. *See* Framework Convention on Climate Change; Convention on Biological Diversity, arts. 3, 6, 8. Compliance with the Climate Change Convention

may be the most difficult task for the northern tier CEE states as they are experiencing a rapid rise in motor fuel consumption. *See Oil and Gas Journal* (25 May 1998).

847. Convention on Environmental Impact Assessment in a Transboundary Context, arts. 2, 3.
848. *See* Convention on the Control of Transboundary Movements of Hazardous Wastes and Their Disposal, arts. 1, 4, 6.
849. Convention on the Transboundary Effects of Industrial Accidents.
850. *See* Convention on Nuclear Safety, arts. 1, 7, 14, 16, 8, 19, 20; Convention on Early Notification of a Nuclear Accident, arts. 1, 2, 3, 10.
851. *See generally* Cole(b).
852. Cassese, at 109. *See also* Kozhevnikov ed.
853. Schöpflin, at 265; Fletcher, at 152–3. For review of the political roots of CEE legal nihilism, *see* Cole(b).
854. For a general review of CEE municipal environmental law prior to 1989, *see* Madar(c). For an examination of pre-1989 environmental regulation in Czechoslovakia, *see* Kiesewetter; Kordik; Silar; Leden; Madar(a) and (b); Survey of the Most Important Czechoslovak Legal Regulations in the Sphere of Environmental Care. For an examination of pre-1989 environmental regulation in Poland, *see* Cole(e), at 303–49; Cummings; Starzewska. For an examination of pre-1989 environmental regulation in Hungary, *see* Bándi; Enycdi & Zentai. For an examination of pre-1989 environmental regulation in Romania, *see* Romania Environment Strategy Paper 66–78. For an examination of pre-1989 environmental regulation in Bulgaria, *see* Djolov & Dimitrova; Hunter(a).
855. Haigh *et al.*, at 26. By the late 1970s, the former Czechoslovakia enacted over 350 environmental statutes, few of which, however, were actually enforced. Kiesewetter, at 79. In cases where fines were applied, their amount was generally insufficient to affect the behavior of polluters. Environmental Action Programme, at III-24. *See also* Juergensmeyer *et al.*, at 836–7.
856. For a review of the early efforts to reform environmental law in CEE, *see* Hunter(b); Salvo. For an illustration of the recent developments in Czech environmental law, *see* Environmental Policy of the Czech Republic.
857. Hartwig, at 450. For a review of contemporary environmental law in Poland, *see* Cole(e), at 349–57; Schlickman *et al.*; Sommer(b). For a review of contemporary environmental law in the Czech Republic, *see* Environment of the Czech Republic; P. Kočíková ed. For a review of contemporary environmental law in Slovakia, *see* Slovak Environmental Programme for the Danube River Basin; Krchnak; Enviro Guide Slovakia 1993. For a review of contemporary environmental law in Hungary, *see* Bándi *et al.*; Bándi. For a review of contemporary environmental law in Bulgaria, *see* Hunter *et al.*, eds.
858. In the recent years, all CEE states have adopted substantially reformed constitutions. Hartwig, at 450, 469; *see also* 6 European Constitutional Review (1997). For an examination of the question of whether international legal principles in fact embody justice and fairness, see Franck(b, c).
859. For a review of recent efforts to implement environmental reform in Poland, *see* Brown *et al.*; Cole(f). For a review of recent efforts to implement

environmental reform in Romania, *see* International Market Insight Reports (18 December 1997).
860. Strategic Environmental Issues, at 31.
861. Environmental Needs Assessment, at 59.
862. ODD, No. 168 (29 August 1995).
863. Fidas & Gershwin, at 16–17.
864. Environmental Action Programme, at IV-7.
865. *See* Cole(f), at 190–213.
866. *See* Article 18 of the Constitution of the Republic of Hungary; Article 55 of the Constitution of the Republic of Bulgaria. *See* Casey-Lefkowitz, at 7.
867. *See generally* Krchnack; Salvo.
868. *See* Poland's 1990 amendment to its 1980 Environmental Protection Act, art. 99 and Act on Division of Duties and Competencies in Certain Acts Between Local Self-Government Agencies and State government Administration Agencies (1990); Acts of the Czech National Council Nos. 114/1992 S.B., 244/1992 S.B., 334/1992 S.B., 466/1992 S.B.; Slovak Act on State Environmental Administration, No. 595/1990; No. 453/1992; Hungarian Act on Local Self-Governments, art. 8, 18 (1990); Bulgarian Law on the Management and Administration of Municipalities (1991); Romanian Leg. Decision 264/1991.
869. *See also* Poland's Nature Conservation Act of 1991, and Executive Order on Air Pollution Control of 1990.
870. *See* Czech Republic Air Protection Act 309/1991 S.B, amended in 1992 and 1993, and amendments to Czech Water Act 138/1973 in 1992; *see also* Czech Republic State Administration for Protection of the Air and Fees Levied for Pollution Act 389/1991 S.B.; Clean Water Act, as amended by Act No. 238/1993.
871. *See* Ministry of Environmental, Natural Resources and Forestry Directive 12/1990 (Poland); Slovak Parliamentary Decree No. 242/1993; Slovak Waste Management Act No. 238/1991 paras. 11, 12; Slovak Clean Air Act No. 309/1991.
872. Bándi *et al.*, at 12; Krchnak, at 62–4. For a review of EU efforts to assist CEE states with the enforcement of environmental legislation, *see Business Insurance* (31 August 1998).
873. *See* Czech Republic Income Tax Act Nos. 586/1992 S.B., 35/1993 S.B., and 96/1993 S.B.; Slovak Consumer Taxes Act No. 213/1992; Slovak State Environmental Fund Act No. 176/1992.
874. *See* Environmental Protection Law of Bulgaria, sections 19–20 (1991).
875. *See* Poland's 1990 amendment to 1980 Environmental Protection Act, art. 100 and Law on Nature Protection (1992); Czech Republic Legislation No. 17/1992 S.B. and 244/1992 S.B.; Slovak Law on Environmental Impact Assessment No. 127/1994.
876. Slovak Law on Environmental Impact Assessment No. 127/1994.
877. Romanian Leg. Decision 97/1991 and 264/1991.
878. *See* Poland's 1990 amendment to 1980 Environmental Protection Act, art. 100, and Act on Land Planning, arts 38, 42 (amended 1990); Constitution of the Czech Republic, art. 7, and Declaration of Basic Human Rights and Freedoms, art. 35 established by Decree of the Presidium of the Czech National Council 2/1993 S.B.; Bulgarian Constitution, art. 41(2), providing

'Citizens shall be entitled to obtain information from state bodies and agencies on any matter of legitimate interest to them which is not a state or official secret and does not affect the rights of others'; the Environmental Protection Law of Bulgaria, sections 8, 9 (1991); Romanian Constitution, art. 31. For an evaluation of recent CEE state participation in the UN ECE Convention on Access to Information, Public Participation in Decision-Making, and Access to Justice in Environmental Matters, *see* Jendroska.
879. *See* Poland's 1990 amendment to 1980 Environmental Protection Act, art. 100; Czech Republic Act on the Environment No. 17/1992 S.B., art. 15.
880. Bándi *et al.*, at 17–18; Krchnak, at 68.
881. CTK News Wire (10 September 1997); the *Guardian* (26 September 1997).
882. CTK News Wire (30 March 1998).
883. *See* the Czech Republic's Environmental Act 17/1992 S.B.; Act on the Protection of the Air Against Pollutants 309/1991 S.B.; Act on the Protection of Nature and the Landscape 114/1992 S.B.; Slovak Act on Waste Management No. 238/1991 as amended by Act No. 255/1993; Slovak Clean Air Act No. 31/1995.
884. Bándi *et al.*, at 6.
885. *But see* Article 100 of Poland's Environmental Protection Act of 1991.
886. Krchnak, at 73.
887. *See* Bándi *et al.*, at 2–47.
888. *Euro-East* (19 May 1998); June 1995 EU White Paper on CEE States Accession to the EU. For a review of EU environmental law and policy, *see* Baldock & Keene; Brinkhorst; Crockett & Schultz; European Commission, Environmental Legislation; Haagsma; Haigh(c); Johnson & Corcelle; Kiss & Shelton(b); Krämer(a, b); Lammers(b); Macrory; Sands(d, f); Sands & Tarasofsky; Wagenbaur; Wilkinson.
889. For a review of the efforts of the Czech Republic to harmonize its legislation and practice with EU standards, *see* Kočíková. For example, citing a dependence on groundwater for drinking and annual discharges of 4.2 million tons of untreated hazardous waste, especially around the Danube, Hungary's government has introduced a new environmental effort intended to harmonize domestic standards with EU regimes as soon as 2002. FBIS-EEU, (8 July 1997).
890. *See* Adamová & Maračk, at 10. In 1993, Poland amended its Land Use Planning Act to incorporate EIA requirements. 5 YIEL, at 374.
891. Cole(e), at 356–7. For a review of the extent to which Poland has harmonized its legislation with relevant EU Directives, *see* Polish Ministry of Environmental Protection Natural Resources and Forestry, The State of Harmonization of the Polish Law and Environmental Legislation of the European Union. For a review of the enforcement of EU environmental laws, *see* Ercmann.
892. *See also* Fidas & Gershwin, at 5.
893. For a more detailed examination of the relationship between international and domestic law in the socialist legal system, *see* Rubanov.
894. Czapliński; Cassese, at 111. Although a number of Western states require implementing legislation in order to give effect to nonself-executing treaties, these states do not reject the application of, or require implementing legislation for principles of customary international law. *See* Butler(a).

895. Kopal, at 504–5.
896. Cassese, at 114. CEE socialist states, however, did engage in a fairly widespread use of private international law to regulate matters relating to civil law, family law and judicial assistance. Mavi & Gabor, at 98–102.
897. For an analysis of the correlation of Hungarian municipal environmental law with international environmental law, *see* Bándi *et al.*, at 29, 30, 42, 44, 90.
898. Hartwig, at 454.
899. Hartwig, at 463, 467–8.
900. For a listing of these agreements, *see* Bándi *et al.*, at 35–6. For a continuous review of the adoption of bilateral and multilateral environmental treaties by CEE states, *see* the YIEL series.
901. The constitution also permits foreign citizens to access Hungarian courts. Hartwig, at 461, 462, *citing* arts. 1(c), 21 para. 3, of the Hungarian Constitution.
902. Hartwig, at 454, 465, *citing* Czech and Slovak Federal Constitution, art. 6.
903. Hartwig, at 454.
904. Sand(c), at 125, *citing* Bilder(a), at 139; *See also* Bilder(c); Handl(g), at 71.
905. Kopal, at 501–2; Góralczyk, at 479.
906. Yearbook of the International Court of Justice 1993–94 (1994).
907. Góralczyk, at 480, *citing* Lissitzyn, at 63; Kopal, at 502–4. *See also* Bierzanek. For a Romanian view on the use of dispute settlement mechanisms, *see* Malitza.
908. Shinkaretskaia, at 246.
909. Góralczyk, at 482.
910. *See* Hrivnak, at 35.
911. *See* Usenko.
912. Góralczyk, at 483. *See* Movchan, at 131.
913. For a review of the Soviet view of international law during the late 1980s, immediately prior to the collapse of the Soviet Union, *see* the collected essays in Butler(b).
914. *See* Kopal.
915. Yearbook of the ICJ, Poland, at 112–13 (21 September 1990), Hungary, at 95–6 (7 October 1992), Bulgaria, at 86 (26 May 1992). Austria also submits to the compulsory jurisdiction of the Court, at 83 (28 April 1971). *See also* W. Góralczyk, at 491.
916. Yearbook of the ICJ, at 112–13.
917. Góralczyk, at 494.
918. Kopal, at 507–8.
919. Danube River Protection Convention, art. 24.
920. For a comparison of the Danube River Protection Convention with other river-basin accords, *see* Milich & Varady.
921. Convention on Conciliation and Arbitration Within the OSCE. List of signatories provided by the secretariat of the Court of Conciliation and Arbitration Within the OSCE, 12 February 1998. For more information on the OSCE and its various dispute resolution mechanisms, *see generally* United States Institute of Peace, Roundtable on CSCE: Mechanisms for the Peaceful Settlement of Disputes. For a discussion of the historic development of the OSCE *see* Andren & Birnbaum. For a discussion of more recent

developments in the OSCE and its role in promoting regional stability, *see* McGoldrick. For a review of the limited incorporation of environmental issues into OSCE instruments, *see* Organization for Security and Co-operation in Europe, Provisions on Environment in the CSCE Documents 1973–1989.
922. OSCE Convention, arts. 1, 2. The OSCE also adopted on 15 December 1992, Provisions for an OSCE Conciliation Commission, which could be invoked upon the consent of two or more OSCE member states regardless of their adoption of the OSCE Convention. Provisions for an OSCE Conciliation Commission, Annex 3 to the Summary of Conclusions of the Stockholm Council Meeting of December 14–15, 1992.
923. OSCE Convention, art. 18, 20, 26.
924. Personal correspondence with the Secretariat of the Court of Conciliation and Arbitration within the OSCE, dated 26 February 1998.
925. Polish–Ukrainian Treaty on Protection of the Environment, 5 YIEL, at 372–3; Danube River Protection Convention.
926. Valletta Mechanism, sect. IV, as adopted by the OSCE (8 February 1991).
927. Valletta Mechanism, sects. VI–XIII.
928. *See* Yankov; Bowett; Max Planck Institute, Judicial Settlement of International Disputes, which also reviews the capacity of regional and nonpermanent dispute resolution mechanisms. For a critical review of the existing methods of dispute resolution for transboundary water resource disputes, *see* Chauhan, at 321–86.
929. For an examination of the traditional role of the ICJ in dispute resolution, *see* Elias(b); Gross(a, b); Jennings(c); McWhinney; Scott, *et al.*(b); Tiefenbrun. According to a model developed by Gamble and Fischer in 1976, the ICJ was not at that time well suited to resolve environmental and natural resource disputes. Gamble & Fischer, at 119–21. *But see* Ferrante.
930. *See* the Ambatielos Case, where the Court held the United Kingdom and Greece were required to resolve their dispute through a previously designated arbitration mechanism.
931. North Sea Continental Shelf Case, at 6. For a review of the operation of the ICJ in resolving disputes, *see* Jennings(a, b); Schwebel(a, b).
932. Gulf of Maine Case.
933. Aegean Sea Continental Shelf Case, at 13.
934. Military and Paramilitary Activities Case, at 438–9.
935. Aegean Sea Continental Shelf Case, at 13.
936. Military and Paramilitary Activities Case, at 436–7, 439; United States Diplomatic and Consular Staff in Tehran Case, at 19–20.
937. *See* Continental Shelf Case, at 20.
938. *See* Lauterpacht, at 17, rejecting the view the ICJ is not 'technically' qualified to handle legal disputes within the context of complex nonlegal issues.
939. Neuhold, at 225.
940. Damrosch ed.
941. Góralczyk, at 495. For a brief general response to these and similar assertions, *see* Schwebel. For a review of suggestions for improving the capacity of the Court, *see* Steinberger, at 271–7; Partan.
942. Grabowska, at 143–4. *See also* Postiglione.

943. Postiglione.
944. The United States, for instance, has included coordinated management of the Danube Basin, including environmental protection, as one objective of the Southeast Europe Cooperation Initiative (SECI) launched by the US. FBIS-EEU (25 January 1997); FBIS-EEU (12 April 1997).
945. ODD, No. 208 (25 October 1995).
946. Fidas & Gershwin, at 18.
947. Polish Press Agency (21 April 1998). In addition, in 1997, the EIB invested $8.5 billion for projects relating to environmental projection. *Europe Environment* (10 February 1998).
948. *See* ODD, No. 190 (29 September 1995).
949. At least four CEE states have concluded cooperative financial agreements with the U.S. Export Import Bank (Exim Bank), allowing them to receive preferential loans for the purchase of American exports. ODD No. 190 (29 September 1995). In addition, on 11 December 1995, the EU agreed to provide Slovakia financial assistance in the amount of $159 million over the next five years. ODD, No. 240 (12 December 1995).
950. *See also* Environmental Action Programme, at I-3.
951. *Financial Times* (10 February 1997).
952. ODD No. 194 (5 October 1995).
953. RFE/RLN, No. 77 Part II (22 April 1998). In March 1998, the EBRD rejected Ukraine's proposal to fund the construction of two new nuclear reactors, which according to the government of Ukraine would permit the permanent closure of Chernobyl. RFE/RLN, No. 41, Part II (2 March 1998).
954. RFE/RLN, No. 53, Part II (18 March 1998).
955. For a review of the efforts of the EU to assist CEE states in promoting regional environmental protection, *see Europe Environment* (7 April 1998).
956. *Europe Energy* (25 July 1997).
957. Environmental Needs Assessment, at 32–3, 41.
958. *See*, *supra*, Chapter 2, section III, B, 6.
959. For a general evaluation of the capacity of international law to assist in the resolution of international disputes, *see* Higgins(b). For an evaluation of the capacity of international law to promote environmental protection, *see* Sands(h); Birnie & Boyle; Birnie(c); Developments – International Environmental Law. For an evaluation of the capacity of international law to assist in the resolution of transboundary environmental disputes in specific geographical areas, *see* Cohen; Glinka; Rivera.
960. *See* Tunkin(a), at 446.
961. Poland, for instance, established agreements with Czechoslovakia, Russia, Belarus, Ukraine and the German Democratic Republic. *See* the Agreement on Water Management in International Watercourses Between Poland and Czechoslovakia (21 March 1958); the Agreement on Water Management in International Watercourses Between Poland and the USSR (17 November 1964); the Agreement on Co-operation on Water Management in International Watercourses Between Poland and the GDR (11 March 1965), *cited in* Sommer (c), at 231, 217.
962. On 28 May 1986, Hungary, for example, signed a Convention on the Protection Against Pollution of the River Tisza System, specifically addressing the issue of water quality in the drainage basin of this tributary to the

Danube River. The convention, however, has had little effect on improving water quality as its provisions are deemed to be vague and lack uniformity. Szilagyi.
963. See, e.g., the Agreement on Protection of the Atmosphere Against Pollution Between Poland and Czechoslovakia (24 September 1974); the Agreement on Co-operation in Environmental Protection Between Poland and the GDR (4 June 1973), cited in Sommer(a), at 219, 224.
964. Presentation of Blenesi, Cambridge, 3 March 1995.
965. For a review of the general principles of international law, see Brownlie(b); Jennings & Watts; O'Connell(b); Rousseau; Higgins(a); Schachter(a, c); Schaffer & Snyder.
966. See also Schultz & Crockett, at 55.
967. FBIS-EEU (2 September 1996). The Czech Republic and Germany concluded a similar agreement on cooperation with respect to transboundary waters on 12 December 1995.
968. For identification of these conventions, see Environmental Action Programme, at I-2, I-11.
969. EYBCR 1992, at 329; Poland, the Czech Republic, Germany, and the European Union are parties to the Elbe and Oder River Conventions. See Convention on the International Commission for the Protection of the Elbe, Official Journal of the European Communities No. L 321/25/1991; Convention on the International Commission for the Protection of the Oder (1996).
970. See Danube River Protection Convention, arts. 2, 4, 5, 11, 14.
971. Elbe River Convention, art. 2; Oder River Convention, art. 2.
972. See North Sea Continental Shelf Case; Lac Lanoux Arbitration; Corfu Channel Case. See also Convention on Transboundary Movements of Hazardous Wastes; Convention on Long-Range Transboundary Air Pollution; Convention on the Protection of the Marine Environment of the Baltic Sea Area (1992); Convention on the Protection of the Marine Environment of the Baltic Sea Area (1974); Danube River Protection Convention; Convention on the Protection of the Black Sea Against Pollution. OECD, Application of Information and Consultation Practices. See further Cole(c), at 313; ZumBrunnen, at 400. The Czech Republic maintains bilateral water cooperation agreements with Poland, Germany and Austria. Kordik, at 109. See generally Murphy; Pallemaerts(a); Partan; Smets; Weber.
973. See Trail Smelter Arbitration, at 716; Statement of the Rules of International Law Applicable to Transfrontier Pollution. See also Framework Convention on Climate Change; Convention on Long Range Transboundary Air Pollution and its subsequent protocols. See further Agreement on Protection of the Atmosphere Against Pollution Between Poland and Czechoslovakia (24 September 1974); Agreement on Cooperation in Environmental Protection Between Poland and the GDR (24 June 1973), discussed in Sommer(a), at 205, 219, 224. See generally Gündling(c); Haigh(b); Merrill.
974. See Lac Lanoux Arbitration; Diversion of Water from the Meuse Case; Convention on the Non-Navigational Uses of International Watercourses; Helsinki Rules on the Uses of the Waters of International Rivers. See also the Black Sea Convention; the 1974 and 1992 Baltic Sea Conventions;

Convention on the Protection and Use of Transboundary Watercourses and International Lakes. *See further* Danube River Convention; the Danube River Protection Convention, the Elbe River Convention and the Oder River Convention. *See generally* Benvenisti; Caponera(b); Caponera(c); Gleick; Florio; Handl(a); Kiss(a); Korbut & Baskin; Lammers(a, b); McCaffrey(a, b, d–g); McCaffrey & Sinjela; Sette-Camara; Schwabach; Sergent; Szilgyi; Teclaff; Teclaff & Teclaff(a); Wouters; Zavadsky.

975. Szilagyi, at 42.
976. FBIS-EEU (15 July 1993).
977. For a review of the more general barriers to the inherent functionality of international air pollution law, *see* Vukas; Rest(a).
978. Environment of the Czech Republic, at 44.
979. *See* dissenting opinion of Judge Koroma, at 6–7, for an acknowledgment of the difficult of balancing the various factors relevant to determining an equitable utilization.
980. International law traditionally has recognized the award of damages for riparian interferences adversely affecting domestic water supply, hydroelectric generation, flood control, fishing resources and navigation. Bush, at 315–21.
981. *Euro-East* (23 September 1997). The Czech Republic, which is one of the leading CEE states in terms of environmental investment, has increased its investment from 1 per cent of GDP in 1990 to 2.6 per cent in 1996. Czech Republic Ministry of the Environment, Environmental Policy of the Czech Republic.
982. Convention on Long Range Transboundary Air Pollution, art. 2, par. 3a; principles 20, 21 and 22 of the Declaration of the United Nations Conference on Environment and Development.
983. *See* Declaration of the United Nations Conference on the Human Environment; World Charter for Nature; Declaration on Environment and Development. *See also* Danube River Protection Convention, art. 2; Polish–Ukrainian Treaty on Protection of the Environment, 5 YIEL, at 372–3, art. 2. *See generally* Clark & Munn; Conable; Di Leva; Munro & Holdgate; M'Gonigle.
984. OECD, Managing The Environment: The Role of Economic Instruments, at 42; OECD, Use of Economic Instruments in Environmental Policy; OECD, The Polluter Pays Principle, Definition, Analysis, Implementation; Declaration on Environment and Development, principle 16. *See also* Danube River Protection Convention, art. 2; Polish–Ukrainian Treaty on Protection of the Environment, art. 2. *See generally* Gaines.
985. Nuclear Tests Case, at 342–4 (Judge Gregory Weeramantry dissent); Bergen ECE Ministerial Declaration on Sustainable Development, in Basic Documents of International Environmental Law, at 558–9. *See also* Danube River Protection Convention, art. 2. *See generally* Bodansky(a); Cameron & Abouchar; Gündling(b); Freesonte; Hey(a, b).
986. Nuclear Tests Case, at 344–5; Convention on Environmental Impact Assessment; Convention on the Protection and Use of Transboundary Watercourses and International Lakes, art. 7; Czech Republic statute No. 17/1992 S.B. (16 January 1992), EYBCR 1992, at 309; Sommer(b), at 14–15. *See also* Colombo; Futrell; Kennedy; OECD, Application of Information

278 *Notes*

and Consultation Practices; Pallemaerts(a); Pineschi; Sheate; Stewart; Wathern.
987. Chorzów Factory Case, at 29; ILC, Report on International Liability for Injurious Consequences Arising Out of Acts Not Prohibited by International Law; ILC, Report on State Responsibility; 'Report of the ILC on the work of its forty-eight session, 6 May to 26 July 1996'. *See* Convention on Transboundary Movements of Hazardous Wastes. *See also* Agreement on Protection of the Environment, signed 1 July 1989 by Germany, the Czech Republic and Poland providing in part they agree to discuss developing rules of liability relating to harm caused by transboundary air emissions. Grabowska, at 140. *See also* Polish–Ukrainian Treaty on Protection of the Environment, art. 13 which contains similar provisions. *See generally* Allott; d'Arge & Kneese; Boyle(b); Brownlie(a); de Sola; Doeker & Gehring; P.-M. Dupuy(a, b); European Commission, Communication on Environmental Liability; Francioni & Scovazzi; Gilbert; Handl(a, e); Hoffman; Kiss(c); McCaffrey(c); Murphy; Perkins; Rest(a, b, c); Rosas; Szasz(b); Zemanek.
988. As members of the IAEA, CEE states generally attempt to operate their nuclear power plants according to the nonbinding standards set by the agency and are subject to verification inspections if they receive assistance from the agency. For a review of the operation of the IAEA, *see* Szasz(a). *See also* Convention on Nuclear Safety; Convention on the Early Notification of a Nuclear Accident; Polish–Ukrainian Treaty on Protection of the Environment, requiring joint EIAs for nuclear power development; Birnie & Boyle, at 348–55. Upon admission to the EU, certain CEE states will be required to operate their nuclear power plants in compliance with the Euratom Treaty and safety standards set by Euratom. Treaty Establishing the European Atomic Energy Community. *See generally* Adede; Boyle(a, c); Handl(f); Handl & Lutz; Sands(a).
989. This tension is highlighted by Judge Oda's pronouncement in his Dissenting Opinion, at 7, that, 'Any construction work relating to economic development would be bound to affect the existing environment to some extent but modern technology would, I am sure, be able to provide some acceptable ways of balancing the two conflicting interests.'
990. For a similar criticism of the general applicability of the precautionary principle, *see* Birnie & Boyle, at 98. *See also* Gündling(a). The requirement for joint participation within an international EIA also tends to lack dispositional relevance in that environmental actors frequently litigate over the procedural aspects of a domestic EIA in order to force a negotiation on the substantive matters of the project, yet this option is not available in the context of an international EIA. Goldberg, at 403.
991. Friends of the Earth – Poland, Mochovce Press Release (5 January 1995).
992. *See also* Oral Pleadings of Slovakia, Samuel Wordsworth, at 34–6 (14 April 1997) responding to oral pleadings by Katherine Gorove.
993. For a similar criticism concerning the principles and rules of state responsibility/liability in general, *see* Sands(h), at 633–7.
994. For a general evaluation of the capacity of the international law of state responsibility/liability, *see* Kiss(c).
995. *See* Hungarian Reply, at 178; *See also* Dissenting Opinion of Judge Koroma, at 2.

996. Kramer(f); Birnie & Boyle, 361–2.
997. For an examination of the emerging human right to a clean and safe environment, *see* R. Dupuy(a); Gormely(a, b); Hodkova; Kiss(d); Pathak; Popovic; Shelton; Tamasevski; Taylor; Trindade; Vukasovic. For a review of principles relating to intergenerational equity, which might also be applicable in transboundary environmental disputes, *see* D'Amato(b); Brown Weiss(a, b, c, e); Gündling(a); Supanich.
998. Sommer(a), at 228.
999. *See* The Role of Economic Instruments, at 87–8; Economic Instruments in Environmental Policy, at 2; Goldberg, at 403. For a review of the operation of marketable emissions permits, *see* OECD, Guidelines and Considerations for the Use of Economic Instruments in Environmental Policy 1, 4.
1000. *See also* Schroeer, at 200.
1001. EYBCR 1992, at 323.
1002. Stypka, at 29.
1003. Goldmark & LaRocco, at 77.
1004. Environmental Action Programme, at III-23.
1005. *See* Economic Instruments in Environmental Policy.

Bibliography

Books and academic journals

K. Abbott, 'Modern International Relations Theory: A Prospectus for International Lawyers', 14 *Yale JIL* 335 (1989).

W. Aceves, 'Institutionalist Theory and International Legal Scholarship', 12 *American University JILP* 227 (1997).

E. Adamová & J. Maračk, 'Harmonization of Czech with EC Law', 2 *RECIEL* 9 (1993).

A. Adede, *The IAEA Notification and Assistance Conventions in Case of a Nuclear Accident: Landmarks in the History of the Multilateral Treaty-Making Process* (1987).

V. Aggarwal & P. Allan, 'Preferences, Constraints and Games: Analyzing Polish Debt Negotiations with International Banks', in *Game Theory and International Relations: Preferences, Information and Empirical Evidence* 9 (P. Allan & C. Schmidt eds, 1994).

N. Åkerman ed., *International Environmental Policy* 1 (1990).

C. Albin, 'The Role of Fairness in Negotiation', *Negotiation Journal* (July 1993). Albin(a).

C. Albin, *Negotiating the Acid Rain Problem in Europe: A Fairness Perspective* (IIASA Working Paper 93-050, 1993). Albin(b).

R. Alford & P. Bekker, 'International Courts and Tribunals', 32 *International Lawyer* 499 (Summer 1998).

D. Alheritiere, 'Settlement of Public International Disputes on Shared Resources: Elements of a Comparative Study of International Instruments', in *Transboundary Resources Law* 139 (1987).

P. Allan & C. Schmidt eds, *Game Theory and International Relations: Preferences, Information and Empirical Evidence* (1994).

G. Allison, *Essence of Decision: Explaining the Cuban Missile Crisis* (1971).

P. Allott, 'State Responsibility and the Unmaking of International Law', 29 *Harvard ILJ* 1 (1988). Allott(a).

P. Allott, *Eunomia: A New Order for a New World* (1990). Allott(b).

J. Alvarez, 'Why Nations Behave', 19 *Michigan JIL* 303 (Winter 1998).

J. Anderman, 'Swimming the New Stream: The Disjunction Between and Within Popular and Academic International Law', 6 *Duke JCIL* 293 (Spring 1996).

H. Anderson, III, 'International Environmental Law at the Grassroots', 11 *Tulane ELJ* 109 (1997).

N. Andren & K. Birnbaum eds, *Belgrade and Beyond: The CSCE Process in Perspective* (1980).

A. Arend, 'Do Legal Rules Matter? International Law and International Politics', 38 *Virginia JIL* 107 (Winter 1998).

R. d'Arge & A. Kneese, 'State Liability for International Environmental Degradation: An Economic Perspective', 20 *NRJ* 427 (1980).

A. Aronson, 'From "Cooperator's Loss" To Cooperative Gain: Negotiating Greenhouse Gas Abatement', 102 *Yale LJ* 2143 (June 1993).
J. Ausubel & D. Victor, 'Verification of International Environmental Agreements', 17 *Annual Review of Energy and the Environment* 1 (1992).
E. Azar & J. Burton, *International Conflict Resolution: Theory and Practice* (1986).
D. Baldock & M. Wenning, 'The EC Structural Funds – Environmental Briefing 2' (WWF–UK & IEEP, London 1990).
D. Baldock & E. Keene, 'Incorporating Environmental Considerations in Common Market Arrangement', 23 *Environmental Law* 575 (1993).
G. Bándi, 'The Development of Environmental Law in Hungary', 18 *RCEEL* 57 (1992).
G. Bándi et al., *Environmental Management and Law Association, Environmental Law and Management System in Hungary: Overview, Perspective and Problems* 1 (June 1993).
A. Barnes, 'The Growing International Dimension to Environmental Issues', 13 *Columbia JEL* 389 (1988).
E. Barratt-Brown, 'Building a Monitoring and Compliance Regime Under the Montreal Protocol', 16 *Yale JIL* 519 (1991).
J. Barros & D. Johnston eds, *The International Law of Pollution* (1983).
R. Beck, 'International Law and International Relations: The Prospects for Interdisciplinary Collaboration', 1 *Journal of International Legal Studies* 119 (Summer 1995).
R. Beck et al., *'International Rules: Approaches from International Law and International Relations* (1996).
P. Bekker, 'Gabcikovo–Nagymaros Project (Hungary/Slovakia), Judgement', 92 *American JIL* 273 (April 1998).
J. Bendor & T. Hammond, 'Rethinking Allison's Models', 86 *APSR* 301 (1992).
E. Benvenisti, 'Collective Action in the Utilization of Shared Freshwater: The Challenges of International Water Resources Law', 90 *American JIL* 384 (July 1996).
L. Bergman, 'Some Basic Issues in International Environmental Policy', in *Maintaining a Satisfactory Environment: An Agenda for International Environmental Policy* 55 (N. Åkerman ed., 1990).
L. Bergman et al., 'A Scheme for Sharing the Costs of Reducing Sulphur Emissions in Europe' (IIASA Working Paper 90-005, January 1990).
G. Berrisch, 'The Danube Dam Dispute Under International Law', 46 *Austrian JPIL* 231 (1994).
G. Bethlem & M. Faure, 'Environmental Toxic Torts in Europe: Some Trends in Recovery of Soil Clean-Up Costs and Damages for Personal Injury in the Netherlands, Belgium, England and Germany', 10 *Georgetown IELR* 855 (1998).
P. Bhagwati, 'Environmental Disputes', in *The Effectiveness of International Environmental Agreements* 436 (P. Sand ed., 1992).
R. Bierzanek, 'Some Remarks on the Function of International Courts in the Contemporary World', 7 *Polish YIL* 121 (1975).
R. Bilder, 'The Settlement of Disputes in the Field of the International Law of the Environment', 144 *RCADI* 139 (1975). Bilder(a).
R. Bilder, 'The Role of Unilateral State Action in Preventing International Environmental Injury', 14 *Vanderbilt JTL* 51 (1981). Bilder(b).
R. Bilder, 'An Overview of International Dispute Settlement', 1 *Emory Journal of International Dispute Resolution* 23 (1986). Bilder(c).

J. Binder, 'Daming Evidence: Gabčíkovo, the Case For', 1993 *East European Reporter* 76 (September/October 1993).
P. Birnie, 'The Role of International Law in Solving Certain Environmental Conflicts', in *International Environmental Diplomacy* 95 (J. Carroll ed., 1988). Birnie(a).
P. Birnie, 'Legal Techniques of Settling Disputes: The "Soft Settlement" Approach', in *Perestroika and International Law* 177 (W. Butler ed., 1990). Birnie(b).
P. Birnie, 'International Environmental Law: Its Adequacy for Present and Future Needs', in *The International Politics of the Environment: Actors, Interests, and Institutions* (A. Hurrell & B. Kingsbury eds, 1992). Birnie(c).
P. Birnie & A. Boyle, *International Law and the Environment* (1992).
L. Björkbom, 'Resolution of Environmental Problems: Use of Diplomacy', in *International Environmental Diplomacy* 123 (J. Carroll ed., 1988).
P. Bliss-Guest, 'The Protocol against Pollution from Land-Based Sources: a Turning Point in the Rising Tide of Pollution', 17 *Stanford JIL* 261 (1981).
B. Boczek, 'International Protection of the Baltic Sea Environment Against Pollution: A Study in Marine Regionalism', 72 *American JIL* 782 (1978). Boczek(a).
B. Boczek, 'Global and Regional Approaches to the Protection and Preservation of the Marine Environment', 16 *Case Western Reserve JIL* 39 (1984). Boczek(b).
D. Bodansky, 'Scientific Uncertainty and the Precautionary Principle', 33 *Environment* 4 (1991). Bodansky(a).
D. Bodansky, 'Protecting the Marine Environment from Vessel-Source Pollution: UNCLOS III and Beyond', 18 *Ecology LQ* 719 (1991). Bodansky(b).
E. Bohm & M. Szucsich, 'Developments in Central and Eastern Europe', 18 *Comparative Lab L* 182 (1997).
M. Bos, *A Methodology of International Law* (1984).
I. Bostian, 'Flushing the Danube: The World Court's Decision Concerning the Gabcikovo Dam', 9 *Colorado JIELP* 401 (Summer 1998).
M. Bothe, 'Transfrontier Environmental Management', in *Trends in Environmental Policy and Law* 391 (M. Bothe ed., 1980). Bothe(a).
M. Bothe, 'International Legal Problems of Industrial Siting in Border Areas and National Environmental Policies', in *Transfrontier Pollution and the Role of States* 79 (M. Bothe ed., 1981). Bothe(b).
S. Bowers, 'Ethnic Politics in Eastern Europe', in *Ethnic and Religious Conflicts: Europe and Asia* 29 (P. Janke ed., 1994).
D. Bowett, 'Contemporary Developments in Legal Techniques in the Settlement of Disputes', 180 *RCADI* 169 (1983).
A. Boyle, 'Nuclear Energy and International Law: An Environmental Perspective' 60 *BYIL* 257 (1989). Boyle(a).
A. Boyle, 'State Responsibility and International Liability for Injurious Consequences of Acts Not Prohibited by International Law: A Necessary Distinction?', 39 *ICLQ* 1 (1990). Boyle(b).
A. Boyle, 'Chernobyl and the Development of International Environmental Law', in *Perestroika and International Law* 203 (W. Butler ed.,1990). Boyle(c).
A. Boyle, 'Saving the World: Implementation and Enforcement of International Environmental Law Through International Institutions', 3 *Journal of Environmental Law* 229 (1991). Boyle(d).
F. Boyle, *World Politics and International Law* (1985).
J. Brierly, *The Law of Nations* (1963).

L. Brinkhorst, 'The Road to Maastricht', 20 *Ecology LQ* 7 (1993).
Z. Brodecki, 'Damage to the Baltic Sea: The Future of International Liability', in *Pollution of the Baltic Sea* 16 (1988).
E. Brown Weiss, *In Fairness to Future Generations: International Law, Common Patrimony, and Intergenerational Equity* (1989). Brown Weiss(a).
E. Brown Weiss, 'Our Rights and Obligations to Future Generations for the Environment', 84 *American JIL* 198 (1990). Brown Weiss(b).
E. Brown Weiss ed., *Environmental Change and International Law* (1992). Brown Weiss(c).
E. Brown Weiss, *'Multidisciplinary Perspectives on The Improvement of International Environmental Law and Institutions, Environmental Change and International Law: New Challenges and Dimensions'* (Spring 1994). Brown Weiss(d).
E. Brown Weiss, 'A Reply to Barresi's "Beyond Fairness to Future Generations"', 11 *Tulane ELJ* 89 (Winter 1997). Brown Weiss(e).
I. Brownlie, *System of the Law of Nations: State Responsibility* (1983). Brownlie(a).
I. Brownlie, 'The Relations of Law and Power', in *Contemporary Problems of International Law: Essays in Honour of Georg Schwarzenberger on his Eightieth Birthday* 19 (B. Cheng & E. Brown eds, 1988). Brownlie(b).
I. Brownlie, *Principles of Public International Law* (4th ed., 1990). Brownlie(c).
J. Brunnée, *Acid Rain and Ozone Layer Depletion: International Law and Regulation* (1988). Brunnée(a).
J. Brunnée, 'The Jigsaw Puzzle of International Environmental Protection: International Approaches to Atmospheric Pollution and the Baltic Sea Area', 20 *International Journal of Legal Information* 1 (1992). Brunnée(b).
J. Brunnée & S. Toope, 'Environmental Security and Freshwater Resources: Ecosystem Regime Building', 91 *American JIL* 26 (January 1997).
J. Bryden, 'Environmental Rights in Theory and Practice, 62 *Minnesota LR* 163 (1978).
M. Bulla & I. Juhasz eds, *State of the Environment in Hungary and Environmental Policy* (1989).
W. Burhenne ed., *International Environmental Soft Law: Collection of Relevant Instruments* (1993).
O. Von Busekist, 'A Bridge Between Two Conventions on Civil Liability for Nuclear Damage: The Joint Protocol Relating to the Application of the Vienna Convention and the Paris Convention', 43 *Nuclear Law Bulletin* 10 (1990).
W. Bush, 'Compensation and the Utilization of International Rivers and Lakes: The Role of Compensation in the Event of Permanent Injury to Existing Uses of Water', in *The Legal Regime of International Rivers and Lakes* 315 (R. Zacklin & L. Caflisch ed., 1981).
J. Busterud, 'International Environmental Relations', 7 *NRL* 325 (1974).
W. Butler, 'International and Municipal Law: Some Reflections on British Practice', in *International Law and the International System* 67 (ed., 1987). Butler(a).
W. Butler ed., *Perestroika and International Law* (1990). Butler(b).
A. Button, 'Prerequisite to Peace: An International Environmental Ethos', 59 *Tennessee LR* 681 (1992).
M. Byers, 'Custom, Power, and the Power of Rules', 17 *Michigan JIL* 109 (1995). Byers(a).
M. Byers, 'Response: Taking the Law Out of International Law: A Critique of the "Iterative Perspective"', 38 *Harvard ILJ* 201 (Winter 1997). Byers(b).

D. Caldwell, *The Dynamics of Domestic Politics and Arms Control: The SALT II Treaty Ratification Debate* (1991).
L. Caldwell, *International Environmental Policy* (1984). Caldwell(a).
L. Caldwell, 'Beyond Environmental Diplomacy: The Changing Institutional Structure of International Cooperation', in *International Environmental Diplomacy* (J. Carroll ed., 1988). Caldwell(b).
L. Caldwell, 'The Geopolitics of Environmental Policy: Transnational Modification of National Sovereignty', 59 *RJUPR* 693 (1990). Caldwell(c).
J. Cameron & J. Abouchar, 'The Precautionary Principle: A Fundamental Principle of Law and Policy for the Protection of the Global Environment', 14 *Boston College ICLR* 1 (1991).
A. Cancado Trindade ed., *Human Rights, Sustainable Development and the Environment* (1992).
D. Caponera, 'Patterns of Cooperation in International Water Law: Principles and Institutions', in *Transboundary Resources Law* 1 (1987). Caponera(a).
D. Caponera, *Principles of Water Law and Administration, National and International* (1992). Caponera(b).
D. Caron, 'Protection of the Stratospheric Ozone Layer and the Structure of International Environmental Lawmaking', 14 *Hastings ICLR* 755 (1991).
E. Carr, *The Twenty Years' Crisis 1919–1939* (2d ed. 1961).
J. Carroll ed., *International Environmental Diplomacy: The Management and Resolution of Transfrontier Environmental Problems* (1987).
F. Carter & D. Turnock eds, *Environmental Problems in Eastern Europe* (1993).
A. Cassese, *International Law in a Divided World* (1986).
B. Cassidy, 'Cleaning up Eastern Europe: Proposals for Coordinated European Hazardous Waste Management Regime', 12 *Virginia ELJ* 185 (1993).
L. Cederman, 'Unpacking the National Interest: An Analysis of Preference Aggregation in Ordinal Games', in *Game Theory and International Relations: Preference Aggregation and Empirical Evidence* 50 (P. Allan & C. Schmidt eds, 1994).
S. Charnovitz, 'Two Centuries of Participation: NGOS and Internaitonal Governance', 18 *Michigan JIL* 183 (Winter 1997).
S. Charnovitz, 'Participation of Nongovernmental Organizations in the World Trade Organization', 17 *University of Pennsylvania Journal of International Economic Law* 331 (Spring 1996).
B. Chauhan, *Settlement of International Water Law Disputes in International Drainage Basins* (1981).
A. Chayes & A. Chayes, 'Adjustment and Compliance Processes in International Regulatory Regimes', in *Preserving the Global Environment* 280 (J. Mathews ed., 1991).
A. Chayes & A. Chayes, 'The New Sovereignty: Compliance with International Regulatory Agreements', 37 *Virginia JIL* 789 (Spring 1997).
L. Chen, *An Introduction to Contemporary International Law: A Policy-Oriented Perspective* (1989).
D. Chenevert, 'Application of the Draft Articles on the Non-Navigational Uses of International Watercourses to the Water Disputes Involving the Nile River and the Jordan River', 6 *Emory ILR* 495 (1992).
C. Chinkin, 'The Challenge of Soft Law: Development and Change in International Law', 38 *ICLQ* 850 (1989).

C. Chinkin & R. Sadurska, 'The Anatomy of International Dispute Resolution', 7 *Ohio St. Journal on Dispute Resolution* 39 (1991).
H. Chodosh, 'An Interpretive Theory of International Law: The Distinction Between Treaty and Customary Law', 28 *Vanderbilt JTL* 973 (November 1995).
M. Choo, 'An Institutionalist Perspective on Resolving Trade–Environmental Conflicts', 12 *Journal of Environmental Law on Litigation* 433 (1997).
W. Clark & R. Munn eds, *Sustainable Development of the Biosphere* (1986).
J. Cohen, 'International Law and the Water Politics of the Euphrates', 24 *ILP* 502 (1991).
R. Cohen, Justice and Negotiation, *3 Research on Negotiation in Organizations* (1991).
D. Cole, 'Cleaning up Kraków: Poland's Ecological Crisis and the Political Economy of International Environmental Assistance', 2 *Colorado IELP* 205 (1991). Cole(a).
D. Cole, 'Marxism and the Failure of Environmental Protection in Eastern Europe and the USSR', 27 *Legal Studies Forum* 35 (1993). Cole(b).
D. Cole, 'Poland's Progress: Environmental Protection in a Period of Transition', 2 *The Parker School Journal of East European Law* 279 (1995). Cole(c).
D. Cole, 'Environmental Protection and Economic Growth: Lessons from Socialist Europe', in *Law and Economics: New and Critical Perspectives* (R. Malloy & C. Braun eds, 1995). Cole(d).
D. Cole, 'An Outline History of Environmental Law and Administration in Poland', *Hastings ICLR* 297 (1995). Cole(e).
D. Cole, *Instituting Environmental Protection: From Red to Green in Poland* (1998). Cole(f).
R. Collin, 'Where did all the blue skies go?', 9 *Journal of Environmental Law and Litigation* 399 (1994).
A. Colombo ed., *Environmental Impact Assessment* (1992).
Comment, 'The Convention on Long-Range Transboundary Air Pollution: Meeting the Challenge of International Cooperation', 39 *Harvard ILJ* 447 (1989).
B. Conable, 'Development and the Environment: A Global Balance', 5 *American University JILP* 217 (1990).
C. Cooper, 'The Management of International Environmental Disputes in the Context of Canada–United States Relations', 1986 *CYIL* 247 (1986).
A. Craik, '"Recalcitrant Reality and Chosen Ideas": The Public Function of Dispute Settlement in International Environmental Law', 10 *Georgetown ILR* 551 (1998).
M. Crane, 'Diminishing Water Resources and International Law: U.S.–Mexico, A Case Study', 24 *Cornell ILR* 299 (1991).
T. Crockett & C. Schultz, 'The Integration of Environmental Policy and the European Community: Recent Problems in Implementation and Enforcement' 29 *Columbia JTL* 169 (1991).
S. Cummings, 'Polish Environmental Regulation: The State of Poland's Environment, Governmental Authorities and Policy', 16 *Suffolk TLR* 379 (1993).
C. Curtis, 'Legality of Seabed Disposal of High-level Radioactive Waste under the London Dumping Convention', 14 *Ocean Development & IL* 383 (1985).
W. Czapliński, 'Relations Between International Law and the Municipal Legal Systems of European Socialist States', 14 *Review of Socialist Law* 105 (1988).
K. Dahlberg *et al.* eds, *Environment and the Global Arena: Actors, Values, Policies, Futures* (1983).

A. D'Amato, 'Is International Law Part of Natural Law?', 9 *Vera Lex* 8 (1989). D'Amato(a).
A. D'Amato, 'Agora: What Obligation Does Our Generation Owe to the Next? An Approach to Global Environmental Responsibility', 84 *AJIL* 190 (1990). D'Amato(b).
L. Damrosch ed., *The International Court of Justice at a Crossroads* (1987). Damrosch(a).
L. Damrosch, 'Multilateral Disputes', in *The International Court of Justice at a Crossroads* 376 (L. Damrosch ed., 1987). Damrosch(b).
R. d'Arge & A. Kneese, 'State Liability for International Environmental Degradation: An Economic Perspective', 20 *NRJ* 427 (1980).
J. Darmstadter & R. Fri, 'Interconnections Between Energy and the Environment: Global Challenges', 17 *Annual Review of Energy and the Environment* 45 (1992).
J. DeBardeleben, *The Environment and Marxism-Leninism: The Soviet and East German Experience* (1985).
L. de La Fayette, 'Nuclear Liability Revisited', 1 *RECIEL* 443 (1992).
R. Delaney, *This is Communist Hungary* (1958).
A. DeMestral, 'The Prevention of Pollution of the Marine Environment Arising from Offshore Mining and Drilling', 20 *Harvard ILJ* 469 (1979).
A. DeRouw, 'Emergency Response to Maritime Pollution Incidents: Legal Aspects', 25 *Proceedings of the Law of the Sea Institute* 325 (1993).
C. de Sola, 'The Council of Europe Convention on Environmental Damage', 1 *RECIEL* 411 (1992).
K. Deutsch & S. Hoffmann eds, *The Relevance of International Law* (1971).
'Developments – International Environmental Law', 104 *Harvard LR* 1487 (1990).
J. De Yturriaga, 'Regional Conventions on the Protection of the Marine Environment', 162 *RCADI* 319 (1979).
C. Di Leva, 'International Environmental Law and Development', 10 *Georgetown IELR* 501 (1998).
G. Djolov & D. Dimitrova, 'The People's Republic of Bulgaria', in *Environmental Policies in East and West* 53 (G. Enyedi *et al.* eds, 1987).
G. Doeker & T. Gehring, 'Private or International Liability for Transnational Environmental Damage – the Precedent of Conventional Liability Regimes', 2 *JEL* (1990).
G. Downs, 'Enforcement and the Evolution of Cooperation', 19 *Michigan JIL* 319 (Winter 1998).
M. Doyle & G. Ikenberry, *New Thinking in International Relations Theory* (1997).
R. Duncan, 'The 1972 Convention on the Prevention of Marine Pollution by Dumping of Wastes at Sea', 5 *Journal of Maritime Law & Commerce* 299 (1974).
J. Dunoff, 'Institutional Misfits: The GATT, The ICJ & Trade–Environment Disputes', 15 *Michigan JIL* 1043 (Summer 1994).
C. Dupont, 'The Rhine: A Study of Inland Water Negotiations', in *International Environmental Negotiation* 135 (G. Sjöstedt ed., 1993). Dupont(a).
C. Dupont, 'Domestic Politics and International Negotiations: A Sequential Bargaining Model', in *Game Theory and International Relations: Preferences, Information and Empirical Evidence* 156 (P. Allan & C. Schmidt eds, 1994). Dupont(b).
P.-M. Dupuy, 'International Liability for Transfrontier Pollution', in *Trends in Environmental Policy and Law* (1980). Dupuy(a).

P.-M. Dupuy, 'The International Law of State Responsibility: Revolution or Evolution?', 11 *Michigan JIL* 105 (1989). Dupuy(b).
P.-M. Dupuy, 'Overview of Existing Customary Legal Regime Regarding International Pollution', in *International Law Pollution* (D. Magraw ed., 1991). Dupuy(c).
P.-M. Dupuy, 'Soft Law and the International Law of the Environment', 12 *Michigan JIL* 420 (1991). Dupuy(d).
R. Dupuy ed., *The Rights to Health and Human Right* (1979). R. Dupuy(a).
R. Dupuy, *The Future International Law of the Environment* (1985). R. Dupuy(b).
G. Eckstein, 'Application of International Water Law to Transboundary Groundwater Resources, and The Slovak–Hungarian Dispute Over Gabickovo–Nagymaros', 19 *Suffolk TLR* 67 (Winter 1995).
'Economic Instruments in Environmental Policy', 20 *Environmental Policy and Law* 140 (1990).
C. Edwards, 'In Search of Legal Scholarship Strategies for the Integration of Science into the Practice of Law', 8 *Southern California Interdisciplinary LJ* (Winter 1998).
M. Efinger & M. Zürn, 'Explaining Conflict Management in East–West Relations', in *International Regimes in East–West Politics* 73 (V. Rittberger ed., 1990).
M. Ehrlich, 'Towards a New Dialogue Between International Relations Theory and International Trade Theory', 2 *UCLA JIL and Foreign Affairs* 249 (Fall/Winter 1997–1998).
T. Elias, *The Modern Law of Treaties* (1974). Elias(a).
T. Elias, *The International Court of Justice and Some Contemporary Problems* (1983). Elias(b).
J. Endicott, 'The 1975–76 Debate Over Ratification of the Nuclear Non-Proliferation Treaty in Japan', 17 *Asian Survey* 275 (1977).
G. Enyedi & V. Zentai, 'The Hungarian Peoples Republic', in *Environmental Policies in East and West* 213 (G. Enyedi *et al.* eds, 1987).
G. Enyedi *et al.* eds, *Environmental Policies in East and West* (1987).
Equipe Cousteau, *The Gabčíkovo Dam: A Textbook Case* (September 1992). Equipe Cousteau(a).
Equipe Cousteau, *The Danube: For Whom and for What?* (March 1993). Equipe Cousteau(b).
D. Esty, 'Linkages and Governance: NGOs at The World Trade Organization', 19 *University of Pennsylvania Journal of International Economic Law* 709 (Fall 1998).
S. Evans, 'Control of Marine Pollution Generated by Offshore Oil and Gas Exploration and Exploitation', 10 *Marine Policy* 82 (1986).
R. Falk, *The Status of Law in International Society* (1970).
M. Falkernmark, 'Fresh Waters as a Factor in Strategic Policy and Action', in *Global Resources and International Conflict* 85 (1986).
B. Fassbender, 'International Law and International Relations Theory: Building Bridges', 86 *Proceed. ASIL* 167 (1–4 April 1992).
J. Fawcett & A. Parry, *Law and International Resource Conflicts* 167 (1981).
A. Ferrante, 'The Dolphin/Tuna Controversy and Environmental Issues: Will the World Trade Organization's "Arbitration Court" and the International Court of Justice's Chamber for Environmental Matters Assist the United States and the World in Furthering Environmental Goals?', 5 *JTL & Policy* 279 (Spring 1996).

G. Fidas & L. Gershwin, *Intelligence Community Assessment: The Environmental Outlook in Central and Eastern Europe* (National Intelligence Council, December 1997).

D. Fisher, *Paradise Deferred: Environmental Policymaking in Central and Eastern Europe* (1992).

R. Fisher, 'Bringing Law to Bear on National Governments', 74 *Harvard LR* 1130 (1961).

J. Fitzmaurice, *Damming the Danube: Gobčíkovo and Post-communist Politics in Europe* (1998).

M. Fitzmaurice, *International Legal Problems of the Environmental Protection of the Baltic Sea* (1991). Fitzmaurice(a).

M. Fitzmaurice, 'The 1992 Convention on the Baltic Sea Environment', 2 *RECIEL* 24 (1993). Fitzmaurice(b).

T. Fleischer, 'Jaws on the Danube: Water Management, Regime Change and the Movement Against the Middle Danube Hydro-electric Dam', 17 *IJ of Urban and Regional Research* 429 (1993).

G. Fletcher, 'Searching for the Rule of Law in the Wake of Communism', 1992 *Bringham Young University LR* 145 (1992).

C. Flinterman *et al.* eds, *Transboundary Air Pollution* (1986).

F. Florio, 'Water Pollution and Related Principles of International Law', 17 *CYIL* 134 (1979).

A. Flournay, 'Legislative Inaction: Asking the Wrong Questions in Protective Environmental Decisionmaking', 15 *Harvard ELR* 327 (1991).

M. Forster, 'Hazardous Waste: Towards International Agreement', 12 *Environmental Policy and Law* 64 (1984).

A. Fraenkel, 'The Convention on Long-Range Transboundary Air Pollution: Meeting the Challenge of International Cooperation', 30 *Harvard ILJ* 447 (1989).

F. Francioni, 'International Co-operation for the Protection of the Environment: The Procedural Dimension', in *Environmental Protection and International Law* 203 (W. Lang *et al.* eds, 1991).

F. Francioni & T. Scovazzi eds, *International Responsibility for Environmental Harm* (1991).

T. Franck, *The Power of Legitimacy Among Nations* (1990). Franck(a).

T. Franck, 'Fairness in the International Legal and Institutional System', 240 *RCADI* 13 (1993). Franck(b).

T. Franck, 'Is International Law Fair?', 17 *Michigan JIL* 615 (Spring 1996). Franck(c).

D. Freestone, 'The Precautionary Principle', in *International Law and Global Climate Change* 21 (R. Churchill & D. Freestone eds, 1991).

H. French, *Green Revolutions: Environmental Reconstruction in Eastern Europe and the Soviet Union* (1990).

Friends of the Earth – *Poland, Mochovce Press Release* (5 January 1995).

R. Frydman *et al.*, *Capitalism with a Comrade's Face* (1998).

M. Führ & G. Roller eds, *Participation and Litigation Rights of Environmental Associations in Europe: Current Legal Situation and Practical Experience* (1991).

W. Futrell, 'Environmental Assessment: The Necessary First Step in Successful Environmental Strategies', 10 *UCLA Pacific Basin LJ* 234 (1991).

I. Gadourek, *The Political Control of Czechoslovakia: A Study in Social Control of a Soviet Satellite State* (1953).

S. Gaines, 'International Principles for Transnational Environmental Liability: Can Developments in Municipal Law Help Break the Impasse?', 30 *Harvard ILJ* 311 (1989). Gaines(a).

S. Gaines, 'The Polluter-Pays Principle: From Economic Equity to Environmental Ethos', 26 *Texas ILJ* 463 (1991). Gaines(b).

J. Galambos, 'Political Aspects of an Environmental Conflict: The Case of the Gabcikovo–Nagymaros Dam System', in *Perspectives on Environmental Conflict and International Relations* 79 (J. Käkönen ed., 1992).

J. Gamble & D. Fischer, *The International Court of Justice: An Analysis of a Failure* (1976).

M. Garner, 'Transnational Alignment of Non-governmental Organisations for Global Environmental Action', 24 *Vanderbilt JTL* 653 (1991).

A. Geddes, 'Locus Standi and EEC Environmental Measures', 4 *JEL* 29 (1992).

T. Gehring, 'International Environmental Regimes: Dynamic Sectoral Legal Systems', 1 *YIEL* 35 (1990).

M. Gelfand, 'Practical Application of International Environmental Law: Does It Work At All?', 29 *Case Western Reserve JIL* 73 (1997).

A. George & R. Keohane, *The Concept of National Interest: Uses and Limitations, in Presidential Decisionmaking in Foreign Policy: The Effective Use of Information and Advice* (A. George ed., 1980).

G. Gilbert, 'The Criminal Responsibility of States', 39 *ICLQ* 345 (1990).

P. Gleick, 'Water Resources: A Long-Range Global Evolution', 20 *Ecology LQ* 141 (1991).

C. Glinka, 'Global Imperative – An Effective System of Resolution Techniques for International Environmental Disputes: The Canadian–United States Example', 14 *Suffolk TLR* 127 (1990).

M. Glode & B. Glode, 'Transboundary Pollution: Acid Rain and United States–Canada Relations', 20 *Environmental Affairs* 1 (1993).

L. Goldie, 'Equity and the International Management of Transboundary Resources', in *Transboundary Resources Law* 103 (1987).

D. Goldberg & D. Hunter, 'EBRD's Environmental Promise: A Bounced Check?', *Center for International Environmental Law Brief* 1 (December 1994).

S. Goldberg *et al.*, *Dispute Resolution* (1986).

G. Goldenman, 'Adapting to Climate Change: A Study of International Rivers and Their Legal Arrangements', 17 *Ecology LQ* 741 (1990). Goldenman(a).

G. Goldenman, 'Environmental Barriers to Foreign Investment in Eastern Europe: Myths and Mistakes', 2 *RECIEL* 1 (1993). Goldenman(b).

G. Goldenman *et al.* eds, *Environmental Liability and Privatization in Central and Eastern Europe* (1994).

P. Goldmark Jr. & P. LaRocco, 'Global Energy Bargaining', 17 *Annual Review of Energy and the Environment* 77 (1992).

G. Goodpaster, 'Rational Decision-Making in Problem-Solving Negotiation: Compromise, Interest-Valuation, and Cognitive Error', 8 *Ohio State Journal on Dispute Resolution* 299 (1993).

W. Góralczyk, 'Changing Attitudes of Central and Eastern European States Towards the Judicial Settlement of International Disputes', in *The Peaceful Settlement of International Disputes in Europe: Future Prospects* 477 (D. Bardonnet ed., 1991).

P. Gormley, *Human Rights and the Environment* (1976). Gormley(a).

P. Gormley, 'The Legal Obligation of the International Community to Guarantee a Pure and Decent Environment: The Expansion of Human Rights Norms', 3 *Georgetown IELR* 85 (1990). Gormley(b).

M. Gottlieb *et al.*, 'The Czechoslovak Socialist Republic', in *Environmental Policies in East and West* 69 (G. Enyedi *et al.* eds, 1987).

G. Grabowska, 'Environmental Conflicts in Border Areas', in *The Peaceful Settlement of International Disputes in Europe: Future Prospects* 137 (D. Bardonnet ed., 1991).

C. Gray, *Judicial Remedies in International Law* (1987).

J. Gregor, 'Interview with Vice Premier of the Czechoslovak Socialist Republic and Chairman of the Commission for Human Environment of the Government of the Czechoslovak Socialist Republic', 14 *BCL* 1 (1975).

F. Grieves, 'Regional Efforts at International Environmental Protection', 12 *International Lawyer* 309 (1978).

L. Gross, 'Settlement of Disputes by the International Court of Justice', in *Essays on International Law and Organization* 677 (1984). Gross(a).

L. Gross ed., *The Future of the International Court of Justice* (1976). Gross(b).

T. Gruchalla-Wisierski, 'A Framework for Understanding "Soft Law"', 30 *McGill LJ* 37 (1984).

L. Guerra, 'Symposium: Peaceful Transitions to Democracy: The Application of the Spanish Model in the Constitutional Transitions in Central and Eastern Europe', 19 *Cardozo LR* 1937 (1998).

L. Gündling, 'Our Responsibility to Future Generations', 84 *American JIL* 207 (1990). Gündling(a).

L. Gündling, 'The Status in International Law of the Principle of Precautionary Action', 5 *International Journal of Estuarine & Coastal Law* 23 (1990). Gündling(b).

L. Gündling, 'Protection of the Environment by International Law: Air Pollution', in *Environmental Protection and International Law* 91 (W. Lang *et al.* eds, 1991). Gündling(c).

L. Guruswamy, 'The Promise of the United Nations Convention on the Law of the Sea (UNCLOS): Justice in Trade and Environment Disputes', 25 *Ecology LQ*, (1998).

E. Haagsma, 'The European Community's Environmental Policy: A Case Study in Federalism', 12 *Fordham ILJ* 311 (1989).

E. Haas, *Beyond the Nation-State: Functionalism and International Organization* (1964). E. Haas(a).

E. Haas, 'Words Can Hurt You: or, Who Said What to Whom About Regimes', 36 *International Organization* 207 (1982). E. Haas(b).

E. Haas, 'Do Regimes Matter? Epistemic Communities and Mediterranean Pollution Control', 43 *International Organization* 377 (1989). E. Haas(c).

P. Haas, *Saving the Mediterranean: the Politics of International Environmental Co-operation* (1990). P. Haas(a).

P. Haas, 'Epistemic Communities and the Dynamics of International Environmental Co-operation', in *Regime Theory and International Relations* 169 (V. Rittberger ed., 1993). P. Haas(b).

P. Haas *et al.*, *Institutions for the Earth: Sources of Effective International Environmental Protection* (1994).

D. Hackett, 'An Assessment of the Basel Convention on the Control of Transboundary Movements of Hazardous Wastes and Their Disposal', 5 *American University JILP* 295 (1990).

S. Haggard & B. Simmons, 'Theories of International Regimes', 41 *International Organization* 491 (1987).
R. Hahn & K. Richards, 'The Internationalisation of Environmental Regulation', 30 *Harvard ILJ* 421 (1989).
R. Hahn & R. Stavins, 'Incentive-Based Environmental Regulation: A New Era from an old Idea?', 18 *Ecology LQ* 1 (1991).
N. Haigh, 'Collaborative Arrangements for Environmental Protection in Western Europe', in *Environmental Policies in East and West* 366 (G. Enyedi *et al*. eds, 1987). Haigh(a).
N. Haigh, 'New Tools for European Air Pollution Control', 1 *International Environmental Affairs* 26 (1989). Haigh(b).
N. Haigh, 'The European Community and International Environmental Policy', in *The International Politics of the Environment: Actors, Institutions and Interests* 228 (A. Hurrell & B. Kingsbury eds, 1992). Haigh(c).
N. Haigh *et al*., 'The Background to Environmental Protection in Market and Planned Economy Countries', in *Environmental Policies in East and West* 22 (G. Enyedi *et al*. eds, 1987).
M. Halperin, *Bureaucratic Politics and Foreign Policy* (1974).
G. Handl, 'Balancing of Interests and International Liability for the Pollution of International Watercourses: Customary Principles of Law Revisited', 13 *CYIL* 156 (1975). Handl(a).
G. Handl, 'Territorial Sovereignty and the Problem of Transnational Pollution', 69 *AJIL* 50 (1975). Handl(b).
G. Handl, 'The Principle of Equitable Use as Applied to International Shared Natural Resources: Its Role in Resolving Potential International Disputes Over Transfrontier Pollution', 14 *Revue Belge de Droit International* 40 (1977–8). Handl(c).
G. Handl, 'International Legal Perspective on the Conduct of Abnormally Dangerous Activities in Frontier Areas: The Case of Nuclear Power Plant Siting', 7 *ELQ* 1 (1978). Handl(d).
G. Handl, 'State Liability for Accidental Transnational Environmental Damage by Private Persons', 74 *AJIL* 525 (1980). Handl(e).
G. Handl, 'Transboundary Nuclear Accidents: The Post-Chernobyl Multilateral Legislative Agenda', 15 *Ecology LQ* 203 (1988). Handl(f).
G. Handl, 'Environmental Security and Global Change: The Challenge of International Law', in *Environmental Protection and International Law* 59 (W. Lang *et al*. eds, 1991). Handl(g).
G. Handl & R. Lutz, 'An International Policy Perspective on the Trade of Hazardous Materials and Technologies', 30 *Harvard ILJ* 351 (1989).
G. Handl *et al*. eds, 1 *Yearbook of International Environmental Law* (1990). 1 *YIEL*.
G. Handl *et al*. eds, 2 *Yearbook of International Environmental Law* (1991). 2 *YIEL*.
G. Handl *et al*. eds, 3 *Yearbook of International Environmental Law* (1992). 3 *YIEL*.
G. Handl *et al*. eds, 4 *Yearbook of International Environmental Law* (1993). 4 *YIEL*.
G. Handl *et al*. eds, 5 *Yearbook of International Environmental Law* (1994). 5 *YIEL*.
G. Handl *et al*. eds, 6 *Yearbook of International Environmental Law* (1995). 6 *YIEL*.
P. Hardi *et al*. eds, *The Hardi Report: The Bös-Nagymaros Barrage System* (September 1989).
P. Hardi & J. Galambos, 'Environmental Legislative Issues in Hungary', 2 *RECIEL* 16 (1993).
H. Hartnell, 'Subregional Coalescence in European Regional Integration', 16 *Wisconsin ILJ* 115 (Winter 1997).

M. Hartwig, 'The Institutionalization of the Rule of Law: The Establishment of Constitutional Courts in the Eastern European Countries', 7 *American University JILP* 449 (1992).

V. Haufler, 'Crossing the Boundary between Public and Private: International Regimes and Non-State Actors', in *Regime Theory and International Relations* 94 (V. Rittberger ed., 1993).

P. Hayward, 'Environmental Protection: Regional Approaches', 8 *Marine Policy* 106 (1984).

G. Height, 'Unfulfilled Obligations: The Situation of the Ethnic Hungarian Minority in the Slovak Republic', 4 *ILSA JICL* 27 (Fall 1997).

H. Heinrich, *Hungary: Politics, Economics, and Society* (1986).

L. Helfer & A. Slaughter, 'Toward a Theory of Effective Supranational Adjudication', 107 *Yale LJ* 273 (Winter 1997).

J. Helland-Hansen, 'The Global Environment Facility', 3 *International Environmental Affairs* 137 (1991).

J. Henckaerts & S. Van der Jeught, 'Human Rights Protection under the New Constitutions of Central Europe', 20 *Loyola ICLJ* 475 (1998).

L. Henkin, *How Nations Behave: Law and Foreign Policy* (1979).

A. Herbert, 'Cooperation in International Relations: A Comparison of Keohane, Haas and Franck', 14 *Berkeley JIL* 222 (1996).

E. Hey, 'The Precautionary Approach: Implications of the Revision of the Oslo and Paris Conventions', 15 *Marine Policy* 1441 (1991). Hey(a).

E. Hey, 'The Precautionary Concept in Environmental Policy and Law: Institutionalizing Caution', 4 *Georgetown IELR* 303 (1992). Hey(b).

R. Higgins, *Conflict of Interests: International Law in a Divided World* (1965). Higgins(a).

R. Higgins, 'International Law and the Avoidance, Containment and Resolution of Disputes', 230 *RCADI* 23 (1991). Higgins(b).

D. Hinrichsen & G. Enyedi eds, *State of the Hungarian Environment* (1990).

I. Hodkova, 'Is There a Right to a Healthy Environment in the International Legal Order?', 7 *Connecticut JIL* 65 (1991).

K. Hoffman, 'State Responsibility in International Law and Transboundary Pollution Injuries', 25 *ICLQ* 509 (1976).

H. Hohmann, *Precautionary Legal Duties and Principles in Modern International Environmental Law* (1994).

M. Holley, 'Central America: Translating Regional Environmental Accords into Domestic Enforcement Action', 25 *Ecology LQ* 89 (1998).

K. Horohoe, 'Theoretical Perspectives on International Institutions', 89 *Proceed. ASIL* 79 (5–8 April 1995).

J. Hrivnak, 'Czechoslovak Regulations Governing Arbitration in International Trade', 24 *BCL* 19 (1985).

L. Hughes, 'The Role of International Environmental Law in the Changing Structure of International Law', 10 *Georgetown IELR* 243 (1998).

D. Hunter *et al.* eds, *State of Environmental Law: Bulgaria* (1992). Hunter(a).

D. Hunter, 'Environmental Reforms in Post-Communist Central Europe: From High Hopes to Hard Reality', 13 *Michigan JIL* 921 (Summer 1992). Hunter(b).

D. Hurlbut, 'Beyond the Montreal Protocol: Impact on Nonparty States and Lessons for Future Environmental Protection Regimes', 4 *Colorado JIELP* 344 (Summer 1993).

A. Hurrell, 'International Society and the Study of Regimes', in *Regime Theory and International Relations* 49 (V. Rittberger ed., 1993).
A. Hurrell & B. Kingsbury eds, *The International Politics of the Environment: Actors, Interests, and Institutions* (1992).
K. Iida, *Two-Level Games with Uncertainty: An Extension of Putnam's Theory* (1991).
Improving Environmental Negotiations, Options 5 (June 1993).
A. Inotai, 'Past, Present and Future of Federalism in Central and Eastern Europe', 1 *New Europe LR* 505 (Spring 1993).
International Legal Problems of the Environmental Protection of the Baltic Sea (1992).
J. Irvin, 'The Role of International Law in Negotiated Settlement of International Disputes', 3 *The Vanderbilt International* 58 (1969).
T. Iwama, 'Emerging Principles and Rules for the Prevention and Mitigation of Environmental Harm', in *Environmental Change and International Law* 107 (1992).
H. Jacobson, 'Conceptual, Methodological and Substantive Issues Entwined in Studying Compliance', 19 *Michigan JIL* 569 (Winter 1998).
M. Jakubowski, 'COMECON and Environmental Protection', 1 *Yearbook on Socialist Legal Systems* 115 (1986).
K. Janics, *Czechoslovak Policy and the Hungarian Minority, 1945–1948* (1982).
M. Janis ed., *International Courts for the Twenty-First Century* (1992).
P. Janke ed., *Ethnic and Religious Conflicts: Europe and Asia* (1994).
J. Jendroska, 'Compliance Monitoring in Poland: Current State and Development', 1 *International Conference on Environmental Enforcement* 351 (1992). Jendroska(a).
J. Jendroska, 'UN ECE Convention on Access to Information, Public Participation in Decision-Making, and Access to Justice in Environmental Matters: Towards More Effective Public Involvement in Monitoring Compliance and Enforcement in Europe', 13 *National Environmental Enforcement Journal* 34 (July 1998). Jendroska(b).
Sir R. Jennings, 'The World Court is Necessarily a Regional Court', in *The Peaceful Settlement of International Disputes in Europe: Future Prospects* 305 (D. Bardonnet ed., 1991). Jennings(a).
Sir R. Jennings, 'The Role of the International Court of Justice in the Development of International Environment Protection Law', 1 *RECIEL* 240 (1992). Jennings(b).
Sir R. Jennings, 'The International Court of Justice after Fifty Years', 89 *American JIL* 493 (July 1995). Jennings(c).
Sir R. Jennings & Sir A. Watts eds, *Oppenheim's International Law* (9th ed., 1992).
B. Johnson, 'The Baltic Conventions', 25 *ICLQ* 1 (1976).
S. Johnson & G. Corcelle, *The Environmental Policy of the European Communities* (1989).
C. Joyner & G. Little, 'It's Not Nice to Fool Mother Nature! The Mystique of Feminist Approaches to International Environmental Law', 14 *Boston University ILJ* 223 (Fall 1996).
J. Juergensmeyer *et al.*, 'Environmental Protection in Post-Socialist Eastern Europe: The Polish Example', 14 *Hastings ICLR* 831 (1991).
T. Judt, 'Radical Politics in a New Key', 81 *Northwestern University LR* 817 (1987).
R. Juelke, 'The Economic Causes and Consequences of Constitutional Reform in Eastern Europe', 34 *William & Mary LR* 1367 (1993).
M. Kaplan, *System and Process in International Politics* (1957).

J. Kara, 'Geo-politics and the Environment, the Case of Central Europe', *Environmental Politics* 1 (1992).
A. Kaswan, 'Environmental Justice: Bridging the Gap Between Environmental Laws and "Justice"', 47 *American University LR* 221 (1997).
D. Kay & H. Jacobson eds., *Environmental Protection: The International Dimension* (1982).
M. Keck & K. Sikkink, *Activists Beyond Borders: Advocacy Networks in International Politics* (1998).
H. Kelsen, *Principles of International Law* (1966).
J. Kenedi, 'Why the Gypsy is the Scapegoat?', 2 *East European Reporter* (1986).
G. Kennan, *American Diplomacy, 1900–1950* (1951).
W. Kennedy, 'The Directive on Environmental Impact Assessment', 8 *Environmental Policy & Law* 84 (1988).
R. Keohane, *After Hegemony: Cooperation and Discord in the World Political Economy* (1984). Keohane(a).
R. Keohane, 'Neoliberal Institutionalism: A Perspective on World Politics', in *International Institutions and State Power* 1 (1989). Keohane(b).
R. Keohane, 'The Demand for International Regimes', in *International Institutions and State Power* 101 (1989). Keohane(c).
R. Keohane, 'International Relations and International Law: Two Optics', 38 *Harvard ILJ* 487 (Spring 1997). Keohane(d).
R. Keohane, 'Commentary: When Does International Law Come Home?', 35 *Houston LR* 699 (Fall 1998). Keohane(e).
R. Keohane & J. Nye, *Transnational Relations and World Politics* (1977). Keohane & Nye(a).
R. Keohane & J. Nye, *Power and Interdependence: World Politics in Transition* (1977). Keohane & Nye(b).
R. Keohane & J. Nye, 'Power and Interdependence Revisited', 41 *International Organization* 725 (1987). Keohane & Nye(c).
Z. Kiesewetter, 'Exploitation and Protection of Natural Resources and Their Impact on Human Environment', 18 *BCL* 77 (1979).
L. Kimball, 'International Law and Institutions: The Oceans and Beyond', 21 *Ocean Development and ILJ* 147 (1990).
J. Kindt, 'The Law of the Sea: Offshore Installations and Marine Pollution', 12 *Pepperdine LR* 381 (1984).
G. King *et al.*, *Designing Social Inquiry: Scientific Inference in Qualitative Research* (1994).
B. Kingsbury, 'A Symposium on Implementation, Compliance and Effectiveness: The Concept of Compliance as a Function of Competing Conceptions of International Law', 19 *Michigan JIL* 345 (Winter 1998).
R. Kinnear & B. Rhode, 'Europe: Its Environmental Identity', in *Environmental Policies in East and West* 5 (G. Enyedi *et al.* eds, 1987).
M. Kirkpatrick, *Environmental Problems and Policies in Eastern Europe and the USSR* (1978).
A. Kiss, 'The Protection of the Rhine Against Pollution', in *Transboundary Resources Law* 51 (1987). Kiss(a).
A. Kiss, 'The International Control of Transboundary Movement of Hazardous Waste', 26 *Texas ILJ* 521 (1991). Kiss(b).
A. Kiss, 'Present Limits to the Enforcement of State Responsibility for Environmental Damage', in *International Responsibility for Environmental Harm* 3 (F. Francioni & T. Scovazzi eds, 1991). Kiss(c).

A. Kiss, 'An Introductory Note on a Human Right to Environment', in *Environmental Change and International Law* 199 (1992). Kiss(d).
A. Kiss & D. Shelton, 'Systems Analysis of International Law: A Methodological Inquiry', 17 *Netherlands YIL* 45 (1986). Kiss & Shelton(a).
A. Kiss & D. Shelton, *Manual of European Environmental Law* (1991); Supplement (1994). Kiss & Shelton(b).
G. Klaasen *et al.*, *Strategies for Reducing Sulphur Dioxide Emissions in Europe Based on Critical Sulphur Deposition Values* (IIAS 1992).
G. Klein & M. Reban eds, *The Politics of Ethnicity in Eastern Europe* (1981).
P. Kočíková ed., *The Czech Republic Environmental Handbook for Industry* (1993).
H. Koh, 'Address: The 1998 Frankel Lecture: Bringing International Law Home', 35 *Houston LR* 623 (Fall 1998).
V. Kopal, 'Evolution of the Position of Czechoslovakia in Relation to the Obligatory Jurisdiction of the World Court', in *The Peaceful Settlement of International Disputes in Europe: Future Prospects* 497 (D. Bardonnet ed., 1991).
L. Korbut & I. Baskin, 'Protection of Water Resources under International Law', in *Perestroika and International Law* 17 (W. Butler ed., 1990).
Z. Kordik, 'Legal Protection of Waters', 18 *BCL* 102 (1979).
A. Korula, *Post-Negotiation Impasses in the Environmental Domain: The Influence of Some Political and Economic Factors on Environmental Treaty Acceptance* (IIASA Working Paper 92-86, 1992).
D. Kosary & S. Vardy, *History of the Hungarian Nation* (1969).
M. Koskenniemi, 'Peaceful Settlement of Environmental Disputes', 60 *Nordic JIL* 73 (1991). Koskenniemi(a).
M. Koskenniemi ed., *International Law* (1992). Koskenniemi(b).
F. Kozhevnikov ed., *Role of Law in Socialist Countries, in International Law* (1962).
J. Kramer, 'The Environmental Crisis in Eastern Europe', 42 *Slavic Review* 204 (1983). Kramer(a).
J. Kramer, *The Energy Gap in Eastern Europe* (1990). Kramer(b).
J. Kramer, 'Eastern Europe and the Energy Shock of 1990–91', *Problems of Communism* (June 1991). Kramer(c).
J. Kramer, 'The Nuclear Power Debate in Eastern Europe', 35 *Radio Free Europe Research Report* 59 (September 1992). Kramer(d).
J. Kramer, 'Energy and Environment in the Czech Republic and Poland', presented during the *V World Congress for Central and East European Studies* (Warsaw, 10 August 1995). Kramer(e).
J. Kramer, 'Nuclear Power in Central and Eastern Europe', *Problems of Communism* 41 (August 1995). Kramer(f).
L. Krämer, 'The Implementation of Community Environmental Directives within Member States: Some Implications of Direct Effect Doctrine', 3 *JEL* 39 (1991). Krämer(a).
L. Krämer, *Focus on European Environmental Law* (1992). Krämer(b).
S. Krasner ed., *International Regimes* (1983). Krasner(a).
S. Krasner, 'Structural Causes and Regime Consequences: Regimes as Intervening Variables', in *International Regimes* (S. Krasner ed., 1983). Krasner(b).
F. Kratochwill, *Rules, Norms, and Decisions: On the Conditions of Practical and Legal Reasoning in International Relations and Domestic Affairs* (1989).
F. Kratochwill & J. Ruggie, 'International Organization: A State of the Art on an Art of the State', 40 *International Organization* 753 (1986).

K. Krchnak, *Environmental Law Institute, Implementation and Enforcement of Environmental Laws in the Slovak Republic* (1995).
V. Kremenyuk ed., *International Negotiation: Analysis, Approaches, Issues* (1991).
V. Kremenyuk & W. Lang, 'The Political, Diplomatic, and Legal Background', in *International Environmental Negotiation* 3 (G. Sjöstedt ed., 1993).
D. Kritsiotis, 'Conceiving the Lawyer as Creative Problem Solver: The Power of International Law as Language', 34 *California Western LR* 397 (Spring 1998).
Neil Kritz ed., *Transitional Justice: How Emerging Democracies Reckon With Former Regimes* (1995).
K. Kummer, 'The International Regulation of Transboundary Traffic in Hazardous Wastes: The 1989 Basel Convention', 41 *ICLQ* 530 (1992).
K. Kummer & I. Rummel-Bulska, *The Basel Convention on the Control of Transboundary Movement of Hazardous Wastes and Their Diposal* (1990).
M. Kurkela, 'International Regulation of Transfrontier Hazardous Waste Shipments: A New EC Environmental Directive', 21 *Texas ILJ* 85 (1996).
B. Kwiatowska, 'Marine Pollution from Land-Based Sources: Current Problems and Prospects', 14 *Ocean Development & ILJ* 315 (1984).
J. Lammers, *Pollution of International Watercourses: A Search for Substantive Rules and Principles of Law* (1984). Lammers(a).
J. Lammers, 'International and European Community Law Aspects of Pollution of International Watercourses', in *Environmental Protection and International Law* 115 (W. Lang *et al.* eds, 1991). Lammers(b).
W. Lang *et al.* eds, *Environmental Protection and International Law* (1991).
W. Lang, 'Diplomacy and International Environmental Law-Making: Some Observations', 3 *YIEL* 108 (1992).
D. Lasok & J. Bridge, *An Introduction to the Law and Institutions of the European Communities* (3rd ed., 1982).
H. Lasswell & M. McDougal, *Jurisprudence for a Free Society* (1992).
E. Lauterpacht, *Aspects of the Administration of International Justice* (1991).
J. Leden, 'Legal Regulation of Air Pollution Control', 14 *BCL* 64 (1975).
A. Levin, *Protecting the Human Environment* (1977).
T. Ljalev, 'Development of Environmental Legislation', in *Development of Social Legislation in Bulgaria* (J. Radev ed., 1984).
B. Lindwall, 'The Ecological Situation in the Baltic Sea', in *Pollution of the Baltic Sea* 4 (1988).
J. Linnerooth, 'The Danube River Basin: Negotiating Settlements to Transboundary Environmental Issues', 30 *NRJ* 629 (1990).
E. Lisitzin, 'Collaborative Arrangements for Environmental Protection in European Socialist Countries', in *Environmental Policies in East and West* 352 (G. Enyedi *et al.* eds, 1987).
M. Liška, 'The Gabčíkovo–Nagymaros Project – Its Real Significance and Impacts', 6 *Europa Vincet* 7 (November 1992).
M. List, 'Cleaning up the Baltic: A Case Study in East–West Environmental Cooperation', in *International Regimes in East–West Politics* 90 (V. Rittberger ed., 1990).
M. List & V. Rittberger, 'Regime Theory and International Environment and Management', in *The International Politics of the Environment: Actors, Interests, and Institutions* (A. Hurrell & B. Kingsbury eds, 1992).
O. Lissitzyn, *The International Court of Justice* (1951).

W. Long, 'Economic Aspects of Transport and Disposal of Hazardous Wastes', 14 *Marine Policy* 199 (1990).
P. Lowe & J. Goyder, *Environmental Groups in Politics* (1983).
E. Luard, *Types of International Society* (1976).
T. Lundmark, 'Systemizing Environmental Law on a German Model', 7 *Dickenson JELP* 1 (1998).
S. McCaffrey, 'Trans-Boundary Pollution Injuries: Jurisdictional Considerations in Private Litigation Between Canada and the United States', 3 *California Western LR* 191 (1972). McCaffrey (a).
S. McCaffrey, 'Pollution of Shared Natural Resources', 71 *Proceed ASIL* 56 (1977). McCaffrey (b).
S. McCaffrey, 'The Work of the International Law Commission Relating to Trans-frontier Environmental Harm', 20 *New York University JILP* 715 (1988). McCaffrey (c).
S. McCaffrey, 'The Law of International Watercourses: Some Recent Developments and Unanswered Questions', 17 *Denver JILP* 505 (1989). McCaffrey (d).
S. McCaffrey, 'The Law of International Watercourses: Ecocide or Ecomanagement?', 59 *RJUPR* 1001 (1990). McCaffrey (e).
S. McCaffrey, 'International Organizations and the Holistic Approach to Water Problems', 31 *NRJ* 139 (1991). McCaffrey (f).
S. McCaffrey, 'The International Law Commission Adopted Draft Articles on International Watercourses', 89 *AJIL* 395 (1995). McCaffrey (g).
S. McCaffrey & M. Sinjela, 'The 1997 United Nations Convention on International Watercourses', 92 *American JIL* 97 (January 1998).
A. McDonald, *Fairness in River Basin Agreements* (IIASA Working Paper, June 1993).
J. McDonald & D. Bendahmane eds, *Conflict Resolution: Two Track Diplomacy* (1987).
M. McDougal, 'International Law, Power and Policy: A Contemporary Conception', 82 *RCADI* 137 (1953).
M. McDougal & Associates, *Studies in World Public Order* (1960).
D. McGoldrick, 'The Development of the Conference on Security and Co-operation in Europe (CSCE) After the Helsinki 1992 Conference', 42 *ICLQ* 411 (1993).
Lord A. McNair, 'The Place of Law and Tribunals in International Relations', in *Lord McNair: Selected Papers and Bibliography* 295 (G. Fitzmaurice & R. Jennings eds, 1974). McNair(a).
Lord A. McNair, *The Law of Treaties* (1986). McNair(b).
J. MacNeill *et al.*, *Beyond Interdependence: The Meshing of the World's Economy and the Earth's Ecology* (1991).
R. Macrory, 'European Community Water Law', 20 *Ecology LQ* 119 (1993).
E. McWhinney, *Judicial Settlement of International Disputes: Jurisdiction, Justiciability and Judicial Law-making on the Contemporary International Court* (1991).
Z. Madar, 'General Problems of the Legal Regulation of Environmental Protection in Czechoslovakia', 14 *BCL* 5 (1975). Madar(a).
Z. Madar, 'Protection Against Air Pollution in Czechoslovakia', 18 *BCL* 120 (1979). Madar(b).
Z. Madar, 'Legal Regulations of Environmental Protection in CMEA Countries', 59 *RJUPR* 779 (1990). Madar(c).

D. Madeo, 'Environmental Contamination and World Trade Integration: The Case of the Czech Republic', 26 *Law and Policy in International Business* 945 (Spring 1995).

P. Magocsi, *Historical Atlas of East Central Europe* (1993).

D. Magraw, 'Transboundary Harm: The International Law Commission's Study of "International Liability"', 80 *AJIL* 305 (1986). Magraw(a).

D. Magraw ed., *International Law and Pollution* (1991). Magraw(b).

M. Malitza, 'The Role of Medium-Sized Countries in the Resolution of Inter-State Conflicts', 47 *Revue Roumaine D'Etudes Internationales* 35 (1980).

L. Malone, 'The Chernobyl Accident: A Case Study in International Law Regulating State Responsibility for Transboundary Nuclear Pollution', 12 *Columbia JEL* 203 (1987).

R. Manser, *Failed Transitions: The Eastern European Economy and Environment Since the Fall of Communism* (1993).

Z. Maoz, *National Choices and International Processes* (1990).

J. Marcinkiewicz, *Pollution in the Heart of Europe* (1987).

M. Marvin, *Beyond Sovereignty: The Challenge of Global Policy* (1986).

V. Mavi & A. Gabor, 'Harmonization of Private International Law in the Soviet Union and Eastern Europe: A Comparative Law Survey', 10 *Review of Socialist Law* 97 (1984).

R. Mekota, 'Legal Protection of Nature', 14 *BCL* 90 (1975).

T. Merrill, 'Golden Rules for Transboundary Pollution', 46 *Duke LJ* 931 (March 1997).

J. Merrills, 'The Role and Limits of International Adjudication', in *International Law and the International System* 169 (W. Butler ed., 1987). Merrills(a).

J. Merrills, *International Dispute Settlement* (1991). Merrills(b).

R. M'Gonigle, '"Developing Sustainability" and the Emerging Norms of International Environmental Law: The Case of Land-based Marine Pollution', 128 *CYIL* 169 (1990).

S. Mines, 'Are International Institutions Doing Their Job? Environment: New Institutional Challenges', 90 *Proceed. ASIL* 488 (27–30 March 1996).

S. Mloch, 'Silesia and the Black Triangle', 2 *RECIEL* 43 (1993).

W. Mock, 'Game Theory, Signalling, and International Legal Relations', 26 *George Washington Journal of International Law and Economics* 33 (1992).

I. Mocsy, *The Effects of World War I: The Uprooted* (1983).

J. Moermond & E. Shirley, 'A Survey of the International Law of Rivers', 16 *Denver JILP* 139 (1987).

E. Moisé, *International Regulations on Radioactive and Toxic Wastes; Similarities and Differences* (1991).

J. Montville ed., *Conflict and Peacemaking in Multiethnic Societies* (1990).

H. Morgenthau, *Politics Among Nations* (3rd ed., 1965).

A. Movchan, 'The Concept and Meaning of Modern International Law and Order', in *International Law and the International System* 123 (W. Butler ed., 1987).

R. Müllerson, 'The History of the Soviet Science of International Law', 3 *Yearbook on Socialist Legal Systems* 321 (1989). Müllerson(a).

R. Müllerson, *International Law, Rights and Politics: Developments in Eastern Europe and the CIS* (1994). Müllerson(b).

D. Munro, *International Environmental Law* (1990).

R. Munro & M. Holdgate eds, *Caring for the Earth: A Strategy for Sustainable Development* (1991).
R. Munro & J. Lammers, *Environmetnal Protection and Sustainable Development: Legal Principles and Recommendations* (1987).
S. Murcott, 'The Danube River Basin: International Cooperation or Sustainable Development', 36 *NRJ* 521 (Summer 1996).
S. Murphy, 'Prospective Liability Regimes for the Transboundary Movement of Hazardous Wastes', 88 *American JIL* 24 (January 1994).
E. Nadelman, 'Global Prohibition Regimes: The Evolution of Norms in International Society', 44 *International Organization* 479 (1990).
V. Nanda, *World Climate Change – The Role of International Law and Institutions* (1983).
J. Nalven, 'Transboundary Environmental Problem Solving: Social Process, Cultural Perception', 26 *NRJ* 793 (1986).
H. Neuhold, 'Commentary to F. Francioni, International Co-operation for the Protection of the Environment: The Procedural Dimension', in *Environmental Protection and International Law* 203 (W. Lang *et al.* eds, 1991).
M. Nicholson, *Rationality and the Analysis of International Conflict* (1992). Nicholson(a).
M. Nicholson, 'Interdependent Utility Functions: Implications for Co-operation and Conflict', in *Game Theory and International Relations: Preferences, Information and Empirical Evidence* 77 (P. Allan & C. Schmidt eds, 1994). Nicholson(b).
Note, 'Economic Implications of European Transfrontier Pollution: National Prerogative and Attribution of Responsibility', 11 *Georgia JICL* 519 (1981).
J. Nye, 'Nuclear Learning and U.S.–Soviet Security Regimes', 41 *International Organization* 371 (1987).
D. P. O'Connell, 'Independence and Succession to Treaties', 38 *British YIL* 84 (1962). O'Connell(a).
D. P. O'Connell, *International Law* (2nd ed., 1970). O'Connell(b).
P. Ordeshook, *Game Theory and Political Theory: An Introduction* (1986).
J. O'Reilly & L. Cuzze, 'Hazardous Wastes', 22 *Iowa Journal of Corporate Law* 507 (1997).
K. Oye ed., *Cooperation Under Anarchy* (1986). Oye(a).
K. Oye, 'Explaining Cooperation Under Anarchy', in *Cooperation Under Anarchy* (ed., 1986). Oye(b).
M. Pallemaerts ed., *The Right to Environmental Information* (1991). Pallemaerts(a).
M. Pallemaerts, 'International Environmental Law from Stockholm to Rio: Back to the Future', 1 *RECIEL* 254 (1992). Pallemaerts(b).
M. Pallemaerts, 'The North Sea Ministerial Declarations from Bremen to the Hague: Does the Process Generate any Substance?', 7 *IJ of Estuarine & Coastal Law* 1 (1992). Pallemaerts(c).
D. Partan, 'Increasing the Effectiveness of the International Court', in *International Law: A Contemporary Perspective* 279 (R. Falk *et al.* eds, 1988). Partan(a).
D. Partan, 'The "Duty to Inform" in International Environmental Law', 6 *Boston University ILJ* 43 (1988). Partan(b).
R. Pathak, 'The Human Rights System as a Conceptual Framework for Environmental Law', in *Environmental Change and International Law* 205 (1992).
J. Pehe, 'The Choice Between Europe and Provincialism', 1 *Transition* 14 (July 1995).

J. Perkins, 'The Changing Foundations of International Law: From State Consent to State Responsibility', 15 *Boston University ILJ* (Fall 1997).
F. Perrez, 'The Efficiency of Cooperation: A Functional Analysis of Sovereignty', 15 *Arizona JICL* 515 (1998).
K. Piddington, 'The Role of the World Bank', in *The International Politics of the Environment* (A. Hurrell & B. Kingsbury eds, 1992).
L. Pineschi, 'Environmental Impact Assessment', in *World Treaties for the Protection of the Environment* 485 (T. Scovazzi & T. Treves eds, 1992).
D. Pirages, 'Ecological Theory and International Relations', 5 *Indiana Journal of Global Legal Studies* (Fall 1997).
R. Pisillo-Mazzeschi, 'Forms of International Responsibility for Environmental Harm', in, *International Responsibility for Environmental Harm* 15 (F. Francioni & T. Scovazzi eds, 1991).
P. Poborski, *Ecological Problems in Upper Silesia* (report prepared by the Ecology Department of the Katowice Local Authority, 1994).
M. Politi, 'The Impact of Chernobyl Accident on States' Perception of International Responsibility for Nuclear Damage', in *International Responsibility for Environmental Harm* 473 (F. Francioni & T. Scovazzi eds, 1991).
N. Popovic, 'In Pursuit of Environmental Human Rights: Commentary on the Draft Declaration of Principles on Human Rights and the Environment', 27 *Columbia Human Rights LR* 487 (Spring 1996).
A. Postiglione, 'A More Efficient International Law on the Environment and Setting up an International Court for the Environment within the United Nations', 20 *Environmental Law* 321 (1990).
M. Potier, 'Towards a Better Integration of Environmental, Economic, and Other Governmental Policies', in *Maintaining a Satisfactory Environment: An Agenda for International Environmental Policy* 69 (N. Åkerman ed., 1990).
A. Potrykowska, *Polish Academy of Sciences, The Consequences of Environmental Pollution for the Population in Poland* (1995).
R. Powell, 'Absolute and Relative Gains in International Relations Theory', 85 *APSR* 303 (1991).
V. Prittwitz, 'Several Approaches to the Analysis of International Environmental Policy', in *Maintaining a Satisfactory Environment: An Agenda for International Environmental Policy* 5 (N. Åkerman ed., 1990).
R. Putnam, 'Diplomacy and Domestic Politics: The Logic of Two-Level Games', 42 *International Organization* 427 (1988).
M. Radinsky, 'Retaliation: The Genesis of a Law and the Evolution Toward International Cooperation: An Application of Game Theory to Modern International Conflicts', 2 *George Mason University LR* 53 (Fall 1994).
K. Raustiala, 'The "Participatory Revolution" in International Environmental Law', 21 *Harvard ELR* 537 (1997).
K. Read, 'Debt-for-Nature Swaps in Central and Eastern Europe: An Answer to Environmental and Economic Troubles in the New Market Economies', 6 *International Legal Perspectives* 69 (Fall 1994).
M. Reisman, 'The View from the New Haven School of International Law', 86 *Proceed. ASIL* 118 (1992).
A. Reitze, Jr., 'A Century of Air Pollution Control Law: What's Worked; What's Failed; What Might Work', 21 *Environmental Law* 1549 (1991).

A. Rest, 'Responsibility and Liability for Transboundary Air Pollution Damage', in *Transboundary Air Pollution* (1986). Rest(a).
A. Rest, 'New Tendencies in Environmental Responsibility/Liability Law: The Work of the UN/ECE Task Force on Responsibility and Liability Regarding Transboundary Water Pollution', 21 *Environmental Policy & Law* 135 (1991). Rest(b).
A. Rest, 'Ecological Damage in Public International Law', 22 *Environmental Policy & Law* 31 (1992). Rest(c).
R. Reuben, 'Public Justice: Toward a State Action Theory of Alternative Dispute Resolution', 85 *California LR* 577 (May 1997).
B. Rice, 'The Emperor's New Clothes: The World Bank and Environmental Reform', 7 *The World Policy Journal* 305 (Spring 1990).
V. Rittberger ed., *International Regimes in East–West Politics* (1990). Rittberger(a).
V. Rittberger ed., *Regime Theory and International Relations* (1993). Rittberger(b).
V. Rittberger & M. Zürn, 'Towards Regulated Anarchy in East–West Relations: Causes and Consequences of East–West Regimes', in *International Regimes in East–West Politics* (V. Rittberger ed., 1990).
L. Rivera, 'Resolving Air Resource Disputes on a Transfrontier Basis: El Paso and Ciudad Juarez', 10 *Houston JIL* 133 (1987).
B. Rhode, 'The United Nations Economic Commission for Europe', in *Environmental Policies in East and West* 388 (G. Enyedi et al. eds, 1987).
N. Robinson, 'Problems of Definition and Scope', in *Law, Institutions and the Global Environment* 43 (1972). Robinson(a).
N. Robinson, 'Introduction: Emerging International Environmental Law', 17 *Stanford JIL* 229 (1981). Robinson(b).
N. Robinson, 'Soviet Environmental Protection: The Challenge for Legal Studies', 7 *Pace ELR* 117 (1989). Robinson(c).
N. Robinson ed., *International Protection of the Environment: Agenda 21 & the UNCED Proceedings* (1992). Robinson(d).
K. Rogers, *Ecological Security and Multinational Corporations* (Woodrow Wilson International Center, Spring 1997).
A. Rosas, 'Issues of State Liability for Transboundary Environmental Damage', 60 *Nordic JIL* 5 (1991).
S. Rosenne, *The World Court: What it is and How it Works* (1989).
L. Ross, *Barriers to Conflict Resolution, Negotiation Journal* (1991).
G. Rottem, 'Compliance with International Standards: Environmental Case Studies', 89 *Proceed. ASIL* 206 (5–8 April 1995).
C. Rousseau, *Droit International Public* (10th ed., 1984).
A. Rubanov, 'The Development of Generally Recognized Rules of International Law Mentioning National Law', in *International Law and the International System* 85 (W. Butler ed., 1987).
J. Ruben, 'Third Party Roles: Mediation in International Environmental Disputes', in *International Environmental Negotiation* 275 (G. Sjöstedt ed., 1993).
P. Rubin, 'Growing a Legal System in the Post-Communist Economies', 27 *Cornell ILJ* 1 (Winter 1994).
J. Rubin & B. Brown, *The Social Psychology of Bargaining and Negotiations* (1975).
J. Ruhl, 'The Seven Degrees of Relevance: Why Should Real-World Environmental Attorneys Care Now About Sustainable Development Policy?', 8 *Duke Environmental Law and Policy Forum* 273 (Spring 1998).

S. Samuels, 'The Soviet Position on International Arbitration: A Wealth of Choices or Choices for the Wealthy', 26 *Virginia JIL* 417 (1986).
A. Sanborn ed., *Transylvania and the Hungarian Romanian Problem: A Symposium* (1979).
F. Sandbach, *Environment, Ideology & Policy* (1980).
P. Sand, *Lessons Learned in Global Environmental Governance* (1990). Sand(a).
P. Sand, *World Resources Institute, Lessons Learned in Global Environmental Governance* (June 1990). Sand(b).
P. Sand, 'Transnational Environmental Disputes', in *The Peaceful Settlement of International Disputes in Europe: Future Prospects* 123 (D. Bardonnet ed., 1991). Sand(c).
P. Sand, 'International Cooperation: The Environmental Experience', in *Preserving the Global Environment* 236 (J. Mathews ed., 1991). Sand(d).
P. Sand, 'New Approaches to Transnational Environmental Disputes', 3 *International Environmental Affairs* 193 (1991). Sand(e).
P. Sand ed., *The Effectiveness of International Environmental Agreements – A Survey of Existing Legal Instruments* (1992). Sand(f).
P. Sands ed., *Chernobyl: Law and Communication: Transboundary Nuclear Air Pollution – The Legal Instruments* (1988). Sands(a).
P. Sands, 'The Environment, Community and International Law', 30 *Harvard ILJ* 393 (1989). Sands(b).
P. Sands, *International Law of Liability for Nuclear Damage* (1990). Sands(c).
P. Sands, 'European Community Environmental Law: Legislation, the European Court of Justice and Common Interest Groups', 53 *Modern LR* 685 (1990). Sands(d).
P. Sands, 'Present at the Creation: A New Development Bank for Europe in the Age of Environment Awareness', 84 *Proceed. AJIL* 77 (1990). Sands(e).
P. Sands, 'European Community Environmental Law: The Emergence of a Regional Regime of International Environmental Protection', 100 *Yale LJ* 2511 (1991). Sands(f).
P. Sands ed., *Greening International Law* (1993). Sands(g).
P. Sands, *Principles of Environmental Law* (1995). Sands(h).
P. Sands & R. Tarasofsky, *Documents in EC Environmental Law* (1994).
Y. Sasamura, 'Prevention and Control of Marine Pollution from Ships', 25 *Proceed. Law of the Sea Institute* 306 (1993).
L. Satterfield, 'The Bi-National Program to Restore and Protect Lake Superior Basin: Talk or Substance?', 4 *Colorado JIELP* 251 (1993).
H. Saunders, 'We Need a Larger Theory of Negotiation: The Importance of Pre-Negotiating Phases', 1 *Negotiation Journal* 249 (1985).
O. Schachter, 'International Law in Theory and Practice', 178 *RCADI* 21 (1982). Schachter(a).
O. Schachter, 'The Emergence of International Environmental Law', 44 *Journal of International Affairs* 457 (1991). Schachter(b).
O. Schachter, *International Law in Theory and Practice* (1991). Schachter(c).
O. Schachter et al., *Toward a Wider Acceptance of UN Treaties* (1971).
E. Schaffer & R. Snyder eds, *Contemporary Practice of Public International Law* (1997).
T. Schelling, *The Strategy of Conflict* (1960).
J. Schlickman et al., *International Environmental Law and Regulation*, Pol-6 (1991).

R. Schmidt, Jr., 'International Negotiations Paralyzed by Domestic Politics: Two-Level Game Theory and the Problem of the Pacific Salmon Commission', 26 *Environmental Law* 95 (Spring 1996).
A. Schneider, 'Symposium Democracy and Dispute Resolution: Individual Rights in International Trade Organizations', 19 *University of Pennsylvania Journal of International Economic Law* 587 (Summer 1998).
J. Schneider, *World Public Order of the Environment* (1979).
G. Schneider, 'Getting Closer at Different Speeds: Strategic Interaction in Widening European Integration', in *Game Theory and International Relations: Preferences, Information and Empirical Evidence* 125 (P. Allan & C. Schmidt eds, 1994).
G. Schöpflin, *Politics in Eastern Europe 1945–1992* (1993).
W. Schroeer, 'Progress Toward Canadian–U.S. Acid Rain Control', in *Nine Case Studies in International Environmental Negotiation* 183 (L. Susskind *et al.* eds, 1990).
C. Schultz & T. Crockett, 'Economic Development, Democratization, and Environmental Protection in Eastern Europe', 18 *Environmental Affairs* 53 (1990).
A. Schwabach, 'The Sandoz Spill: The Failure of International Law to Protect the Rhine from Pollution', 16 *Ecology LQ* 443 (1989). Schwabach(a).
A. Schwabach, 'Diverting The Danube: The Gabcikovo–Nagymaros Dispute and International Freshwater Law', 14 *Berkeley JIL* 290 (1996). Schwabach(b).
A. Schwabach, 'The United Nations Convention on the Law of Non-Navigational Uses of International Watercourses, Customary International Law and the Interest of Developing Upper Riparians', 33 *Texas ILJ* 257 (Spring 1998). Schwabach(c).
G. Schwarzenberger, *The Frontiers of International Law* (1961).
S. Schwebel, 'Reflections on the Role of the International Court of Justice', 61 *Washington LR*, 1061. (1986). Schwebel(a).
S. Schwebel, *Justice in International Law* (1994). Schwebel(b).
G. Scott *et al.*, 'Success and Failure Components of Global Environmental Cooperation: The Making of International Environmental Law', 2 *ILSA JICL* 23 (Fall 1995). Scott *et al.*(a).
G. Scott *et al.*, 'Recent Activity Before the International Court of Justice: Trend or Cycle?', 3 *ILSA JICL* 1 (Fall 1996). Scott *et al.*(b).
M. Sergent, '1997 Comparison of the Helsinki Rules to the 1994 U.N. Draft Articles: Will the Progression of International Watercourse Law be Dammed?', 8 *Villanova ELJ* 435 (1997).
J. Setear, 'An Iterative Perspective on Treaties: A Synthesis of International Relations Theory and International Law', 37 *Harvard ILJ* 139 (Winter 1996). Setear(a).
J. Setear, 'Law in the Service of Politics: Moving Neo-Liberal Institutionalism from Methaphor to Theory by Using the International Treaty Process to Define "Iteration"', 37 *Virginia JIL* 641 (Spring 1997). Setear(b).
J. Setear, 'Responses to Breach of a Treaty and Rationalist International Relations Theory: The Rules of Release and Remediation in the Law of Treaties and the Law of State Responsibility', 83 *Virginia LR* 1 (February 1997). Setear(c).
J. Sette-Camara, 'Pollution of International Rivers', 186 *RCADI* 117 (1984).
R. Shaw, 'Acid-Rain Negotiations in North America and Europe: A Study in Contrast', in *International Environmental Negotiation* 84 (G. Sjöstedt ed., 1993).
W. Sheate, 'Public Participation: the Key to Effective Environmental Assessment', 21 *Environmental Policy & Law* 156 (1991).

G. Shell, 'Trade Legalism and International Relations Theory: An Analysis of the World Trade Organization', 44 *Duke LJ* 829 (March 1995). Shell(a).
G. Shell, 'The Trade Stakeholders Model and Participation by Nonstate Parties in the World Trade Organization', 17 *University of Pennsylvania Journal of International Economic Law* 359 (Spring 1996). Shell(b).
D. Shelton, 'Human Rights, Environmental Rights, and the Right to the Environment', 28 *Stanford JIL* 103 (1991). Shelton(a).
D. Shelton, 'The Participation of Nongovernmental Organizations in International Judicial Proceedings', 88 *American JIL* 611 (October 1994). Shelton(b).
I. Shihata, 'The World Bank and the Environment: A Legal Perspective', 16 *Maryland JIL & Trade* 1 (1992).
G. Shinkaretskaia, 'International Adjudication Today in the View of a Soviet International Lawyer', in *Perestroika and International Law* 245 (W. Butler ed., 1990).
R. Sidman et al., 'Big Bangs and Decision-Making: What Went Wrong?', 13 *Boston University ILJ* 435 (Fall 1995).
C. Siegal, 'Rule Formation in Non-Hierarchical Systems', 16 *Temple Environmental Law and Technology Journal* 173 (1998).
P. Sieghart, *The International Law of Human Rights* (1983).
J. Silar, 'Legal Protection of Water as it Relates to Environmental Protection', 14 *BCL* 75 (1975). Silar(a).
J. Silar, 'The Forestry Act', 18 *BCL* 110 (1979). Silar(b).
Sir I. Sinclair, *The Vienna Convention on the Law of Treaties* (2nd ed., 1984).
J. Singer ed., 'Control of Power-Plant Stack Emissions', in *Combustion, Fossil Power*, 15–1 (1994).
G. Sjöstedt, *International Environmental Negotiation* (1993).
A. Slaughter, 'International Law and International Relations Theory: A Dual Agenda', 87 *AJIL* 205 (1993). Slaughter(a).
A. Slaughter, 'The Liberal Agenda for Peace: International Relations Theory and the Future of the United Nations', 4 *Transnational Law & Contemporary Problems* 377 (Fall 1994). Slaughter(b).
A. Slaughter, 'Liberal International Relations Theory and International Economic Law', 10 *American University JILP* 717 (Winter 1995). Slaughter(c).
A. Slaughter et al., 'International Law and International Relations Theory: A New Generation of Interdisciplinary Scholarship', 92 *American JIL* 367 (July 1998).
R. Smart & G. Murray, 'International Drug Treaties: The Connection Between Ratification and Social and Economic Conditions', 13 *Drug and Alcohol Dependence* 107 (1984).
H. Smets, 'The Right of Information on the Risks Created by Hazardous Installations at the National and International Levels', in *International Responsibility for Environmental Harm* (F. Francioni & T. Scovazzi eds, 1991).
N. Smith, 'The Real Challenge to the Polish Revolution: Cleaning the Polish Environment Through Privatization and Preventive Market-Based Incentives', 19 *Pepperdine LR* 553 (1992).
D. Snidal, 'The Game Theory of International Politics', in *Cooperation Under Anarchy* (K. Oye ed., 1986). Snidal(a).
D. Snidal, 'Relative Gains and the Pattern of International Cooperation', 85 *APSR* 701 (1991). Snidal(b).

D. Snidal, 'Political Economy and International Institutions', 16 *International Review of Law and Economics* 121 (March 1996). Snidal(c).
G. Snyder & P. Diesing, *Conflict Among Nations: Bargaining, Decision Making, and System Structure in International Crises* (1977).
A. Springer, 'Towards a Meaningful Concept of Pollution in International Law', 26 *ICLQ* 531 (1977). Springer(a).
A. Springer, *The International Law of Pollution: Protecting the Global Environment in a World of Sovereign States* (1983). Springer(b).
E. Somers, 'The Role of the Courts in the Enforcement of Environmental Rules', 5 *International Journal of Estuarine and Coastal Law* 195 (1990).
J. Sommer, 'Transboundary Co-operation Between Poland and its Neighboring States', in *Transboundary Air Pollution* 205 (C. Flinterman *et al*. eds, 1986). Sommer(a).
J. Sommer, 'Polish Environmental Law Developments from an EC Perspective', 2 *RECIEL* 11 (1993). Sommer(b).
J. Sommer, 'Poland', in *International Encyclopedia of Laws* (R. Blanpain ed., 1993). Sommer(c).
R. Soni, *Control of Marine Pollution in International Law* (1985).
M. Soto, 'General Principles of International Environmental Law', 3 *ILSA JICL* 193 (Fall 1996).
B. Spector, *Post-Negotiation: Is the Implementation of Future Negotiated Environmental Agreements Threatened?* (IIASA Working Paper, 1992). Spector(a).
B. Spector, *International Environmental Negotiation: Insights for Practice* (IIASA Working Paper No. 92-22, 1992). Spector(b).
B. Spector & A. Korula, *The Post-Agreement Negotiation Process: The Problems of Ratifying International Environmental Agreements* (IIASA Working Paper, December 1992).
A. Springer, *The International Law of Pollution* 142 (1983).
D. Sprinz & T. Vaahtoranta, 'The Interest-Based Explanation of International Environmental Policy', 48 *International Organization* 77 (1994).
B. Stark, 'Conceptions of International Peace and Environmental Rights: "The Remains of the Day",' 59 *Tennessee LR* 651 (1992).
A. Starzewska, 'The Polish People's Republic', in *Environmental Policies in East and West* 294 (G. Enyedi *et al*. eds, 1987).
J. Stein ed., *Getting to the Table: The Processes of International Pre-Negotiation* (1989).
R. Stein, 'The Settlement of Environmental Disputes: Towards a System of Flexible Dispute Settlement', 12 *Syracuse JIL & Commerce* 283 (1985).
R. Stein & B. Johnson, *Banking on the Biosphere? Environmental Procedures and Practices of Nine Multilateral Development Agencies* (1979).
H. Steinberger, 'The International Court of Justice', in *Judicial Settlement of International Disputes* 194 (Max Planck Institute for Comparative Public Law and International Law ed., 1974).
N. Stephen, 'The Growth of International Environmental Law', 8 *Environmental & Planning LJ* 183 (1991).
K. Steven, 'Hard Law, Soft Law and Diplomacy: The Emerging Paradigm for Intergovernmental Cooperation in Environmental Assessment', 31 *Alberta LR* 644 (1992).
A. Stewart, 'Environmental Risk Assessment: The Divergent Methodologies of Economist, Lawyers and Scientists', 10 *Environment & Planning LJ* 10 (1993).

B. Stewart & K. Wilshusen, 'U.S.–Canadian Negotiations on Acid Rain', in *Acid Rain and Friendly Neighbors: The Policy Dispute Between Canada and the United States* 61 (J. Schmandt & H. Roderick eds, 1985).

A. Strauss, 'From Gattzilla to the Green Giant: Winning the Environmental Battle for the Soul of the World Trade Organization', 19 *University of Pennsylvania Journal of International Economic Law* 769 (Fall 1998).

L. Strauss & J. Cropsey, *History of Political Philosophy* (1973).

T. Stypka, *Cambridge Global Security Fellows' Initiative, Air Pollution from Large Industrial Sources: An International Perspective* (Occasional Paper No. 1, 1997).

P. Sugar ed., *Ethnic Diversity and Conflict in Eastern Europe* (1980).

M. Sullivan, 'Transboundary Pollution from Mexico: Is Judicial Relief Provided by International Principles of Tort Law?', 10 *Houston JIL* 105 (1987).

G. Supanich, 'The Legal Basis of Intergenerational Responsibility: An Alternative View – the Sense of Intergenerational Identity', 3 *YIEL* 94 (1992).

'Survey of the Most Important Czechoslovak Legal Regulations in the Sphere of Environmental Care', 14 *BCL* 117 (1975).

E. Susman, 'Regulation of Ocean Dumping by the European Economic Community', 18 *Ecology LQ* 559 (1991).

J. Symonides, 'International Legal Problems of the Fight Against the Pollution of Rivers', *Polish YIL* 99 (1972–73).

P. Szasz, *The Law and Practices of the IAEA* (1970); Supplement (1993). Szasz(a).

P. Szasz, 'International Responsibility for Manmade Disasters', 81 *Proceed. ASIL* 320 (1987). Szasz(b).

P. Szasz, 'International Norm-Making', in *Environmental Change and International Law* 41 (1992). Szasz(c).

P. Szilagyi, 'Hungarian Perspective on the Protection of Transboundary Rivers', 2 *RECIEL* 40 (1993).

K. Tamasevski, 'Monitoring Human Rights Aspects of Sustainable Development', 8 *American University JILP* 1 (1992).

L. Teclaff, 'Fiat or Custom: The Checkered Development of International Water Law', 31 *NRJ* 45 (1991). Teclaff(a).

L. Teclaff, 'Evolution of the River Basin Concept in National and International Water Law', 36 *NRJ* 359 (Spring 1996). Teclaff(b).

L. Teclaff & E. Teclaff, 'Transboundary Toxic Pollution and the Drainage Basin Concept', 25 *NRJ* 589 (1985). Teclaff & Teclaff(a).

L. Teclaff & E. Teclaff, 'International Control of Cross Media Pollution – An Ecosystem Approach', in *Transboundary Resources Law* 253 (1987). Teclaff & Teclaff(b).

R. Teitel, 'Transitional Jurisprudence: The Role of Law in Political Transformation 106 *Yale LJ* 2009 (May 1997).

'Ten Years of Environmental Co-operation in the Baltic Sea: An Evaluation and Look Ahead', 14 *Agua Fenica* 3 (1989).

F. Teson, 'Realism and Kantianism in International Law', 86 *Proceed. ASIL* 113 (1992).

Y. Tharpes, 'International Environmental Law: Turning the Tide on Marine Pollution,' 20 *Univeristy of Miami Inter-American LR* 579 (1989).

S. Tiefenbrun, 'The Role of the World Court in Settling International Disputes: A Recent Assessment', 20 *Loyola of Los Angeles ICLJ* 1 (November 1997).

G. Timagenis, *International Control of Marine Pollution* (1980).

A. Timoshenko, 'International Environmental Law: Fundamental Aspects', 59 *RJUPR* 653 (1990).
C. Tomuschat, 'International Liability for Injurious Consequences Arising out of Acts not Prohibited by International Law: The Work of the International Law Commission', in *International Responsibility for Environmental Harm* 15 (F. Francioni & T. Scovazzi eds, 1991).
J. Trachtman, 'Reflections on the Nature of the State: Sovereignty, Power and Responsibility', 20 *Canada–United States LJ* (1994).
G. Tsebelis, *Nested Games: Rational Choice in Comparative Politics* (1990).
G. Tunkin, *Theory of International Law* (1974). Tunkin(a).
G. Tunkin, 'On the Primacy of International Law in Politics', in *Perestroika and International Law* 5 (W. Butler ed., 1990). Tunkin(b).
E. Usenko, 'The Norm-Creating Activity of COMECON and Soviet Law', in *International Law and the International System* 113 (W. Butler ed., 1987).
A. Utton, 'The Transfer of Water from an International Border Region: A Tale of Six Cities and the All American Canal', 16 *North Carolina Journal of International Law and Comparative Regulation* 477 (1991).
A. Utton & L. Teclaff, eds, *Transboundary Resources Law* (1987).
A. Vamvoukos, *Termination of Treaties in International Law* (1985).
M. van de Kerdrove & F. Oss, 'Breaking Frames: The Global Interplay of Legal and Social Systems', 45 *American JCL* 149 (Winter 1997).
J. Vavrousek *et al.*, *The Environment in Czechoslovakia* (1990).
F. Vicuna, 'State Responsibility, Liability, and Remedial Measures Under International Law: New Criteria for Environmental Protection', in *Environmental Change and International Law* 124 (1992).
O. Vidlakova, 'Guest Editorial', 2 *RECIEL* v (1993).
M. Villiger, *Customary International Law and Treaties* (1985).
B. Vitanyi, *The International Regime of River Navigation* (1979).
I. Volgyes ed., *Environmental Deterioration in the Soviet Union and Eastern Europe* (1974).
B. Vukas, 'Concluding Observations on General International Law and New Challenges in the Field of Transboundary Air Pollution', in *Transboundary Air Pollution* (1986).
V. Vukasovic, 'Protection of the Environment: One of the Key Issues in the Field of Human Rights', 59 *RJUPR* 889 (1990).
R. Wagenbaur, 'Regulating the European Environment: The EC Experience', 1992 *University of Chicago Legal Forum* 17.
H. Wagner, 'Theory of Games and the Problem of International Cooperation', 77 *APSR* 330 (1983).
S. Wajda & J. Sommer, 'Environmental Liability in Property Transfer in Poland', in *Environmental Liability and Privatization in Central and Eastern Europe* 179 (G. Goldenman *et al.* eds, 1994).
W. Wallace, *Czecho-Slovakia* (1976).
K. Waltz, *Theory of International Politics* (1979).
H. Ward, 'Game Theory and the Politics of the Global Commons', 37 *Journal of Conflict Resolution* 203 (1993). Ward(a).
H. Ward, 'Game Theory and the Politics of Global Warming: The State of Play and Beyond', 54 *Political Studies* 850 (1996). Ward(b).
P. Wathern ed., *Environmental Impact Assessment: Theory and Practice* (1988).

S. Weber, 'Environmental Information and the European Convention on Human Rights', 12 *Human Rights LJ* 177 (1991).

G. Weeramantry, *Nauru: Environmental Damage Under International Trusteeship* (1992).

A. Weinstein, 'Environmental Problems in Eastern Europe: Learning from the United States' Mistakes', 21 *Cumbria LR* 535 (1991).

B. Weintraub, 'Environmental Security, Environmental Management, and Environmental Justice', 12 *Pace Environmental LR* 533 (Spring 1995).

D. Weissbrodt, 'A New United Nations Mechanism for Encouraging the Ratification of Treaties', 4 *Human Rights Quarterly* 333 (1982).

A. Westing ed., *Global Resources and International Conflict* (1986). Westing(a).

A. Westing, 'An Expanded Concept of International Security', in *Global Resources and International Conflict* 183 (1986). Westing(b).

G. Westone & A. Rosencranz, 'Transboundary Air Pollution: The Search for an International Response', 8 *Harvard ELR* 89 (1984).

J. Wettestad & S. Andresen, *The Effectiveness of International Resource Cooperation: Some Preliminary Findings* (1991).

L. Wicke, 'Environmental Damage Balance Sheets', in *Maintaining a Satisfactory Environment: An Agenda for International Environmental Policy* 34 (N. Åkerman ed., 1990).

D. Wilkinson, 'Maastricht and the Environment: the Implications for the EC's Environment Policy of the Treaty on European Union', 4 *JEL* 222 (1992).

P. Williams, 'International Environmental Dispute Resolution: The Dispute Between Slovakia and Hungary Concerning Construction of the Gabčíkovo and Nagymaros Dams', 19 *Columbia JEL* 1 (1994).

J. Wilson, *Bureaucracy* (1990).

G. Winter, 'The Implementation of the Oslo Convention for the Prevention of Marine Pollution by Dumping from Ships and Aircraft', 3 *Zeitschrift fur Umweltpolitik* 707 (1980).

C. Wojatsek, *From Trianon to the First Vienna Arbitral Award: The Hungarian Minority in the First Czechoslovak Republic – 1918–1938* (1980).

P. Wouters, 'An Assessment of Recent Developments in International Watercourse Law through the Prism of the Substantive Rules Governing Use Allocation', 36 *NRJ* 417 (Spring 1996).

D. Wyatt & A. Dashwood, *The Substantive Law of the EEC* (2nd ed., 1987).

A. Yankov, 'The Integration of Central and Eastern European States into a Common European System for the Peaceful Settlement of Disputes', in *The Peaceful Settlement of International Disputes in Europe: Future Prospects* 447 (D. Bardonnet ed., 1991).

Yearbook of the International Court of Justice 1993–94 (1994).

K. Youel, 'Theme Plenary Session: Implementation, Compliance and Effectiveness', 91 *Proceed ASIL* 50 (9–12 April 1997).

O. Young, *International Cooperation: Building Regimes for Natural Resources and the Environment* (1989). Young(a).

O. Young, 'The Effectiveness of International Institutions: Hard Cases and Critical Variables', in *Governance Without Government: Order and Change in World Politics* (1992). Young(b).

H. Young & A. Wolf, 'Global Warming Negotiations: Does Fairness Matter?', 10 *The Brookings Review* 46 (1992).

R. Zacklin & L. Caflisch eds, *The Legal Regime of International Rivers and Lakes* (1981).
K. Zahariev, 'European Bank for Reconstruction and Development: Environmental Aspects of Operations', 2 *RECIEL* 31 (1993).
D. Zalob, 'Approaches to Enforcement of Environmental Law: An International Perspective', 3 *Hastings ICLR* 299 (1980).
W. Zartman, 'International Environmental Negotiation: Challenges for Analysis and Practice', 8 *Negotiation Journal* 113 (1992).
I. Závadský, 'Environmental Management of the Danube', 2 *RECIEL* 36 (1993).
K. Zemanek, 'State Responsibility and Liability', in *Environmental Protection and International Law* 187 (W. Lang *et al.* eds,1991).
M. Zeppetello, 'National and International Regulation of Ocean Dumping: The Mandate to Terminate Marine Disposal of Contaminated Sewage Sludge', 12 *Ecology LQ* 619 (1985).
C. ZumBrunnen, 'The Environmental Challenges in Eastern Europe', 19 *Millennium Journal of International Studies* 389 (1990).
M. Zürn, 'Intra-German Trade: An Early East–West Regime', in *International Regimes in East–West Politics*, 151 (V. Rittberger ed., 1990). Zürn(a).
M. Zürn, 'Bringing the Second Image (Back) in: About the Domestic Sources of Regime Formation', in *Regime Theory and International Relations* 282 (V. Rittberger ed., 1993). Zürn(b).
C. Zvrosec, 'Environmental Deterioration in Eastern Europe', *Survey* (Winter 1984).

Publications by governments, international organizations and nongovernmental organizations

Bulgaria, Protection and Reproduction of the Environment, in Documents of the State Council of the People's Republic of Bulgaria (1980).
Czech Republic, Ministry of Environment, Environment of the Czech Republic (1990).
Czech Republic, Ministry of Environment, Environmental Year-Book of the Czech Republic 1990 (1991). EYBCR 1990.
Czech Republic, Ministry of Environment, Environmental Year-Book of the Czech Republic 1991 (1992). EYBCR 1991.
Czech Republic, Ministry of Environment, Environmental Year-Book of the Czech Republic 1992 (1993). EYBCR 1992.
Czech Republic, Water Research Institute, Elbe Project (Bulletin, December 1993).
The Danube Defense Action Committee, The Danube Blues: Questions and Answers About the Bös (Gabčíkovo)–Nagymaros Hydroelectric Station System (Budapest, October 1992).
The Danube Task Force, Environmental Programme for the Danube River Basin: Strategic Action Plan (1994).
Ecologia, The Bös-Nagymaros Barrage Study: Program Operations and Impacts (May 1989).
Environment for Europe, Environmental Action Programme for Central and Eastern Europe (29 March 1993).

Environmental Management and Law Association, Environmental Law and Management System in Hungary: Overview, Perspective and Problems (June 1993).
European Commission, Environmental Legislation (1992).
European Commission, Regional Environmental Program Evaluation (1992).
European Commission, Communication on Environmental Liability (1993).
European Commission, White Paper on CEE States Accession to the EU (June 1995).
Euroregion Neisse–Nisa–Nysa, Basic Facts (1992).
Friends of the Earth – Poland, Mochovce Press Release (5 January 1995).
Greenpeace, Poland Suffers from Severe Pollution: Krakow is Not a Nice Place Even to Visit (15 August 1990).
Ground Water Consulting Ltd., Gabčíkovo–WWF: the Pros and Cons (April 1994) – prepared by I. Mucha.
Republic of Hungary, Ministry for Environment and Regional Policy, Review of the Environment, Nature Conservation, Building and Regional Policy in Hungary (1991).
Republic of Hungary, The Danube Story: Educational Summary (Budapest, February 1991).
Republic of Hungary, Information about the Scope of Problems Connected with the System of Gabčíkovo–Nagymaros Waterworks (25 February 1992).
Republic of Hungary, The Prime Minister's Office, Gabčíkovo–Nagymaros File (October 1992) (Secretariat of Mr Ferenc Madl, Minister without Portfolio).
Republic of Hungary, Ministry of Foreign Affairs, Declaration of the Government of the Republic of Hungary on the Termination of the Treaty Concluded Between the People's Republic of Hungary and the Socialist Republic of Czechoslovakia on the Construction and Joint Operation of the Gabčíkovo–Nagymaros Barrage System, Signed in Budapest on 16 September 1977 (19 May 1992).
Republic of Hungary, Application of the Republic of Hungary v. The Czech and Slovak Federal Republic on the Diversion of the Danube River, submitted to the Registrar of the International Court of Justice (signed Budapest, 22 October 1992).
Republic of Hungary, Memorial of the Republic of Hungary in the Case Concerning the Gabčíkovo–Nagymaros Project (Hungary/Slovakia) (2 May 1994).
Republic of Hungary, Counter-Memorial of the Republic of Hungary in the Case Concerning the Gabčíkovo–Nagymaros Project (Hungary/Slovakia) (5 December 1994).
Republic of Hungary, Reply of the Republic of Hungary in the Case Concerning the Gabčíkovo–Nagymaros Project (Hungary/Slovakia) (20 June 1995).
International Law Association, Rules of International Law Applicable to Transfrontier Pollution, Report of the 60th Conference 158 (1982).
International Maritime Organization, The London Dumping Convention: The First Decade and Beyond (1991).
International Law Commission, Report on Draft Articles on the Law of the Non-Navigational Uses of International Watercourses, 45 U.N. GAOR Supp. (No. 10), U.N. Doc. A/45/10 (1990).
International Law Commission, Report on International Liability for Injurious Consequences Arising Out of Acts Not Prohibited by International Law (1990) 2 YILC 90 (1993).

International Law Commission, Report on State Responsibility (1990) 2 YILC 68 (1993).
International Law Commission, Report on Draft Articles on the Non-Navigational Uses of International Watercourses, UN Doc. A/CN.4/L492, Add.1 (1994).
International Law Commission, Report of the International Law Commission on the work of its forty-eighth session, 6 May to 26 July 1996, Official Records of the General Assembly, Fifty-first Session, Supplement No. 10 (A/51/10).
ISTER, Future of the Danube (1991).
International Union for the Conservation of Nature – East European Programme, Environmental Status Reports 1988/89, Vol. I., Czechoslovakia, Hungary, Poland (1990).
Max Planck Institute for Comparative Public Law and International Law, Judicial Settlement of International Disputes (1974).
OECD, The Polluter Pays Principle, Definition, Analysis, Implementation (1975).
OECD, Nuclear Third Party Liability: Nuclear Legislation (1976).
OECD, Responsibilities and Liability of States in Relation to Transfrontier Pollution (1979).
OECD, Application of Information and Consultation Practices for Preventing Transfrontier Pollution, in Transfrontier Pollution and the Role of States (1981).
OECD, Transfrontier Pollution and the Role of States (1981).
OECD, Compensation for Pollution Damage (1981).
OECD, Report by the Environment Committee on Responsibility and Liability of States in Relation to Transfrontier Pollution (1984).
OECD, Economic Instruments for Environmental Protection (1989).
OECD, Guidelines for the Application of Economic Instruments in Environmental Policy (1991).
OECD, Pollution Insurance and Compensation Funds for Accidental Pollution (1991).
OECD, Use of Economic Instruments in Environmental Policy (Recommendation of 31 January 1991).
OECD, Guidelines and Considerations for the Use of Economic Instruments in Environmental Policy (Annex to Use of Economic Instruments in Environmental Policy, 31 January 1991).
OECD, Managing the Environment: The Role of Economic Instruments (1994).
OSCE, Provisions on Environment in the CSCE Documents 1973–1989 (1989).
Republic of Poland, Polish Ministry of Environmental Protection, Natural Resources and Forestry, the State of the Environment in Poland: Damage and Remedy (1992).
Republic of Poland, Polish Agency for Regional Development, Presentation of Selected Regions (May 1995).
Republic of Poland, Polish Ministry of Environmental Protection Natural Resources and Forestry, The State of Harmonization of the Polish Law and Environmental Legislation of the European Union (15 July 1998).
Polish Ecological Club, Environment and Development for Poland: Declaration of Sustainable Development (1990).
Regional Environmental Center for Central and Eastern Europe, Strategic Environmental Issues in Central and Eastern Europe: Regional Report (August 1994).
Regional Environmental Center for Central and Eastern Europe, Strategic Environmental Issues in Central and Eastern Europe: Environmental Needs Assessment in Ten Countries (August 1994).

Regional Environmental Center for Central and Eastern Europe, Use of Economic Instruments in Environmental Policy in Central and Eastern Europe (1995).
Report of the European Commission Fact Finding Mission Concerning Variant C (31 October 1992).
Report of the Working Group of Independent Experts (November 1992).
Report of the Working Group of Water Management Experts for Gabčíkovo System of Locks (November 1993).
Report on Temporary Water Management Regime from the Working Group of Monitoring and Water Management Experts for Gabčíkovo System of Locks (December 1993).
Romania, Ministry of Water, Forestry and Environmental Protection, List of Projects to be Taken into Account for Inclusion in the EC Project Preparation Facility (1992).
Slovak National Agency for Foreign Investment and Development, Press Release: The Gabčíkovo Dam Q&A Fact Sheet (March 1993).
National Council of the Slovak Republic, Resolution on the Gabčíkovo–Nagymaros Hydroelectric Project adopted 24 March 1993, reprinted in FBIS-EEU 15–16 (30 March 1993).
Republic of Slovakia, Ministry of Environment, Enviro Guide Slovakia 1993 (1993).
Republic of Slovakia, Ministry of Environment, Background for Environmental Policy: An Overview of the Environmental Situation (1993).
Republic of Slovakia, Government Office, Press and Information Department, The Gabčíkovo Water Works Project: Conserving the Danube's Inland Delta (1993).
Republic of Slovakia, Memorial of the Slovak Republic in the Case Concerning the Gabčíkovo–Nagymaros Project (Hungary/Slovakia) (2 May 1994).
Republic of Slovakia, Counter-Memorial of the Slovak Republic in the Case Concerning the Gabčíkovo–Nagymaros Project (Hungary/Slovakia) (5 December 1994).
Republic of Slovakia, Reply of the Slovak Republic in the Case Concerning the Gabčíkovo–Nagymaros Project (Hungary/Slovakia) (20 June 1995).
Republic of Slovakia, Ministry of Environment, Environmental Programme for the Danube River Basin (1995).
Slovak Union of Nature and Landscape Protectors & Slovak Rivers Network, Damming the Danube: What Dam Builders Don't Want You To Know (March 1993).
Statement of the Slovak Non-Governmental Organizations Concerning the Mochovce Nuclear Power Plant Project Documentation Submitted for the Public Discussion (17 February 1995).
United Nations, Handbook on the Peaceful Settlement of Disputes Between States (1995).
United Nations ECE, The State of Transboundary Air Pollution, 1992 Update (1992).
United Nations ECE, Impacts of Long-Range Transboundary Air Pollution (1992).
United States EPA, Air Pollution Study in Northern Bohemia and Silesia: Concept Paper (8 August 1991).

United States General Accounting Office, East European Energy, Romania's Energy Needs Persist (August 1992).
United States Institute of Peace, Roundtable on CSCE: Mechanisms for the Peaceful Settlement of Disputes (6 October 1992).
United States Office of Technology Assessment, Energy Efficiency Technologies for Central and Eastern Europe (May 1993).
Vodohospodarska Vystavba, Gabčíkovo–Nagymaros Project – Basic Information about its Actual State and Perspectives (Bratislava, June 1991).
Vodohospodarska Vystavba, Gabčíkovo–Nagymaros Project: Standpoint of the Czecho–Slovak Side and Answers to Questions (Bratislava, April 1992).
Vodohospodarska Vystavba, Gabčíkovo–Nagymaros Project Counter-Proposal of Operation Gabčíkovo (Bratislava, 24 October 1992).
Vodohospodarska Vystavba, The Gabčíkovo–Nagymaros Project – Latest Developments 1991 (Bratislava, no date).
Vodohospodarska Vystavba, The Gabčíkovo–Nagymaros Project: the Temporary Solution on the Territory of the CSFR-Slovakia (Bratislava, no date).
Vodohospodarska Vystavba, The Gabčíkovo–Nagymaros Project: Part Gabčíkovo, Solution According to the 1977 Treaty (Bratislava, no date).
World Bank, The World Bank and the Environment – A Progress Report (1991).
World Bank, World Bank and the Environment (1992).
World Bank, Environment and Development in Bulgaria: A Blueprint for Transition (September 1992).
World Bank, Bulgaria Environment Strategy Study (17 March 1992).
World Bank, Romania Environment Strategy Paper (31 July 1992).
World Bank, Bulgaria: Crisis and Transition to a Market Economy Volume II (1992) (A World Bank Country Study).
World Bank, Czech and Slovak Federal Republic Joint Environmental Study Volume I (22 January 1992).
World Bank, Czech and Slovak Federal Republic Joint Environmental Study: Technical Study Volume II (22 January 1992).
The World Bank, Czechoslovakia: Transition to a Market Economy (1992) (A World Bank Country Study).
World Commission on Environment and Development – Experts Group on Environmental Law, Report on Environmental Protection and Sustainable Development (1987).
World Wildlife Fund, Repercussions of the Power Station (1991).
World Wildlife Fund, A New Solution for the Danube: WWF Statement on the EC Mission Reports of the 'Working Group of Monitoring and Management Experts' and on the Overall Situation of the Gabčíkovo Hydrodam Project (1993).

Newspaper articles, popular journals and electronic news summaries

By author

W. Broad, *IHT* 1 (24 July 1995).
D. Hinrichsen, On a Slow Trip Back From Hell: The Infamous Black Triangle had the Worst Pollution Ever Recorded in the Industrialized World. Now a Sick and

Weary People are Tackling Their Nightmare, *International Wildlife* (11 January 1998).
M. Jordon, 'Environmentalists Decry Europe Dam Decision', *Christian Science Monitor* 6 (29 September 1997).
N. King, 'Cleaning Up Eastern Europe is a Tidy Business', *Wall Street Journal* (5 April 1996).
M. Ledford, Safety Czech: Trying to contain Soviet Reactors, *Multinational Monitor* (1 September 1995).
A. MacLachlan, Czech Utility Now Wants to Expand On-Site Spent Fuel Capacity, *Nuclear Fuel* (19 May 1997).
A. MacLachlan, EU-sponsored Tests to Establish Whether V-213 Confinements Work, *Nucleonics* (30 April 1998).
M. Milner, 'Nuclear Bill Adds up to 100 Million Pounds', The *Guardian* 35 (3 September 1994).
J. Pehe, 'The Choice Between Europe and Provincialism', 1 *Transition* 14, 14–16 (July 1995).
A. Steele, New Focus on Old Threat to Security: Nuclear Power, *Jane's Intelligence Review* (1 December 1997).
J. Thompson, East Europe's Dark Dawn, *National Geographic* 36 (June 1991).
C. Woodard, Fighting for the Scraps: Western Nuclear Companies are on the Prowl in Eastern Europe, *Bulletin of Atomic Scientists* (15 May 1996).

By publisher

AFP, Green Groups Protest 'Stalinist' Hungarian Dam Decision (6 February 1998).
BBC World Broadcasts, Premier Urges Completing Second Nuclear Power Station (2 June 1996).
BBC World Broadcasts, Caretaker Cabinet Not to Make Decision on Belene Nuclear Plant Project (28 February 1997).
BBC World Broadcasts, Russian Gas Giant Secures 25-Year Agreement on Supplies to Turkey (1 April 1997).
BBC World Broadcasts, Environment Ministry Opposes Black Sea Gas Pipeline (13 November 1997).
BBC World Broadcasts, Safety Recommendations for Kozloduy Nuclear Plant Implemented (4 December 1997).
BBC World Broadcasts, Slovakia Rejects Hungary's 'One-Sided Approach' to Dam Dispute (23 March 1998).
BBC World Broadcasts, Slovakia to Ask EU to Make Hungary Fulfil Disputed Dam Obligations (3 April 1998).
BBC World Broadcast, Government Approves Three-year Public Investment Programme (23 July 1998).
BBC World Broadcasts, Government Officially Endorses Transit of Bulgarian Nuclear Fuel (17 August 1998).
BBC World Broadcasts, Foreign Minister Optimistic About Relations with Czech Republic (20 November 1998).
BNA, Environmental Protection Issues in Eastern Europe, 13 *IER* 258 (1990).
BNA, Polish Environmentalists Say Economic, Environmental Recovery Linked, 13 *IER* 330 (1990).

BNA, Pact Reached with IMF on Reform Plan; Loans for $171 Million Pledged by EU, 4 *Eastern Europe Reporter* 304 (11 April 1994).
BNA, Air Pollution Control in the Czech Republic, 4 *Eastern Europe Reporter* 362 (24 April 1994).
Business Insurance, EU Initiative Helps Countries Enforce Environmental Laws (31 August 1998).
Christian Science Monitor, An Unsafe Nuclear Plant in Eastern Europe Appears Unclosable (6 December 1996).
CTK News Wire, Most Czechs Approve of Nuclear Energy – Poll (21 June 1995).
CTK News Wire, Czech and American Armies Cooperate on Environment (6 February 1997).
CTK News Wire, Rainbow Movement Organizes Radiation Monitoring Around Dukovany (18 March 1997).
CTK News Wire, Slovak Power Plants Denounce Temelín Environmentalists (7 July 1997).
CTK News Wire, Euroregion Money to be Used Against Floods (2 September 1997).
CTK News Wire, Austrian Public Resents Czech, Slovak Nuclear Plants (5 September 1997).
CTK News Wire, Environmentalists Sue Temelín Nuclear Plant (10 September 1997).
CTK News Wire, MPs Want Surveys of Temelín's Impact to be Disclosed (14 November 1997).
CTK News Wire, Czech Parliamentary Delegation Rejects Revisions of Benes Decrees (25 November 1997).
CTK News Wire, Austrians to Continue with Anti-Nuclear Campaign (6 January 1998).
CTK News Wire, Austria to Oppose Plans for Expanded Nuclear Waste Site (24 March 1998).
CTK National News Wire, Slovakia Has Right, But No Reason to Apply to The Hague (25 March 1998).
CTK National News Wire, Meciar Intends to Apply to the Hague, Rejects Neutrality (25 March 1998).
CTK News Wire, Dukovany Waste Storage Site is International Issue (30 March 1998).
CTK News Wire, Bursik Looking for Ways of State to do Without Temelín (1 April 1998).
CTK News Wire, Czech Nuclear Power Must Not be Linked to EU Membership (2 April 1998).
CTK News Wire, Czech, Austrian Environment Ministers Discuss Nuclear Power (2 April 1998).
CTK National News Wire, Way Open for Gabcikovo Referendum in Hungary (2 April 1998).
CTK News Wire, Austria Concerned About Safety of Czech Nuclear Power (3 April 1998).
CTK News Wire, State Nuclear Safety Body Finds Shortcomings in Temelín (14 April 1998).
CTK News Wire, EU Easing up over Czech Nuclear Reactors (16 April 1998).
CTK News Wire, Czechs to invest Kc500 billion in Environmental Laws (16 April 1998).

CTK News Wire, Czech Environment Better, But Still Among the Worst (20 April 1998).
CTK News Wire, Launch of 'Slovak Temelín' Source of Many Doubts (2 May 1998).
CTK News Wire, Profile of Mochovce, Thorn in Side of Slovak–Austrian Relations (8 June 1998).
CTK News Wire, Austria Welcomes Temelín Decision (2 July 1998).
CTK News Wire, Kuzvart, Prammer Discuss Nuclear Plant (9 October 1998).
CTK News Wire, Prammer Satisfied with Talks in Prague (9 October 1998).
CTK News Wire, Hungarians Ready to Discuss Gabcikovo with Slovakia (18 November 1998).
Deutsche Presse-Agentur, World Bank Supports Baltic Sea Clean-Up Programmes (17 May 1995).
East European Energy Report, Black Triangle Progress Report (19 December 1995).
East European Markets, EU Push on Baltics (26 April 1996).
The Economist 54 (16 September 1995).
The Economist, Emerging-Market Indicators, 120 (2 March 1996).
Euro-East, The Other Five Accession Partnerships Agreed by European Commission (19 May 1998).
Euro-East, Environment Looked at in Terms of Eastern Enlargement (23 September 1997).
Europe Energy, Agenda 2000: Environment to Remain High Priority into 21st Century (25 July 1997).
Europe Energy, Nuclear Safety: Commission Asked to Step in to prevent Disaster at Slovakian Plant (5 June 1998).
Europe Environment, EIB: 23 per cent Increase in Loans for Environmental Actions in 1997 (10 February 1998).
Europe Environment, European Commission: DG XI Director-General James Currie Sets out the Commission's Priorities (7 April 1998).
FBIS-EEU, Russian Offer for Ecological Damage Unacceptable (20 February 1992).
FBIS-EEU, Dispute Over Nuclear Waste Storage Reported (20 April 1992).
FBIS-EEU, Views on the Gabčíkovo–Nagymaros Project (20 April 1992).
FBIS-EEU, Ecoglasnost Sees Kozloduy as 'Threat' to Europe (22 April 1992).
FBIS-EEU, Ministers Discuss Nuclear Waste Storage (31 May 1993).
FBIS-EEU, Minister on Pollution From Romanian Plant (12 July 1993).
FBIS-EEU, Kovac on Decommissioning Nuclear Plant (14 July 1993).
FBIS-EEU, Slovakia Protests Nagymaros Restoration Project (14 July 1993).
FBIS-EEU, Talks Continue on Danube Dam with EC Mediation (14 July 1993).
FBIS-EEU, Bank Supports Environmental Project with Romania (15 July 1993).
FBIS-EEU, World Bank Supports Environmental Project with Romania (15 July 1993).
FBIS-EEU, Ethnic Slovaks on 'Deteriorating Situation' (16 July 1993).
FBIS-EEU, Article Ponders Border Cooperation Issues (16 September 1994).
FBIS-EEU, Dukovany to Choose Nuclear Fuel Supplier (30 September 1994).
FBIS-EEU Environment Protection Document Signed with Czech Republic (21 October 1994).
FBIS-EEU, Austrian View of Temelín Power Station Explained (2 November 1994).
FBIS-EEU, Euroregion NYSA Three-Year Activities Examined (14 November 1994).
FBIS-EEU, Pollution Damages Forests in Republic (16 November 1994).

FBIS-EEU, Ecology International Cooperation Results Reported (1 December 1994).
FBIS-EEU, Euroregion NYSA Three-Year Activities Examined (14 December 1994).
FBIS-EEU, Czech Republic Negotiates with Russia on Nuclear Waste (3 January 1995).
FBIS-EEU, Mayors Reject Storage Idea for Dukovany Plant (4 January 1995).
FBIS-EEU, Havel: State Lacks Clear Idea of Power Industry (5 February 1995).
FBIS-EEU, PHARE Environmental Project in Black Triangle Region (9 March 1995).
FBIS-EEU, Independent Expert on Environment Protection (31 March 1995).
FBIS-EEU, Official Discusses Future of Kozloduy Reactors (5 April 1995).
FBIS-EEU, Support Shown for Nuclear Power Plant Usage (17 April 1995).
FBIS-EEU, Melescanu, Bulgarian Counterpart Examine Ties (24 April 1995).
FBIS-EEU, Strategies for Rehabilitating Silesia Examined (13 May 1995).
FBIS-EEU, Japan Offers Safety Equipment for Kozloduy Plant (26 June 1995).
FBIS-EEU, Oil Terminals Worsening Baltic Pollution (22 July 1995).
FBIS-EEU, Videnov Comments on Energy Accords with Russia (20 September 1995).
FBIS-EEU, Hungary Ratifies International Water Management Agreement (27 October 1995).
FBIS-EEU, Bulgaria to Receive Untied Loan for Pollution Control (25 November 1995).
FBIS-EEU, Czech Clippings (31 May 1996).
FBIS-EEU, Hungary: Draft Text of Hungarian–Romanian Basic Treaty (2 September 1996).
FBIS-EEU, Czech Republic: Fire Breaks out at Temelín Nuclear Power Station (17 October 1996).
FBIS-EEU, Nysa Euroregion Sets Up Tripartite Parliamentary Group (30 November 1996).
FBIS-EEU, Croatia: Commentary Views Pros, Cons of SECI (25 January 1997).
FBIS-EEU, Czech Republic: Temelín Nuclear Power Station Not to be Ready Before 1999 (2 April 1997).
FBIS-EEU, Romania: Danube Basin Alert Emergency Warning System Inaugurated (12 April 1997).
FBIS-EEU, Slovakia: Czech Spent Nuclear Fuel Shipped Back to Czech Republic (12 April 1997).
FBIS-EEU, Romania: Council of Europe Discusses European Danube Charter (28 May 1997).
FBIS-EEU, Hungary: EU Environmental Standards to be Met by 2002 (8 July 1997).
FBIS-EEU, Romania: Nuclear Plant Authorized to Operate Provisionally (5 August 1997).
FBIS-WEU, Bundesbericht Forschung Report on German Research Policy, Programs, Funding: Main Areas of Federal R&D Support (16 August 1994).
FBIS-WEU, East Europeans Likely to Sign (20 September 1994).
FBIS-WEU, Klaus Views Temelín as Grandiose Project (19 October 1994).
FBIS-WEU, Independent Expert on Environment Protection (31 March 1995).
FBIS-WEU, Vranitzky Asks Klaus for Information on Temelín (7 June 1995).
Financial Times, West's Policies on Eastern Nuclear Plants 'Misguided' (10 February 1997).
Financial Times, Asia Intelligence Wire, Black Sea Becoming 'Dead Sea' (15 December 1997).

318 Bibliography

The Green Voice (6 August 1995).
Greenwire, Worldview Poland: Government Shifts Focus from Industry to Consumers (25 March 1998).
Greenwire, EU: Enviros Warn Ministers On Union Expansion (19 August 1997).
The *Guardian*, Germany and Austria Sued for Polluting Black Sea Via Danube (26 September 1997).
Hungarian News Agency, Government Orders New Impact Study on Danube Dam (24 July 1998).
Hungarian News Agency, Report on Costs of Environmental Development (10 August 1998).
Hungarian News Agency, Talks on Danube Dam to Resume on 27 November (17 November 1998).
International Market Insight Reports, Environmental Market in Estonia (11 September 1997).
International Market Insight Reports, Romania: Environmental Technologies Market Overview (18 December 1997).
Lloyds List, St. Petersburg Leads Baltic Revival: Port Development Initiative Acting as Investment Magnet for Environmental Clean-Up (19 May 1997).
Los Angeles Times, A Still-Clouded Pollution Issue Environment: An International Government–Business Effort Gave a Czech City Clean Air (1 November 1998).
NYT, Eastern Europeans Protest Hydroelectric Project (4 February 1990).
NYT, Gabčíkovo Journal; On the Danube, Unbuilt Dams But Pent-Up Anger (5 December 1990).
NYT, Hungarians Ease Stand Over Dam (29 August 1991).
NYT, Slovaks Finish Much-Criticized Dam on Danube (3 November 1992).
NYT, Diverted, Danube Fades, and the Rage Rises (8 November 1992).
NYT, Exported for Decades, Canadian Reactors are Plagued by Operating Problems (3 December 1997).
Nuclear Engineering International, World Survey: Looking Beyond the Short-Term (30 June 1998).
Nucleonics Week, Other Delays at Temelín Take Regulators off the Critical Path (10 September 1998).
ODD, No. 171 (1 September 1995).
ODD, No. 176 (11 September 1995).
ODD, No. 180 (15 September 1995).
ODD, No. 181, Part I (18 September 1995).
ODD, No. 183 (20 September 1995).
ODD, No. 185 (22 September 1995).
ODD, No. 187 (26 September 1995).
ODD, No. 188 (27 September 1995).
ODD, No. 190 (29 September 1995).
ODD, No. 193 (4 October 1995).
ODD, No. 194 (5 October 1995).
ODD, No. 196 (9 October 1995).
ODD, No. 208 (25 October 1995).
ODD, No. 210 (27 October 1995).
ODD, No. 213 (1 November 1995).
ODD, No. 219 (9 November 1995).
ODD, No. 233 (1 December 1995).

ODD, No. 234 (4 December 1995).
ODD, No. 235 (5 December 1995).
ODD, No. 239 (11 December 1995).
ODD, No. 240 (12 December 1995).
ODD, No. 241 (13 December 1995).
ODD, No. 244 (18 December 1995).
ODD, No. 248 (22 December 1995).
ODD, No. 7 (10 January 1996).
ODD, No. 24 (2 February 1996).
ODD, No. 31 (13 February 1996).
ODD, No. 32 (14 February 1996).
ODD, No. 42 (28 February 1996).
ODD, No. 47 (6 March 1996).
ODD, No. 49 (10 March 1996).
ODD, No. 51 (12 March 1996).
ODD, No. 55 (18 March 1996).
Offshore, Black Sea Afflicted with many Biological Problems (July 1998).
Pipeline and Gas Journal, Planned Pipeline Between Russia, Turkey, To Be World's Deepest (1 September 1998).
Polish News Bulletin, Weekly Supplement: Pollution (12 November 1998).
Polish Press Agency, World Bank Credit for Environment Protection Used in Full (21 April 1998).
Polish Press Agency, Czechs to Clear Up Cyanide Pollution of Odra (1 October 1998).
PR Newswire, Westinghouse Wins Contracts for Nuclear Upgrades in Russia, Bulgaria (19 February 1998).
Reuters Limited, Bulgarian Reactor Could Affect EU Ties (18 December 1995).
Reuters Textline, Europe: Cover Story: Nuclear Liability (30 April 1995).
Reuters Textline, Poland: Turów Power Station Announces Modernization Plans (21 February 1995).
RFE/RL Caucasus Report, Cracks Emerge in the Georgian–Turkish Partnership, No. 7 (14 April 1998).
RFE/RLN, More Problems with Kozloduy Nuclear Plant, No. 98, Part II (19 August 1997).
RFE/RLN, Hungarian Parties Cross-Cooperate on Gabčíkovo Ruling, No. 127, Part II (29 September 1997).
RFE/RLN, Slovakia Rejects Hungarian 'Zero-Option' Proposal Following Hague Ruling, No. 134, Part II (8 October 1997).
RFE/RLN, Agreement on Gabčíkovo–Nagymaros Project to be Reached by March, No. 180, Part II (16 December 1997).
RFE/RLN, Hungarian, Slovak Foreign Ministers to Meet Soon, No. 191, Part II (8 January 1998).
RFE/RLN, Czech Government to Continue with Construction of Temelín, No. 192, Part II (7 January 1998).
RFE/RLN, Hungary, Slovakia Discuss Hydropower Dam Proposals, No. 194, Part II (13 January 1998).
RFE/RLN, Gabčíkovo–Nagymaros Settlement Still Hangs in Balance, No. 9, Part II (15 January 1998).
RFE/RLN, Partial Agreement in Slovak–Hungarian Dam Dispute, No. 14, Part II (22 January 1998).

RFE/RLN, Hungary, Slovakia Agree to Speed up Dam Talks, No. 17, Part II (27 January 1998).
RFE/RLN, Hungarian Coalition Divided Over Gabčíkovo–Nagymaros Compromise, No. 25, Part II (6 February 1998).
RFE/RLN, Hungarian Premier Denies Accusations over Dam Decision, No. 27, Part II (10 February 1998).
RFE/RLN, Moldovan President in Romania, No. 36, Part II (23 February 1998).
RFE/RLN, Baltic Sea Regions Sign Accord in Poland, No. 37, Part 1 (24 February 1998).
RFE/RLN, Hungarian Parliament Debates Dam Dispute, No. 39, Part II (26 February 1998).
RFE/RLN, Hungary, Slovakia Sign Preliminary Agreement on Danube Dam, No. 41, Part II (2 March 1998).
RFE/RLN, Chernobyl Director Upset with EBRD Rejection of Funding, No. 41, Part II (2 March 1998).
RFE/RLN, Hungarians Oppose Agreement on Danube Dam, No. 44, Part II (5 March 1998).
RFE/RLN, Slovakia, Hungary Disagree over Dam Dispute Deadline, No. 45, Part II (6 March 1998).
RFE/RLN, Hungarian Rally Against Dam, No. 51, Part II (16 March 1998).
RFE/RLN, EBRD Releases Money to Ukraine for Chernobyl, No. 53, Part II (18 March 1998).
RFE/RLN, Hungarian Cabinet Says No Dam at Nagymaros, No. 60, Part II (27 March 1998).
RFE/RLN, Czech Environment Minister Against Nuclear Power Plant, No. 64, Part II (2 April 1998).
RFE/RLN, Slovak Parliament Passes Resolution on Danube Dam, No. 64, Part II (2 April 1998).
RFE/RLN, Bulgaria Rejects EU Concerns about Nuclear Plant, No. 64, Part II (2 April 1998).
RFE/RLN, Hungarian Cabinet Will Not Dispute Ruling on Dam Referendum, No. 65, Part II (3 April 1998).
RFE/RLN, Hungarian Opposition Leader Details Program, No. 69, Part II (9 April 1998).
RFE/RLN, EBRD Agrees to Later Shutdown of Bulgarian Nuclear Reactors, No. 73, Part II (16 April 1998).
RFE/RLN, Slovakia Estimates Environmental Damage by Former Soviet Troops, No. 77 Part II (22 April 1998).
RFE/RLN, Kyiv Criticizes G-8 for not Abiding by Accord on Chernobyl Closure, No. 77 Part II (22 April 1998).
RFE/RLN, Romanian–French Deal on Black Sea Oil Drilling, No. 8, Part II (23 April 1998).
RFE/RLN, Construction on Controversial Czech Nuclear Plant to Continue, No. 84, Part II (4 May 1998).
RFE/RLN, Bulgaria Premier Refuses EU Request to Close Nuclear Plant, No. 87, Part II (7 May 1998).
RFE/RLN, Expert Team Chief Warns Against Slovak Nuclear Plant, No. 93, Part II (18 May 1998).
RFE/RLN, Bulgarian Nuclear Material to Transit Neighboring Countries (18 November 1997).

RFE/RLN, Polish Parliament Adopts Mining Restructuring Plan (30 November 1998).
Royal Society of Chemistry Newsbrief, Baltic nations Agree to Cut Emissions into Sea (22 April 1997).
Sofia BTA (2 February 1992).
Sofia BTA (3 April 1993).
Turkish Daily News, Black Sea Issues a Distress Call (6 November 1998).

Interviews and presentations

Dr Anna Adamus-Matuszynska, Do Stereotypes Influence Social Conflicts?, presented during the Global Security Fellows Initiative Public Forum Series entitled Regions, States and Peoples in Transformation (Cambridge, 2 March 1995).

Mr Paul Almeida, East European Programs, United States Environmental Protection Agency (Washington, D.C., 14 September 1994).

Mr Bohuslav Bezúch, Office of Waste Management, Ministry of the Environment of the Slovak Republic (Bratislava, 4 October 1994).

Mr Ritt Bjerregaard, Member of the Commission responsible for Environment Baltic Agenda 21 Foreign Ministers' meeting – Council of the Baltic Sea States in Nyborg, Denmark (22 June 1998).

Ms Eva Blenesi, Transnational Mediation, presented during the Global Security Fellows Initiative Public Forum Series entitled Regions, States and Peoples in Transformation (Cambridge, 3 March 1995).

Mr Martin Brenosk, Project Development Officer, USAID Office, Slovakia (Bratislava, 3 October 1994).

Mr James Chamberlin, Counselor for Scientific Affairs, Embassy of the United States, Poland (Warsaw, 11 August 1995).

Mr Miroslav Čomaj, Director, Office of Hydraulic and Dams Construction, Hydroconsult (Bratislava, 7 October 1994).

Ms Darina Dzurjaninová, Department of International Relations, Ministry of the Environment of the Slovak Republic (Bratislava, 4 October 1994).

Dr Ferenc Fehér, Team Leader, The Regional Environmental Center for Central and Eastern Europe, Hungary (Budapest, 12 October 1994).

Ms Gabriela Fischerová, Department of International Relations, Ministry of the Environment of the Slovak Republic (Bratislava, 5 October 1994).

Dr Božena Gašparíková, Director, Legislation Department, Ministry of the Environment of the Slovak Republic (Bratislava, 6 October 1994).

Dr Krzysztof Görlich, Deputy Mayor for the City of Kraków, Poland (Kraków, 9 May 1994).

Ms Katherine Gorove, Consultant, International Law Section, Ministry of Foreign Affairs for Hungary (Budapest, 10 October 1994).

Dr Michael Grodzinski, Perceptual Sources of Environmentally Caused Ethnic Tensions, presented during the Global Security Fellows Initiative Public Forum Series entitled Regions, States and Peoples in Transformation (Cambridge, 2 March 1995).

Mr James Hooper, Deputy Chief of Mission, Embassy of the United States, Poland (Warsaw, 10 August 1995).

Dr Vladimir Ira, The Perception of Potential Ethnic and Religious Conflict, presented during the Global Security Fellows Initiative Public Forum Series entitled Regions, States and Peoples in Transformation (Cambridge, 2 March 1995).

Dr Jacek Jaśkiewicz, Adviser to the Minister in the Ministry of Environmental Protection, Natural Resources and Forestry for Poland (Warsaw, 4 May 1994).

Dr Katarzyna Juda-Rezler, Assistant Professor at the Warsaw University of Technology, Institute of Environmental Engineering Systems (Warsaw, 5 May 1994).

Dr Cecília Kandráčová, Director, Department of International Law, Ministry of Foreign Affairs of the Slovak Republic (Bratislava, 5 October 1994).

Dr Jan Kára, Chief of the Office of International Economic Organizations, Ministry of Foreign Affairs of the Czech Republic (Prague, 13 May 1994).

Dr Andrzej Kassenberg, President of the Institute for Sustainable Development (Warsaw, 5 May 1994).

Dr Vilmos Kiszel, President, Göncöl Alliance, Hungary (Vác, 10 October 1994).

Dr Viliam Klescht, Director, Department of Nature and Landscape Protection, Ministry of the Environment of the Slovak Republic (Bratislava, 5 October 1994).

Mr Peter Kollárik, Director, Department of International Economic Cooperation, Ministry of Foreign Affairs of the Slovak Republic (Bratislava, 5 October 1994).

Ms Zofia Kordela-Borczyk & Ms Izabela Suchanek, Economic Co-operation within the Carpathian Euroregion Association, presented during the Global Security Fellows Initiative Public Forum Series entitled Regions, States and Peoples in Transformation (Cambridge, 3 March 1995).

Dr Juraj Králik, Chairman, Slovak National Centre for Human Rights (Bratislava, 3 October 1994).

Ms Karin Krchnak, Environmental Consultant, Environmental Law Institute, Slovakia (Bratislava, 3 October 1994).

Dr Sláva Kubátová, Czech Environment Management Center (Prague, 22 August 1994).

Mr Milan Kunc, Mayor for the City of Děčín, Czech Republic (Děčín, 18 May 1994).

Ms Adela Ladzianska, Department of International Relations, Ministry of the Environment of the Slovak Republic (Bratislava, 4 October 1994).

Dr Miroslav Liška, Chief Deputy of Public Relations and Expertises, Vodohospodárska Výstavba (Bratislava, 3 October 1994).

Dr Július Maljkovič, Director, Division of Geological Research and Exploration, Ministry of the Environment of the Slovak Republic (Bratislava, 4 October 1994).

Mr Stephen McCormick, Senior Associate, National Civic League, Czech Republic (Děčín, 18 May 1994).

Mr Jaroslav Mráz, Regional Chairman of Euroregion NISA (Liberec, 17 May 1994).

Dr Bedřich Moldan, former Minister of the Environment of the Czech Republic (Prague, 13 May 1994).

Mr Lubomír Novák, Chief Engineer for the City of Děčín, Czech Republic (Děčín, 18 May 1994).

Dr Maciej Nowicki, President of Ecofund and former Minister of Environmental Protection, Natural Resources and Forestry for Poland (Warsaw, 5 May 1994).

Dr Thomas Owen, Environmental and Resource Economics Expert, Harvard Institute for International Development, Slovakia (Bratislava, 3 October 1994).

Dr Lubomír Paroha, Director of the Foundation Project North (Ústí nad Labem, 19 May 1994).

Dr Yaroslav Pilinski, The Ways of Solving Problems of National Minorities Divided by Borders of Different States, presented during the Global Security Fellows Initiative Public Forum Series entitled Regions, States and Peoples in Transformation (Cambridge, 3 March 1995).

Mr René Pisinger, Adviser to the Minister of the Environment of the Czech Republic (Teplice, 19 May 1994).

Mr Jan Pisko, Coordinator, United States Agency for International Development (Prague, 21 August 1994).

Dr Juraj Podoba, The National State as a Historical Source of Ethnic Tension, presented during the Global Security Fellows Initiative Public Forum Series entitled Regions, States and Peoples in Transformation (Cambridge, 2 March 1995).

Dr Václav Poštolka, Professor on the Faculty of Environment at the Jan Evangelista Purkyně University (Liberec, 16 May 1994).

Ambassador Theodore Russell, Embassy of the United States of America, Slovakia (Bratislava, 3 October 1994).

Dr Ivan Rynda, former Member of the Presidium of the Czech Federal Assembly and the Chairman of the Environmental Committee of the Chamber of People (Prague, 12 May 1994).

Mr Loren Schulze, Project Development Officer, USAID Office, Slovakia (Bratislava, 3 October 1994).

Mr Stanislav Sitnicki, Executive Director, The Regional Environmental Center for Central and Eastern Europe, Hungary (Budapest, 12 October 1994).

Dr Jozef Sloboda, Deputy Director, Department of Nature and Landscape Protection, Ministry of the Environment of the Slovak Republic (Bratislava, 5 October 1994).

Dr Jerzy Sommer, Professor of Law and President of the Polish Environmental Law Association (Wrocław, 6 May 1994).

Mr Steve Stec, Regional Coordinator, American Bar Association Central and East European Law Initiative, Hungary (Budapest, 10 October 1994).

Dr Tomasz Stypka, Mechanical Engineer, Technical University of Kraków, Institute of Environmental Protection (Kraków, 7 May 1994).

Mr Steve Taylor, Counselor for Scientific Affairs, Embassy of the United States, Hungary (Budapest, 11 October 1994).

Dr Jindrich Tichy, Professor of Ecology at the University of Ústí nad Labem (Ústí nad Labem, 19 May 1994).

Dr Juraj Tözsér, Director of the Geological State Administration Department, Ministry of the Environment of the Slovak Republic (Bratislava, 4 October 1994).

Mr Josef Vejvoda, Director of the Department of Air Protection, Ministry of the Environment of the Czech Republic (Prague, 13 May 1994).

Ms Anna Violová, Director, Office of Air Pollution, Ministry of the Environment of the Slovak Republic (Bratislava, 6 October 1994).

Dr Roman Vyhnanek, Czech Environment Management Center (Prague, 22 August 1994).

Dr Stanisław Wajda, Adviser to the Minister in the Ministry of Environmental Protection, Natural Resources and Forestry for Poland (Warsaw, 4 & 5 May 1994).

Mr Stanisław Zadrożny, Adviser to the Minister in the Ministry of Environmental Protection, Natural Resources and Forestry for Poland (Warsaw, 4 & 5 May 1994).

Mr Jaroslav Zámečník, Deputy Regional Chairman of Euroregion NISA (Liberec, 17 May 1994).

Mr Ivan Závadský, Director, Division of Water and Air Pollution, Ministry of the Environment of the Slovak Republic (Bratislava, 6 October 1994).

Mr János Zlinszky, Team Leader, The Regional Environmental Center for Central and Eastern Europe, Hungary (Budapest, 12 October 1994).

Index

acid rain, 140, 168, n737, n743
   Black Triangle, 17
Acid Rain Agreement, US–Canada, n80
   Black Triangle, 21
   Silesian Coal Basin, 23
acquifers, 168
   Gabčíkovo–Nagymaros Project, 56, 57, 98, 103
   Soviet military legacy, 32
   threats from coal-mining, 18
adjudication, 4, 5, 11–12, 51, 126, 163, 199–200, 201, 219, 220, 222, n720
   Black Sea, 44
   during implementation phase, 134
   during pre-resolution phase, 130
   during resolution phase, 131, 133, 134
   Gabčíkovo–Nagymaros Project, 67, 71, 107, 118, 119, 224
   Romanian–Bulgarian Nuclear Corridor, 42
   Temelín dispute, 225
   see also International Court of Justice, international dispute resolution mechanism
Aegean Sea Continental Shelf Case, n585, n933, n935
agency structure, 125
agricultural production
   Gabčíkovo–Nagymaros Project, 57, 58–9, 66, 79, 98
air quality
   Western Europe, 24
   see also transboundary air pollution
Air Service Agreement Case, n468
Albania, 183
Ambatielos Case, n588, n930
approximate application, doctrine of, 75
arbitration, 4, 11–12, 126, 153, 163, 201, 219, 220, 222
   Black Sea, 224
   Danube River Protection Convention, 200
   during implementation phase, 134

during pre-resolution phase, 130
during resolution phase, 133, 134
Gabčíkovo–Nagymaros Project, 69, 71, 107, 119
Romanian–Bulgarian Danube River Gauntlet, 224
Romanian–Bulgarian Nuclear Corridor, 42
Temelín dispute, 225
see also international dispute resolution mechanism
Association of Carpathian National Parks, 181
Atomenergoexport, Russia, n135
atomization, 176, 177
Austria, 84, 144, 183, 195
   admission to European Union, 28
   Black Sea, 44
   Bohunice nuclear plant, 29
   Dukovany dispute, 28–9
   energy entitlement owed by Hungary, 53, 62–3
   Environmental Program for the Danube River Basin, 36
   Gabčíkovo–Nagymaros Project, 60, 62, 112
   Mochovce dispute, 29
   Paks dispute, 30
   Temelín dispute, 25–30, 142, 220, 225

Baltic Agenda 21, 12–13
Baltic Sea, 4, 11, 12–16, 45, 131, 140, 143, 144, 151, 158, 209
   attempts to reduce pollution of, 14–15
   environmental degradation, 12–13
   ICJ decision in Gabčíkovo–Nagymaros, 89, 94
   Joint Comprehensive Environmental Action Program, 14, 140, 181
   role of international law, 13–16, 188, 219, 220, 224

Baltic Sea – *continued*
  Silesian Coal Basin, 23
  similarities to Black Sea, 42
Baltic Sea Convention (1974), 14, 131,
    140, 219, 222, n19, n22, n29, n31,
    n974
  Marine Environmental Protection
    Commission (HELCOM), 14,
    45, 134, 220
  prohibitions on pollution, 14
Baltic Sea Convention (1992), 14–15,
    131, 140, 222, n19, n22, n29, n39,
    n40, n974
  Marine Environmental Protection
    Commission (HELCOM), 14
  prohibitions on pollution, 15
Baltic Sea Euroregion, 12
Baltic States, 144
  see also Estonia, Latvia, Lithuania
Basel Convention on the Transboundary
    Movement of Hazardous Waste,
    191, n484, n972, n987
Bavaria, 178–9
Bechtel Corporation, 212, n256
Becva River, n86
Belarus, 6
  Baltic Sea, 12
Belene nuclear power plant, Bulgaria, 39
  ICJ decision in Gabčíkovo–
    Nagymaros, 84, 97
Belgian Concern Powerfin S.A., n638
best environmental practice
  Baltic Sea Convention (1992), 15
bi-polar international system, 3
bilateral cooperation agreements, 208,
    n147, n447, n972, n973
  Czechoslovakia and Hungary, 71
  Hungary and Romania, 208
biodiversity, 11, 31, 43, 45, 141, 181,
    190, 209, n211
  Biodiversity Convention, 190
  Gabčíkovo–Nagymaros Project, 92
Black Sea, 4, 11, 45, 46, 131, 151, 158,
    209, 218
  attempts to reduce pollution of, 43
  basin states, 6, 44
  Black Sea Commission, 43
  Black Sea Convention, 43, n213,
    n216, n974

Black Sea Declaration, 43, n217
  environmental degradation, 42–3
  Environmental Management
    Program, 181
  ICJ decision in Gabčíkovo–
    Nagymaros, 89, 91, 94
  role of international law, 43–4, 188,
    209, 219, 224
Black Triangle, 16–22, 46, 128, 137,
    143, 144, 145, 148, 152, 158, 172,
    204, 205–6, 209, 218
  attempts to reduce pollution of,
    19–20
  cost–benefit analysis, 175
  environmental degradation, 16–19
  ICJ decision in Gabčíkovo–
    Nagymaros, 94, 100, 108
  national minorities, 186
  Polish–Czech–German
    Environmental Program, 181
  possible role for ICJ, 203
  role of international law, 19–22,
    224–5
  Silesian Coal Basin, 25
  sub-state actors, 222
  Temelín dispute, 26
  Working Group for Neighborly
    Cooperation on Environmental
    Issues, 143
Bodiky, Slovakia, 57, 184
Bogatynia Region, Upper Silesia, n89
Bohunice nuclear power plant, 29, 143
  Gabčíkovo–Nagymaros Project,
    relation to, 62
Boundary Waters Convention 1976
    (Czechoslovakia and Hungary)
  Gabčíkovo–Nagymaros Project, 66,
    76, 97–8, n408, n447, n450
boycotts, 124
Bratislava, Slovakia, 54, 110, 113,
    n655
Bucharest Declaration, n160
Budapest, Hungary, 54, 98, 110, 115
Bukowno, Poland, 23
Bulgaria, 6, 144, 145, 148, 171, 190,
    191
  Black Sea, 43
  EIA, 194–5
  environmental investment, 174, 205

environmental regulation, n761,
    n854, n857, n866, n868, n878
  EU membership, 142
  implementing legislation, 198
  nuclear power, 38–42, 84
  public participation, 195
  see also Romanian–Bulgarian
    Danube River Gauntlet;
    Romanian–Bulgarian Nuclear
    Corridor
Bulgarian Atomic Energy Commission
  closure of Kozloduy nuclear power
    plant, 40

Calafat, Romania, 35
Câlâradi, Romania, 35, 37
Canada, n737
  design of Cernavoda nuclear power
    plant, 39, n182
Canadian Atomic Energy Agency
  Cernavoda nuclear power plant, 39
Canadian Export Development
    Corporation
  Cernavoda nuclear power plant, 39
Carpathian region, 186
Central Asian oil, 13
central government actors, 138, 141,
    171, 179, 182, 194, 197, 213, 219
  Gabčíkovo–Nagymaros Project, 69,
    111, 113
  Silesian Coal Basin, 23
  Soviet Military Legacy, 33
Cernavoda nuclear power plant,
    Romania, 39, 128
  ICJ decision in Gabčíkovo–
    Nagymaros, 84, 97
chemical plants
  Black Sea, 42
  Black triangle, 16–17
  Giurgiu–Ruse, 37
chemical weapons containers
  Baltic Sea, 12
Chernobyl nuclear power plant,
    Ukraine, 206
Choana, Poland, 32
Chorzów Factory Cases, 105, n14,
    n419, n481, n495, n579, n987
civil litigation, 195, 196
  Black Sea, 44, 195

Black Sea Convention, 43
  Danube River, 195
  Soviet military legacy, 34
  Temelín dispute, 28, 195
class struggle, 188–9
  see also socialist political and legal
    doctrine
Clean Air Act, United States, n648
clean collapse, 169
  Black Triangle, 18–19
  see also Germany
Climate Change Convention, 190,
    n82, n846, n973
  Black Triangle, 21
CO_2
  Silesian Coal Basin, 23
coal-fired power plants
  Baltic Sea, 12
  Black Triangle 18–20
  Gabčíkovo–Nagymaros Project, 53,
    62
  Silesian Coal Basin, 22
  see also Herschfeld, Turów and
    Tušimice power plants
coal-mining, 138
  open-pit mines in Black Triangle, 17
  operations in Black Triangle, 17
  operations in the Silesian Coal
    Basin, 22–4
  social dislocation, 23–4, 172
cognitive theory
  functionality of international law,
    156
command economy, 53
  see also economic transformation
communism
  opposition to, 35
  see also socialist political and legal
    doctrine
communist ideology
  see socialist political and legal
    doctrine
communist plan of industrialization
  Romanian–Bulgarian Danube River
    Gauntlet, 38
conciliation, 4, 11–12, 126, 163, 201,
    219
  Black Sea, 44
  during pre-resolution phase, 130

conciliation – *continued*
  during resolution phase, 131
  Gabčíkovo–Nagymaros Project, 109, 119
  Romanian–Bulgarian Danube River Gauntlet, 37–8
  Romanian–Bulgarian Nuclear Corridor, 42
  Silesian Coal Basin, 24–5
Confederation of Independent Trade Unions of Silistra, Poland, n157
consistent social preference order, 136
Constanta, Romania, 42
Constitutional Courts, 198
consultation, 128, 209, 219, 224
  Black Triangle, 20
  Gabčíkovo–Nagymaros Project, 75, 97–8, 117, 119
Continental Shelf Case (Libya/Malta), n594, n937
Continental Shelf Case (Tunisia/Libya), n579
Convention on Early Notification of a Nuclear Accident, 191, n115, n850, n988
Convention on Long-Range Transboundary Air Pollution, 190, n454, n840, n973, n982
  Baltic Sea, 15
  Czech efforts at compliance, 210
  First SO_2 Protocol, 190, n841
  NO_x Protocol, 190, n842
  Second SO_2 Protocol, 190, n843
  VOC Protocol SO_2 Protocol, 190, n844
Convention on Marine Pollution from Land-Based Sources
  Baltic Sea, 15
Convention on Nuclear Safety, 191, n115, n850, n988
  Temelín dispute, 27
Convention on the Law of the Non-Navigational Uses of International Watercourses, 104, n454, n457, n470, n490, n974
Convention on the Transboundary Effects of Industrial Accidents, 191, n849
Corfu Channel Case, n459, n724, n972

cost–benefit analysis, 147–8, 172, 175
  Gabčíkovo–Nagymaros project, 86, 112
  Silesian Coal Basin, 24
Council of Europe Parliamentary Assembly
  Romanian–Bulgarian Danube River Gauntlet, 37
Council of Mutual Economic Assistance (CMEA), 55, 118, n803
  Program for Cooperation Concerning Environmental Protection, 71
countermeasures, 91, 92, 97, 100–2
Croatia
  Paks dispute, 30
Čunovo works, 106, 110, 120
customary international law, 125, 160, 210, 213
  Baltic Sea, 15
  Black Sea, 44
  Gabčíkovo–Nagymaros Project, 66, 80, 85, 89, 106
Czech Republic, 6, 46, 144, 147, 148, 179, 194, 203, 204, 213
  admission to European Union, 27–8, 196, 207
  Black Triangle, 16–20, 137, 143, 152, 175
  Constitutional Court, 198
  economic transformation, 147
  EIA, 195
  environmental investment, 174
  environmental regulation, 196, n854, n857, n868, n870, n873, n875, n878, n879, n883
  EU membership, 137, 217, 223, 225
  hazardous waste regulations, 152
  implementing legislation, 198
  Ministry of Industry and Trade
    Temelín dispute, 26, 28
  Ministry of the Environment,
    Temelín dispute, 25–6, 28
  Mochovce dispute, 206
  National Fund for the Environment, 174
  national minorities, 185
  NATO membership, 180
  Parliament, Temelín dispute, 28
  Silesian Coal Basin, 22–5

State Nuclear Safety Authority, 28
Temelín dispute, 25–30, 225
Czechoslovakia, 183
  dissolution of, 69, 70, 72, 75
  Gabčíkovo–Nagymaros Project, 51–120
  national minorities, 186
  Soviet invasion of, 189

damages and reparation, 124, 210, 212
  dispute formation, 127
  Gabčíkovo–Nagymaros Project, 73, 74, 75, 76–7, 83, 109–10, 116;
    Effect of ICJ decision, 102–8
  Soviet military legacy, 32–3
Danube Fisheries Agreement (1958), 97–8, n404, n447
Danube River, 47, 51, 208
  Environmental Program, 181
  erosion, 66
  Gabčíkovo–Nagymaros Project, 51–120
  nuclear contamination, 38
  pollution of, 35, 36, 144
  pollution of Black Sea, 42
  protection of, 37
Danube River Commission, 66, n314
Danube River Convention (1948), 41, n447, n974
  Gabčíkovo–Nagymaros Project, 66, 76, 97–8
Danube River Protection Convention (1994), 36, 37, 38, 134, 200, 201, 203–4, 209, n479, n919, n920, n925, n970, n972, n974, n983, n984, n985
deadlock situation, 158
  see also game theory
Děčín, Czech Republic, n82
democratic transformation, 1, 2, 6, 45–6, 47, 48, 60, 61, 69, 70, 86, 113, 118–19, 139, 150, 151, 167, 173, 177, 189, 192, 215, 222, 225–6
Denmark
  Baltic Sea, 12
descriptive inference, 3, n10
devolution
  environmental decision making, 177, 178, 225; Black Triangle, 22

dilemma situations, 158
  see also game theory
Diplomatic and Consular Staff Case, n594, n936
dispute formation phase, 6–7, 17, 24, 32, 40, 109, 127, 128
  environmental actors, 138
  functions of international law during, 127–9, 162, 218
  Gabčíkovo–Nagymaros Project, 51–9
  public participation, 139, 149
dispute resolution process, 4, 11, 146
  Baltic Sea, 14
  Black Sea, 43
  Black Triangle, 17–18
  Dukovany dispute, 28–9
  Romanian–Bulgarian Danube River Gauntlet, 37
  Romanian–Bulgarian Nuclear Corridor, 40–2
  Silesian Coal Basin, 24–5
  Soviet military legacy, 32–5
  Temelín dispute, 27
  see also dispute formation phase, pre-resolution phase, resolution phase, implementation phase
Djubga, Russia, 42
Dobrohost, Slovakia, 57, 184
domestic legal mechanisms, 124, 169, 177, 197
  Soviet military legacy, 34–5
domestic ratification procedures, 134, 163
domestic resource disputes, 146, 216
  Gabčíkovo–Nagymaros Project, 119
  see also municipal environmental law
Donaukraftwerke, 62–3, 142
Draft Articles on State Responsibility, 78, 81–2, n345, n362, n445, n468, n987
Dukovany nuclear power plant, Czech Republic, 25, 196
Dunakiliti works, 56, 57, 59, 66, 69, 70, 76, 90, 91, 95, 102, 110
duty to cooperate, 92, 130, 134, 208
  Baltic Sea, 12

330 *Index*

duty to cooperate – *continued*
  Gabčíkovo–Nagymaros Project, 75, 105
  Romanian–Bulgarian Danube River Gauntlet, 36
  see also transboundary environmental cooperation
Dzurinda, Mikulas, 111–12

early warning system, 45, 208
  Mochovce and Bohunice nuclear power plants, 29
  Romanian–Bulgarian Danube River Gauntlet, 37–8
ecological circumstances, 46–7, 119, 145–6, 163, 167–70, 220
ecological police, 203
ecological state of necessity
  Gabčíkovo–Nagymaros Project, 66, 74–75, 78–84; as a basis for treaty termination, 95–6
economic actors, 46, 136, 137–8, 146, 163, 168, 170, 171, 172, 178, 221, 222
economic circumstances, 2, 20, 47, 118, 145, 146–8, 163, 167, 170–5, 220
economic inefficiency, 213
  Silesian Coal Basin, 23
economic sanctions, 124
economic transformation, 2, 6, 47, 141, 147, 167, 169, 170, 172, 173, 189, 190, 192, 215, 225–6
  Baltic Sea, 147
  Black Sea, 44, 147
  Black Triangle, 18, 147
  Gabčíkovo–Nagymaros Project, 53, 60, 86, 112, 118–19
  Silesian Coal Basin, 147
  Poland, 13
Elbe River, 17, 19, 144, 181, 211, n58, n657
Elbe River Convention, 19–20, 209, n73, n74, n971, n974
Electricité de France, n638
  Mochovce dispute, 29
  Paks dispute, 30
emergency response plan
  Black Sea Declaration, 43

energy-bargaining, 213
energy entitlement, 53, 112
energy pricing, 22, 172, 225
energy production, 147
  Gabčíkovo–Nagymaros Project, 51, 53, 87, 88, 90, 95, 103, 110
environmental actors, 138, 163, 169, 175, 177, 221
  Black Sea, 44
  Gabčíkovo–Nagymaros Project, 60, 61, 62, 111, 114, 119, 222
  Romanian–Bulgarian Danube River Gauntlet, 38
  Silesian Coal Basin, 24
  Temelín dispute, 26
environmental awareness and activism, 86–7, 176
Environmental Chamber, ICJ, 202
environmental degradation, 48, 130, 144, 145, 146, 169, 170, 177, 212, 215, 218
  American intelligence community report on, 1, 31
  consequences, 1, 145, 148, 175, 218
  costs of remediation/abatement, 148, 150, 185, 210, n664, n779; Silesian Coal Basin, 22; Soviet military legacy, 31, 34
  dispute formation, 127
  economic actors, 137
  economic costs of, 45, 138, 148, 150, n674, n681; Black Sea, 43; Black Triangle, 17
  Gabčíkovo–Nagymaros Project, 57, 58–9, 66, 69, 73, 74, 78–84, 95–6, 97–8, 101, 112
  national minorities, 153, 154, 182
  structural-symmetry, 147
environmental fees, 192, 197
environmental fines, 173, 192, 194, 197, 198
environmental hot spots
  Baltic Sea, 14
Environmental Impact Assessment (EIA), 130, 159, 163, 191, 193, 194, 195, 198, 208, 210–11, 212, n38
Baltic Sea Convention (1992), 15

Black Sea Declaration, 43
Djubga pipeline, 42
Gabčíkovo–Nagymaros Project, 78, 87, 89, 92, 93, 97, 103, 110, 111, 212
Environmental Impact Assessment Convention, 190–1, n38, n455, n986
environmental information, 150, 190
  Silesia Coal Basin, 24
environmental investment, 172, 173, 174, 206
  Baltic Sea, 14
  Black Triangle, 18
environmental organizations
  see environmental actors
Environmental Program for the Danube River Basin, 36
environmental regulation, 173, 177, 179
  Black Triangle, 18
  for specific regulations, see specific states
environmental technology, 144, 146, 147, 150, 170, 173
  cost of, 148, 174
  Gabčíkovo–Nagymaros project, 86, 119
  nuclear technology, 26
  restrictions on dissemination, 194
  Romanian–Bulgarian Danube River Gauntlet, 37
  Silesian Coal Basin, 24
epistemic community, 181, 182
equitable utilization, 127, 209, 210
  Baltic Sea, 15, 224
  Black Sea, 44
  Gabčíkovo–Nagymaros Project, 66, 75, 76, 98, 99, 100, 101, 104, 105–6
  Romanian–Bulgarian Danube River Gauntlet, 38
Estonia, 6
  Baltic Sea, 12
  Defense Cooperation Agreement with Poland, n147
  EU membership, 13
European Bank for Reconstruction and Development (EBRD), 139, 141, 143, n615, n808

Chernobyl, 206, n953
construction of Mochovce nuclear power plant, 29, 206
Environmental Program for the Danube River Basin, 36
Romanian–Bulgarian Nuclear Corridor dispute, 41
European Court of Justice
  Black Sea, 44
European Danube Charter, 37
European geography, 6, 145, 167–8
European integration, 152, 181–2
European Investment Bank (EIB), 139, 141, n947
  Black Triangle, 21
European Parliament
  Gabčíkovo–Nagymaros Project, 70, 111, 116
  Mochovce dispute, 29
European Presidency
  Mochovce dispute, 29
European Trade Union Confederation, n157
European Union, 1, 11, 46, 47, 128, 139, 142–3, 153, 158–9, 163, 180, 193
  air emission standards, 148; Black Triangle, 18
  association agreements, 142, 151–2, 196; Czechoslovakia and Hungary, 71, 196
  Baltic Sea, 12
  Black Sea, 44
  Black Triangle, 18–21; program of inquiry, 19
  Directives, 196, 210; Black Sea, 44; Harmonization with, 71, 143, 152, 196, 205, n888, n889, n891; Relevant for Baltic Sea,13
  Elbe River Convention, 19
  Environmental Program for the Danube River Basin, 36; financial assistance, 140, 143, 144, 205; Structural Funds/Cohesion Funds, 152
  Gabčíkovo–Nagymaros Project, 113, 114, 115–16, 119, 139, 206; European Commission, 60, 63–4, 65, 67, 72, 105, 108, 113, 114,

European Union – *continued*
116; efforts at inquiry and
mediation, 68–70, 71, 73, 119
membership, 181–2, 217, 221, 223;
general linkage with
environmental protection, 137,
140, 142, 144, 158, 206, 226;
linkage between Bulgarian
membership and Kozloduy plant,
40; linkage between Czech
membership and Temelín
dispute, 27–8, 207; linkage
between Hungarian and Slovak
membership and the Gabčíkovo–
Nagymaros Project dispute, 68,
70, 72
Mochovce dispute, European
Commission 29
national minorities, 185
Paks dispute, 30
role in establishing the Baltic Sea Joint
Comprehensive Environmental
Action Program, 14
Romanian–Bulgarian Danube River
Gauntlet, 37–8
Romanian–Bulgarian Nuclear
Corridor, 38, 40, 42
Silesian Coal Basin, 24
Soviet military legacy, 34
sub-state actors, 141
Temelín dispute, 27–8, 225
Treaty Establishing the European
Atomic Energy Community, n988
water quality standards, 148
Euroregions, 182
Euroregion NISA, 19, 46

feminist theory of international law,
n555
Finland, 140
Baltic Sea, 12–13
Silesian Coal Basin, 22
Fisheries Jurisdiction Case, 85, 87,
n361, n384, n385, n458
Fisheries interests
Baltic Sea, 12, 222
Black Sea, 42, 43
Gabčíkovo–Nagymaros Project, 57,
93, 97–8

flood control
Gabčíkovo–Nagymaros Project, 51,
53, 54–5, 66, 69, 87, 88, 90, 95,
103, 105
forced industrialization, 169, 218
foreign direct investment, 174, n781
forestry resources, 195
Black Triangle, 19, 175
Gabčíkovo–Nagymaros Project, 57,
79, 98
Silesian Coal Basin, 22
Framatome
construction of Mochovce nuclear
power plant, 29
France, 143, 144, 195
Mochovce dispute, 142, 206
oil exploration in the Black Sea, 42
Temelín dispute, 26, 28
Free Zones Case, n480
Friendship Treaty, Slovak–Hungarian,
113
fuel switching, 21, 25

Gabčíkovo works, 59, 76–7, 78–84,
102, 103, 106, 110, 114, 131, 132
Gabčíkovo–Nagymaros Case
see Gabčíkovo–Nagymaros Project
Gabčíkovo–Nagymaros Project, 4–5,
47, 48, 51–120, 127, 128, 133,
139, 158, 176, 184, 188, 204, 206,
209, 216, 219, 224
cost–benefit analysis, 175
national minorities, 110, 112, 113,
114, 185, 186
relation to Mochovce and Bohunice
disputes, 30
sub-state actors, 136–7, 220
trilateral experts commission,
64–5, 67
Gabčíkovo–Nagymaros Project,
Agreement (1977 Agreement), 55,
73, 76, 77–108, 186
article 9, 104
article 10, 104
article 14, 83
article 15, 81, 88, 89, 90, 92, 93–4,
97–8, 100, 105
article 19, 81, 88, 89, 90, 92, 93–4,
97–8, 100, 105

article 20, 88, 89, 90, 93–4, 97–8, 100
article 27, 81
London Agreement, 70
1983 Protocol, 58
1989 Protocol, 59
1998 Protocol, 110–11, 116
Game theory, 136, n603, n604, n607
political circumstances, 149, 152
gas pipelines
Black Sea, 42
Gaz de France, n629
General Electric
Cernavoda nuclear power plant, 39
general public, 47, 138, 139, 149, 163, 175, 177, 193, 221
Gabčíkovo–Nagymaros Project, 61, 64, 111, 113, 114, 115
Silesian Coal Basin, 24
support for Temelín nuclear power plant, 26
Georgia
Black Sea, 42, 43, 44
Germany, 144, 147, 181, 195, 203, 204, 213
Baltic Sea, 12
Black Sea, 44
Black Triangle, 16–20, 143, 152; clean collapse, 18–19
Gabčíkovo–Nagymaros Project, support for, 62
hazardous exports, 144
Mochovce dispute, 142, 206
national minorities, 186
Paks dispute, 144
Romania–Bulgaria Nuclear Corridor, 40
Silesian Coal Basin, 22
Temelín dispute, 28
Giurgiu, Romania, 35–8
Global Environmental Facility (GEF), 181, n82, n213, n808
Black Triangle, 21
Environmental Program for the Danube River Basin, 36
fund, 141
good offices, 46, 73, 109, 110, 126, 154, 201, 219, n79
during dispute formation phase, 128

good faith, 80, 89, 91, 105, 107, 111, 116, 119, 133, 140, 141, n29
Görlitz, Germany, 147
government officials, 136–7, 155, 163, 211, 221
Greece, 44
Green Party
Austria, 63, 142
Slovakia, 62
Greifswald/Lubmin nuclear plant, Germany, 144
groundwater, *see* acquifers
Group of Seven Industrialized Nations (G7)
Chernobyl, 206
closure of Kozloduy nuclear power plant, 40, 42
Gulf of Maine Case, n590, n932
Györ, Hungary, 113

Hainburg Hydroelectric Plant, Austria, 62
Hans van der Broek, 142
hard budget constraints, 172
hard law, 126, 160
Haya de la Torre Case, n587, n595
hazardous waste, 11, 31, 45, 152, 191, 218, n150
Baltic Sea, 12
Baltic Sea Convention, 14
Black Sea Convention, 43
Black Sea Declaration, 43
export/import, 144
proposed incineration, 18
see also Basel Convention
heavy industry, 47, 167, 171, 172, 179, 215
Baltic Sea, 12
Black Triangle, 17
Gabčíkovo–Nagymaros Project, 53; shift away from, 86
heavy metals
Baltic Sea, 12
Helsinki Rules on the Uses of the Waters of International Rivers, n454, n974
Henkin, Louis, 126
Herschfeld, Germany
transformation of power plant, 18
Horn, Gyula, 114

Hradcany, Czech Republic, 32
human rights, 139, 140–1, 153, 154, 205, 213
  Covenant on Civil and Political Rights, n458, n701, n702
  Covenant on Economic, Social and Cultural Rights, n458
  Declaration on the Right to Development, n454
  human right to a healthy environment, 79, 92, 98, 101, 193, 213
Hungarian Academy of Sciences, 58, 82
Hungarian national minority in Slovakia, 57–8, 110, 112, 113, 114, 185
Hungarian Socialist Workers' Party, 58
Hungary, 6, 51, 144, 148, 181, 183, 190, 194, 207
  admission to European Union, 207
  admission to OECD, 141–2
  bilateral agreements for environmental protection, 208
  Bohunice dispute, 29
  Constitutional Court, 193
  economic transformation, 147
  environmental investment, 174
  environmental regulation, 196, 198, n854, n857, n868
  EU membership, 223
  Gabčíkovo–Nagymaros Project, 51–120, 137, 175
  implementing legislation, 198
  Mochovce dispute, 29
  national minorities, 185, 186
  NATO membership, 180
  nuclear power, 28–30
  public participation, 195
  Soviet invasion of, 189
  Soviet military legacy, 31–5
Hungary Parliament
  Gabčíkovo–Nagymaros Project, 59, 60, 61, 64
Hydro Quebec International, 212, n256
Hydrocarbon pollution, 222
  Gabčíkovo–Nagymaros Project, 56
  prohibitions on, 14
  threats to the Baltic Sea, 13
hydroelectric facilities, 2, 6
  Gabčíkovo–Nagymaros Project, 52, 53, 110, 118

implementation phase, 6–7, 14, 43, 94, 106, 108, 127, 218
  Danube River Gauntlet, 137
  economic actors, 146
  functions of international law during, 134, 163, 220
  Gabčíkovo–Nagymaros Project, 108–17
independence of the judiciary, 213
industrial accidents
  Black Sea, 42
  Black Triangle, 17
industrial efficiency, 213
  Black Triangle, 20
industrial interests, 46, 138, 213
industrial pollution, 213
  Romanian–Bulgarian Danube River Gauntlet, 35–6, 38
industrial waste
  Silesian Coal Basin, 22
informal enforcement mechanisms, 126
inquiry, 4, 5, 11–12, 46, 51, 126, 162, 181, 219, n68
  Black Triangle, 19
  during dispute formation, 128
  Gabčíkovo–Nagymaros Project, 67–70, 118, 119, 224
  Romanian–Bulgarian Nuclear Corridor, 41
  Silesia Environmental Remediation Project, 24
  Soviet military legacy, 34
  US EPA and Silesian Coal Basin, 24
integration with western institutions, 3, 180, 216, 221
interested third parties, 3, 4, 5–6, 7, 12, 46, 128, 131, 135, 139–45, 146, 157, 163, 164, 167, 170, 218, 221–2, 223, 226, 227
  Black Triangle, 21
  Gabčíkovo–Nagymaros Project, 62–4, 67, 93, 94, 107, 109, 113, 115, 116, 117, 118

incentives, 129, 130, 131, 132, 163, 173, 178, 204, 213–14, 216, 219
Silesian Coal Basin, 25
Romanian–Bulgarian Nuclear Corridor dispute, 41
see also Austria, European Union, Finland, France, Germany, Netherlands, Nordic states, Russia, Scandinavian states, Sweden, Switzerland, United Kingdom, United States
international actors, 1, 179
see also European Union, IFIs and interested third parties
International Atomic Energy Agency (IAEA), 46, 191, 212
economic actors, 222
Kozloduy dispute, 39, 41
Temelín dispute, 26–7
International Bank for Reconstruction and Development (World Bank), 1, 139, 141, 193, 205, n633, n650, n664, n808
Baltic Sea, 143
Black Triangle, 21
Environmental Program for the Danube River Basin, 36
international boundary river, 99
International Court of Justice (ICJ), 47, 48, 157, 199, 200, 202–3, 206
compulsory jurisdiction, 199, 200
during pre-resolution phase, 130
during resolution phase, 131, 132, 133
Gabčíkovo–Nagymaros Project, 63, 65, 67–8, 69, 70–108, 109, 110, 111, 113, 114, 116, 119, 120, 142, 202, 223, 224
operational capacity, 202
Romanian–Bulgarian Nuclear Corridor, 42
Temelín dispute, 27, 220
international dispute resolution mechanisms, 11, 133, 142, 157–8, 189, 200, 216, 226
Black Sea Convention, 43
Gabčíkovo–Nagymaros Project, 60, 119
see also International Court of Justice

international environmental law, 99, 108, 117, 119
evolution, 103, 107
see also duty to cooperate, Environmental Impact Assessment, polluter-pays principle, precautionary principle, prohibition against causing serious harm/injury, transboundary air pollution, transboundary water resources
International Financial Institutions (IFIs), 141, 153, 163, 205, 221, 223–4
see also EBRD, EIB, IMF
international law, 145, 146, 167, 189, 204
application by dispute resolution bodies, 157
capacity, 159–61, 207–14
ecological circumstances, influence of, 145–6, 168
economic circumstances, influence of, 148, 170
function of, 4, 5, 67, 118, 120, 127–34, 135, 162, 188, 218–20
functionality, 3, 5–6, 91, 154–61, 162, 164, 188, 225
general norms, 91
national minority circumstances, 153, 154, 185, 187
operation of, 4, 5–6, 123–6
political circumstances, influence of, 149, 151, 153
role of, 1, 2, 52, 65–6, 70–3, 77, 81–2, 83, 84, 88, 89, 91, 93–4, 95–6, 97, 99–100, 101–2, 104, 105, 106–8, 118, 120, 135, 164, 188, 216, 218, n694
social environment, 152
value of, 71, 72, 226
International Law Association (ILA) Rules of International Law Applicable to Transfrontier Pollution, n973
International Law Commission (ILC), 78
International legal theory/scholarship, 123, 125, n555, n559, n560, n669, n690

International legal theory/scholarship
– *continued*
functionality of international law, 156
International Monetary Fund (IMF),
139, 141, n206, n633
Black Triangle, 21
international organizations, 5,
123, 124
see also interested third parties
international relations theory/
scholarship, 123, 126, n541, n555,
n559, n560, n669, n699
economic actors, 146
functionality of international law,
154, 156, 158
see also cognitive theory, game
theory, liberal approach to legal
theory, policy-oriented approach
to legal theory, realist perspective,
regime theory
Interpretation of Peace Treaties Cases,
n362, n602
inviolability of international
borders, 48
Gabčíkovo–Nagymaros Project, 65,
97–98

J.P. Morgan Bank group, 138
Jablonec River, n52
Japan
Romanian–Bulgarian Nuclear
Corridor, 41
Japanese Overseas Economic
Cooperation Fund
Romanian–Bulgarian Danube River
Gauntlet, 37
Jaworzno, Poland, 23
Jizerskehory mountains
reforestation, 19
Joint Contractual Plan, 52, 78, 93,
97–8, 99
joint investment (ownership) program
Silesian Coal Basin, 25
Gabčíkovo–Nagymaros Project, 52,
86, 87, 97, 99, 104, 106, 107, 108,
110, 111, 113, 114, 116–17, 120
judicial review, 195, 196
jus cogens
Gabčíkovo–Nagymaros Project, 92

Katowice, Poland
Silesian Coal Basin, 22–3
Kelsen, Hans, 126
Klaus, Vaclav, 26
Kozloduy nuclear power plant, Bulgaria,
39, 40, 131, 142, 215
sub-state actors, 137
Kraków, Poland
Silesian Coal Basin, 22–3

Lac Lanoux Arbitration, n454, n972,
n974
land-based pollution, 31, n41
Baltic Sea, 12–13, 15
Black Sea Convention, 43
prohibitions on, 14
Latvia, 6
Baltic Sea, 12
EU membership, 13
Law of the Sea Convention
Baltic Sea, 15
Legality of the Threat or Use of Nuclear
Weapons Advisory Opinions, 82,
82–3, 93, n369
Lenin Steelworks in Nowa Huta
(Kraków), Poland, 170
Lex posterior derogat legi priori, 92
Lex specialis, 92–3, 98, 104
liability, 168, 200, 208, n83
Black Sea Convention, 43
Black Triangle, 21–2
Gabčíkovo–Nagymaros Project, 119
Romanian–Bulgarian Danube River
Gauntlet, 38
Silesian Coal Basin, 25
Soviet military legacy, 32, 33
see also state responsibility/
liability
Liamco Arbitration, n733
liberal approach to legal theory, n690
functionality of international
law, 156
political circumstances, 149, 150
Liberec, Czech Republic, n52
life expectancy
Black Triangle, 17
Silesian Coal Basin, 23
lignite, 171
Gabčíkovo–Nagymaros Project, 53

Lithuania, 6
  Baltic Sea, 12
  EU membership, 13
Lower Silesia
  Black Triangle, 16

marine ecosystem, 195, 218
  Baltic Sea, 12
  Black Sea Declaration, 43
maritime pollution, 218
  regulation of, 14
market economy, 2, 141, 172, 173, 192
  Romanian–Bulgarian Danube River Gauntlet, 38
  see also economic transformation
market-oriented instruments, 194, 213, 215
Marxist ideology
  see socialist political and legal doctrine
Mavrommatis Case, n14
maximin strategy, 158
McNair, Lord, 132
media, 86
mediation, 4, 5, 11–12, 51, 126, 153, 154, 157, 162, 201, 219
  Black Sea, 44
  Black Triangle, 21
  during dispute formation phase, 128
  during pre-resolution phase, 129
  during resolution phase, 131
  Gabčíkovo–Nagymaros Project, 63, 67–70, 71, 73, 109, 116, 118, 224
  Romanian–Bulgarian Danube River Gauntlet, 37–8
  Silesian Coal Basin, 25
  Temelín dispute, 225
Mediocredito Centrale, Italy
  Cernavoda nuclear power plant, 39
Mediterranean Sea
  in comparison to Black Sea, 42
metallurgy factories
  Silesian Coal Basin, 22
Meuse River Case, n974
Milovice, Czech Republic, 32
Mochovce nuclear power plant, Slovakia, 29, 142, 144, 206, 211, 215

ICJ decision in Gabčíkovo–Nagymaros, 84, 97
Moldova
  Cernavoda nuclear power plant, 39
  Romanian–Bulgarian Nuclear Corridor, 41
monitoring compliance, 45, 134, 138, 144, 150, 163, 194, 197, n19
monitoring pollution/environmental degradation, 134, 138, 144, 168, 190, 209
  Black Triangle, 19
  Danube River, 36, n160
  Gabčíkovo–Nagymaros Project, 110, 114
Morava River, n86, n646
Moravians, 185, n816
Mosoni Danube, 70
Most, Czech Republic, 138
municipal environmental law, 150, 155, 169, 172, 177, 178, 191, 192, 210, 216, 217, 226
  current status in CEE, 193–7
  Czechoslovakia, 71
  Gabčíkovo–Nagymaros Project, 119
  Hungary, 71
municipal governments, 138, 144, 176, 177, 178, 179, 194, 222
  nuclear waste, 38
  Silesian Coal Basin, 23
  Soviet military legacy, 32, 33–4
municipal law
  functionality of international law, 155–6, 164
  modification of, 197–8

Nagymaros works, 58, 59, 61, 62–3, 69, 73–4, 76–7, 78–84, 91, 95, 99, 100, 102, 103, 106, 110, 111, 114, 137, 142
nash equilibrium, 158
national identity, 184
national interests, 136, 137, 139, 163, 172
  national minority circumstances, 153
  political circumstances, 149
national monorities, 153, 183–6, 213, 216

national minority circumstances, 145, 153–4, 167, 182–7, 189, 205
NATO, 180, 185
  Committee on the Challenges on Modern Society, NATO, 141
  Gabčíkovo–Nagymaros Project, 115–16
  soviet military legacy, 34
  see also Czech Republic, Hungary, Poland
Nauillaa Arbitration, n465
navigation, 32, 183, 208
  Gabčíkovo–Nagymaros Project, 51, 52, 53, 54, 68, 69, 87, 88, 90, 103, 110, 118
negotiation, 51, 158, 163, 185, 202, 204, 219
  Black Sea Convention, 43
  during implementation phase, 134
  during pre-resolution phase, 130
  during resolution phase, 131, 133
  environmental actors, 145
  economic actors, 146
  Gabčíkovo–Nagymaros Project, 61, 65, 66, 72, 73, 79, 83, 90, 91, 94, 95–6, 97, 99, 100, 101, 105, 106, 107–8, 109, 111, 112, 113, 114, 116, 120
  political circumstances, 150
  role of power, 125
  soviet military legacy, 34
neo-colonialism, environmental, n655
nesting, 152
Netherlands
  Environmental Program for the Danube River Basin, 36
Nicaraguan Military and Paramilitary Case, n465, n468, n586, n593, n934, n936
Nikopol, Bulgaria, 35
Nisa River, 17, 181
  erosion, 19
Nitrate pollution
  Black Sea, 42
  Black Triangle, 17
non-discrimination, principle of, 75, 98
non-governmental organizations, 1, 140–1, 142, 150, 177, 178, 195, 196
  Black Sea dispute, 195
  civil litigation, 195–6
  Gabčíkovo–Nagymaros Project, 60–1, 113, 114
  Mochovce dispute, 211
  Temelín dispute, 25–6, 195
  see also economic actors, environmental actors
Nordic states, 140, 144
  Baltic Sea, 13
  oil exploration, 13
normative-institutional approaches
  functionality of international law, 156
North Sea, 12, 13
North Sea Continental Shelf Case, n492, n589, n931, n972
Northern Bohemia
  Black Triangle, 16–20
  Temelín dispute, 26
Norway
  Baltic Sea, 12–13
notice, 128, 134
  Baltic Sea Convention (1992), 15
  Gabčíkovo–Nagymaros Project, 52, 75, 117
NO_x, 169
  Black Triangle, 17
  see also transboundary air pollution
nuclear energy, 2, 6, 11, 31, 173–4, 211, 218
  Dukovany nuclear power plant, 25, 28–30
  Temelín dispute, 25–30
  see also Belene, Bohunice, Cernavoda, Chernobyl, Dukovany, Kozloduy, Mochovce, Paks and Temelín power plants
Nuclear Tests Case, n986
nuclear waste, 28–9, 38–9, 41

Oder River, 17, 137–8
Oder River Case, n470
Oder River Convention, 20, 209, n75, n969, n971, n974
Olkusz, Poland, 23
Olomouc, Czech Republic, 32
Olse River, n86
Opava river, n86

Orban, Victor, 114
Organization for Economic Cooperation and Development (OECD), 141, 173
Organization for Security and Cooperation in Europe (OSCE), 201
OSCE Convention on Conciliation and Arbitration, 201, n921
organochloride residues
 Baltic Sea, 12
Ostrava, Czech Republic
 Silesian Coal Basin, 22
Ostravice River, n86

Pacta sunt servanda, 78, 84–5, 105
Paks nuclear power plant, Hungary, 30, 144
Pareto-optimality, 158
Paris Peace Treaty (1947), 183
Partnership for Peace, 34
Peace Treaty of Trianon (1920), 183, n817
Petkus, Germany, 32
phosphates
 Black Sea, 42
Pilismarot, Hungary, 110, 111
Ploučnice River, n52
political science research, 138, n613
 economic circumstances, 146
 functionality of international law, 155
 national minority circumstances, 153
 political circumstances, 151
Poland, 6, 140, 145, 147, 148, 152, 179, 190, 194, 197, 203, 204, 212, 213
 Baltic Sea, 12
 Black Triangle, 16–20, 137, 143, 152, 175
 Bohunice dispute, 29
 Constitutional Court, 198
 Defense Cooperation Agreement with Estonia, n147
 economic transformation, 147
 EIA, 195
 environmental investment, 174, 206
 environmental regulation, 198, n854, n857, n868, n869, n871, n872, n875, n878, n879, n885
 EU membership, 13, 137, 217, 223
 Framework Act on the Environment, 152
 implementing legislation, 198
 Mochovce dispute, 29
 NATO membership, 180
 reservation to the compulsory jurisdiction of the ICJ, 200
 Silesian Coal Basin, 22–5
 Soviet military legacy, 31–5
 transformation to market economy, 13
 Treaty on Protection of the Environment, with Ukraine, n799
Poland–Hungary Assistance for the Reconstruction of the Economy (PHARE), 36, 128, 143
Poland Ministry of Environmental Protection
 Baltic Sea, 13
policy-oriented approach to legal theory, n682
political circumstances, 149
Polish EcoFund, 21, 143, n62
political circumstances, 2, 3, 47, 118, 125, 145, 149–53, 163, 167, 220
 domestic, 149–51, 175–9
 international, 151–3, 179–82
political will, 20
 Silesian Coal Basin, 24
polluter-pays principle, 173, 193, 196, 198, 209, 210, 211, 217, 225
 Baltic Sea Convention (1992), 15
 Black Triangle, 21–2
Prague, Czech Republic, 32
pre-resolution phase, 6–7, 14, 17, 24, 27, 29, 32, 37, 38, 40, 47, 109, 127, 130, 131, 134
 economic actors, 146
 environmental actors, 138
 functions of international law during, 129–31, 163, 218, 219
 Gabčíkovo–Nagymaros Project, 60–6
 public participation, 139
 role of ICJ, 202
precautionary principle, 193, 196, 198, 209, 210, 211, 217, 225
 Baltic Sea Convention, 15

precautionary principle – *continued*
 Black Sea Declaration, 43
 Black Triangle, 21–2
 Gabčíkovo–Nagymaros Project, 92, 93, 98, 103
 Romanian–Bulgarian Danube River Gauntlet, 38
private industry
 Soviet military legacy, 32, 34
private investment, 138
 Soviet military legacy, 32
Privatization, 173
 effect on Black Triangle, 18
 former Soviet military sites, 35
 Romanian–Bulgarian Danube River Gauntlet, 38
prohibition against causing serious harm/injury, 127, 209, 210, n45
 Baltic Sea, 15
 Black Sea, 44
 Gabčíkovo–Nagymaros Project, 66, 75, 98
 Temelín dispute, 225
proletarian internationalism, 188–9
protests, diplomatic, 124, 140
protests, public, 176
 Belene nuclear power plant, 39
 Gabčíkovo–Nagymaros Project, 61, 115, 119, 139
 Romanian–Bulgarian Danube River Gauntlet, 35–36
 Soviet military legacy, 32
 Temelín dispute, 25
provincial governments
 Silesian Coal Basin, 23
Provisional Solution, 59, 61, 63–4, 65, 66, 68–9, 71–6, 89, 90, 92, 106, 111, 120, 133, 138
 legality of its implementation, 96–102
Prunéřov, Czech Republic, n51
pseudo regimes, 125
public access to environmental information, 193, 195, 209
 Baltic Sea Convention (1992), 15
public advocacy, 47, 156
 Gabčíkovo–Nagymaros Project, 59, 86, 112, 113

public participation, 3, 139, 143, 149, 150, 151, 175, 176, 177, 195, 210, 215, n177
 Gabčíkovo–Nagymaros project, 86, 176
 national minorities, 153, 154
 Romanian–Bulgarian Danube River Gauntlet, 38
public referendum
 Temelín dispute, 25
punitive measures, 198

radioactive waste, 31, 45, 218, n35
 Baltic Sea, 12
 Black Sea Convention, 43
 Ralsko, Trutnuv and Vesecko municipalities, n52
Rainbow Movement, Czech Republic, 138
realist perspective, 147, n543, n673
regime formation, 134, 158, 188, 224, n683
regime theory, n81, n543, n546, n547, n553, n559
 functionality of international law, 156
 political circumstances, 149
regimes, international, 45, 123, 126, 164, n19, n81, n163, n490, n546, n547, n553, n559, n729, n821, n889
 assessing strength, 125
 constituent elements, 125, 162
 definition, 124
 law as a constituent element, 5, 123, 125–6, 127
 past participation in, 154
regional economic instruments, 213
regional energy programs, 213, 225
 Black Triangle, 20–1
 Silesian Coal Basin, 25
regional environmental policy, 143, 152, 174
 Silesian Coal Basin, 23–4
 Gabčíkovo–Nagymaros Project, 53
regional tradable emissions permit schemes, 171, 213, 214, 225, n81
 Baltic Sea, 13
 Black Triangle, 21
 Silesian Coal Basin, 24

remediation, 102–8, 134, 143, 146, 150
resolution phase, 6–7, 14, 24, 27, 29, 32, 34, 37, 43, 47, 109, 127, 129, 130
 Danube River Gauntlet, 137
 environmental actors, 138
 functions of international law during, 131–4, 163, 218, 219
 Gabčíkovo–Nagymaros Project, 67–108
 role of ICJ, 202
restitution
 see damages and reparation
Rhine River, n89
Rio Declaration, n479
Romania, 6, 144, 148, 190
 adoption of municipal environmental law, 192–3
 Black Sea, 43
 EIA, 195
 environmental regulation, n854, n857, n868, n878
 implementing legislation, 198
 import of hazardous waste, 144
 nuclear power, 36, 38–42, 84
 national minorities, 185
 oil exploration in the Black Sea, 42
 Paks dispute, 30
Romanian–Bulgarian Danube River Gauntlet, 45, 46, 47, 130, 143, 145, 148, 151, 218, 219
 attempts to reduce pollution of, 36–7
 economic actors, 137
 environmental actors, 222
 environmental degradation, 35–6
 Giurgiu–Ruse Environmental Cooperation Commission, 36, 37, 181
 ICJ decision in Gabčíkovo–Nagymaros, 91, 94, 100, 108
 role of international law, 36–8, 188, 209, 224
Romanian–Bulgarian Environmental Cooperation Agreement, 36, 37, 210
Romanian–Bulgarian Nuclear Corridor, 46, 47, 151, 222
 environmental degradation, 38–40

ICJ decision in Gabčíkovo–Nagymaros, 91, 94
role of international law, 40–2, 219
Rühr region, Germany, 23
rule of law, 140, 144, 178, 182, 188, 193, 205, 216, 223, 226
 Black Sea, 44
 Gabčíkovo–Nagymaros Project, 119
Ruse, Bulgaria, 35–8
Russia, 6
 Baltic Sea, 12, 13
 Black Sea, 42, 43, 44
 funding for Mochovce nuclear power plant, 29
 limited relations with CEE states, 180
 Romanian–Bulgarian Nuclear Corridor, 38, 41
 upgrading of Kozloduy nuclear power plant, 39
Russian oil and gas, 171
 Black Triangle, 20–1
 soviet military legacy, 34
Ruthenian region, 181, 186
Ruthenians, n813
RWE-Energie-Versorgung Schwaben, Bayernwerk and Isar Amperwerk Consortium, n638

Samsum, Turkey, 42
Saxony, 178–9
 Black Triangle, 16
 Silesian Coal Basin, 22
Schwarzenberger, Georg, 126
Scandinavian states, 140
sea-based pollution, 31
 Black Sea Convention, 43
 Black Sea Declaration, 43
 prohibitions on, 14, 15
secret treaties, 33
self-determination, 189
Serbia
 Paks dispute, 30
serious injury
 see prohibition against causing serious harm/injury
sharing information, 45, 128, 130, 134, 141, 162, 163, 181, 190, 208, 209, 214, 219
 Baltic Sea, 224

sharing information – *continued*
  Black Triangle, 20
  Gabčíkovo–Nagymaros Project, 52, 59, 64, 75, 97–8, 100, 117, 119
  Romanian–Bulgarian Danube River Gauntlet, 36, 37
  Temelín dispute, 27
Siemens
  construction of Mochovce nuclear power plant, 29
Silesia Environmental Remediation Project, 24
Silesian Coal Basin, 46, 145
  attempts to reduce pollution of, 24–5
  Black Triangle, 16
  environmental degradation, 22–3
  ICJ decision in Gabčíkovo–Nagymaros, 108
  national minorities, 186
  possible role for ICJ, 203
  role of international law, 23–5, 219, 224–5
Silesians, 186, n816
Silistra, Bulgaria, 35, 37
  town council objection to Cernavoda nuclear power plant, 39
situational circumstances 3, 4, 5, 46–7, 120, 135, 145–54, 163, 164, 224
situational bargaining power, 145
Skoda Praha
  construction of Mochovce nuclear power plant, 29
Slovak Ministry of Agriculture, 114
Slovakia, 6, 144, 145, 148, 170, 181, 194, 213
  Baltic Sea, 12
  Commission for the Environment, 62
  Constitutional Court, 198
  EIA, 195
  environmental investment, 174
  environmental regulation, 196, 197, n854, n857, n868, n871, n873, n875, n876, n883
  Gabčíkovo–Nagymaros Project, 51–120, 175
  national minorities, 185, 186
  nuclear power, 28–30
  rule of law, 193

Soviet military legacy, 34
SO_2, 148, 168, 169–70, 210
  Baltic Sea, 13
  Black Triangle, 16–19
  Gabčíkovo–Nagymaros Project, 62
  Scandinavian states, 140
  Silesian Coal Basin, 22–3
  see also transboundary air pollution
social dislocation, 148, 183, 186
  see also coal-mining
social institutions, 124, 125, 126
social norms, 5, 123, 126, 176, 177, 192
social science theory/scholarship, 127, 136, 145, n10, n567, n659, n737
  economic circumstances, 148
  political circumstances, 149–150, 151
social structure, 125, 126
social tension, 182
  Silesian Coal Basin, 23
social transformation, 6
socialist economic system, 48, 178
  environmental degradation, 2, 35
socialist integration, 221
  Gabčíkovo–Nagymaros Project, 51, 55, 86
socialist international law, 159, 188–9, 197, 204, 207–8, 216, 225
socialist internationalism, 188–9
socialist municipal law, 197
socialist perspective on international law, 3
  Soviet military legacy, 34
socialist political and legal doctrine, 176, 179, 183–4, 191–2, 215, 216
  environmental degradation, 2, 170–1, n6
soft budget constraints, 178
soft law, 98, 160
soil contamination,
  Silesian Coal Basin, 22
Southeast Europe Cooperation Initiative, n944
South Africa/Namibia Case, n361
South West Africa Case, n480
sovereignty, 159–60, 189, 192, 197, 199, 207
  and regimes, 124
  Baltic Sea dispute, 15

Gabčíkovo–Nagymaros Project, 66, 72, 75, 76, 98
  Hobbesian view, 220
  Rousseauian view, 220
  transforming notions of, 3
Soviet Black Sea fleet, 54
Soviet-designed nuclear plants, 25–6, 28, 39–40
Soviet hegemony, 2, 55, 86, 180, 188, 189, 204, 221
Soviet military bases
  pollution of the Baltic Sea, 12
Soviet military legacy, 46, 47, 141, 158
  environmental actors
  environmental degradation, 31–2
  ICJ decision in Gabčíkovo–Nagymaros, 88–9
  role of international law, 32–5
Soviet Union, 11, 43, 45, 179, 180, 184, 189, 200, 220
  dissolution of, 2, 118, 188, 189, 225
  Gabčíkovo–Nagymaros Project, 51, 52, 54, 63, 86
  involvement in nuclear disputes, 28–9
Special Agreement, 70, 73, 74, 96, 104, 111
state actors, 3, 5–6, 7, 118, 123, 145, 164
state responsibility/liability, 78, 79, 83, 84, 104, 119, 210–11, 212
  as a basis for treaty termination, 95–6
  see also Draft Articles on State Responsibility, liability
Status of Forces Agreements
  Soviet military legacy, 32–3
Stockholm Declaration, n737
strict liability standard, 196
structural-symmetry, 147, 150, 151
sub-state actor transnational networks, 46, 123, 125, 129, 146
  see also transborder domestic political coalitions
sub-state actors, 3, 4, 5–6, 7, 12, 46, 123, 125, 135–9, 142, 145, 149, 150, 159, 163, 164, 167, 170, 216, 218, 219, 220, 221–2, 226, 227

bureaucratic actors, 69, 137, 151, 156
  domestic litigation, 124, 195, 226
  during implementation phase, 134
  Gabčíkovo–Nagymaros Project, 60–2, 109, 111, 113, 118, 119
  role in negotiations, 129, 130, 138
  see also economic actors, environmental actors, government officials, municipal governments, national minorities
sub-systematic aggregation of a state's national interest, 149
sub-systematic approaches
  functionality of international law, 156
subsequent intervention of international norms
  Gabčíkovo–Nagymaros Project, 65, 105
subsidized state-owned enterprises
Sulphur Triangle
  Silesian Coal Basin, 22
submarine, sunken 32
sustainable development, 48, 172, 173–4, 190, 198, 210, 211, 225
  Baltic Sea, 12–13
  Black Triangle, 21–2
  Gabčíkovo–Nagymaros Project, 80, 95, 98, 103, 105, 107, 108
  Romanian–Bulgarian Danube River Gauntlet, 36, 38
Sweden, 140, 211
  Baltic Sea, 12–13
  Silesian Coal Basin, 22
Swinoujscie, Poland, 32
Switzerland, 143, 205
Szigetköz Region, 98, 101
Szob, Hungary, 110

technical assistance
  Black Triangle, 20
  Gabčíkovo–Nagymaros Project, 63, 68
Temelín nuclear power plant, 46, 144, 158, 215, 222
  Black Triangle, 26
  ICJ decision in Gabčíkovo–Nagymaros, 84, 97
  modification costs, 26

Temelín nuclear power plant – *continued*
  perceived threats to the environment, 25–7
  role of international law, 27–8, 210, 224–5
  territorial integrity
    Gabčíkovo–Nagymaros Project, 66
Tibrek Project, 181
tourism, 147, 148, 182, n55
  Baltic Sea, 12
  Black Sea, 43
  Gabčíkovo–Nagymaros Project, 79
toxic fog, 35
Trail Smelter Arbitration, n454, n734, n973
transaction costs, 124
transborder domestic political coalitions, 139, 163, 170
  see also sub-state actor transnational networks
transboundary air pollution, 6, 11, 31, 45, 47, 169, 173, 181, 190, 203, 218, n67
  bilateral agreements, 208
  Black Triangle, 16–18, 223
  Gabčíkovo–Nagymaros Project, 53, 71
  prohibitions on, 15, 209, 210, 217
  Romanian–Bulgarian Danube River Gauntlet, 35
  Silesian Coal Basin, 22–3, 186
  sulphur pollution: Baltic Sea, 12
transboundary environmental cooperation, 20, 46, 147, 155, 178, 179, 180, 181, 190, 216, 227
  among economic actors, 146
  among local governments, 178
  Silesian Coal Basin, 24
transboundary environmental dispute, definition, 6–7, n13, n14; *see also specific disputes*
transboundary regulatory commissions, 22, 25
transboundary water resources, 210, 218, n67
  Black Triangle, 17–19, 203, 223
  cost of remediation, 148
  efforts to reduce pollution, 20
  Gabčíkovo–Nagymaros Project, 55–6, 71, 79, 82, 83, 89, 93, 97–8, 100
  pollution of, 7, 11, 31, 45, 47, 181
  prohibitions on pollution, 15, 209, 210, 217
  Romanian–Bulgarian Danube River Gauntlet, 35, 181
  Silesian Coal Basin, 22, 186
treaty law, 51, 83, 84
  see also treaty termination
treaty termination, 74, 80, 119
  continuation despite termination, 107, 108
  fundamental change of circumstances, 66, 74–5, 84–9
  impossibility of performance, 66, 74–5, 80–1
  material breach, 66, 74, 78, 84, 89–91, 99
  mitigation, 75, 89, 95
  peremptory norms, 93
  reciprocal breach, 66
  state dissolution, 75
  supervening custom, 74–5, 84, 88, 92–4
Turkey, 6, 11
  Black Sea, 42, 43, 44
Turkish national minority
  Black Sea, 43
Turnu Margurele, Romania, 35
Turon, Poland, 32
Turów power plant, Poland, 18, 20, 203, n62
  modernization, 18
Tušimice power plant, Czech Republic, n51, n656

Ukraine, 6, 181, 186, 201, 206
  Baltic Sea, 12
  Black Sea, 43, 44
  Treaty on Protection of the Environment, with Poland, n799
United Kingdom
  Mochovce dispute, 29
United Nations, 140, 141
United Nations Convention on Transboundary Water Courses and International Lakes, n454, n974, n988
  Baltic Sea, 15

United Nations Environmental
  Programme (UNEP), 36, 37, 140
United Nations Headquarters Case,
  n602
United States, 46, 139, 140, 143,
  144–5, 221, 223, 226
  Black Triangle, 21
  Environmental Program for the
    Danube River Basin, 36, 37
  Gabčíkovo–Nagymaros Project, 60,
    63, 64, 72
  hegemony, 216
  Mochovce dispute, 29
  national minorities, 185
  Romanian–Bulgarian Danube River
    Gauntlet, 37–8
  Silesian Coal Basin, 24
  Soviet Military Legacy, 34
  Temelín dispute, 142
United States Agency for International
  Development (USAID), 205, n808
  Environmental Program for the
    Danube River Basin, 36
United States Environmental
  Protection Agency (EPA)
  Silesian Coal Basin, 24
United States Export–Import Bank
  (EXIM), n949
  Temelín dispute, 26
Upper Austrian Landtag, 27
use of force, 124
Ústí nad Labem, Czech Republic, 19

Valletta mechanism, 201
velvet revolution (1989 revolutions),
  2, 60, 61, 175, 186, 220, 222
verification
  Black Triangle, 21
  economic actors, 146
  functionality of international law,
    156–7
  Gabčíkovo–Nagymaros Project,
    109, 116
Videnov, Zhan, 39
Vidin, Bulgaria, 35
Vienna, Austria, 110
Vienna Convention on Ozone
  Depletion, 190, n845
  Montreal Protocol, 190, n845

Vienna Convention on Treaties
  article 60 (material breach), 80, 89
  article 62 (fundamental change), 80,
    85, 87
  article 64 (preemptory norms), 93
Vistula River
  pollution of, 22–3
Vladimír Mečiar, 113
Vojca, Slovakia, 57, 184
Vranitzky, Franz, 27

Warnemunde, Germany, n19
Warsaw Pact
  dissolution of, 54, 55, 86, 118
  Soviet military legacy, 31–3
water and construction interests,
  61, 114
water pollution, 45, 168, 173
  Black Triangle, 20
  Gabčíkovo–Nagymaros Project,
    66, 100
  Soviet military legacy, 32
  see also transboundary water
    resources
western assistance, 144, 159, 189–90,
  205, 217, n164, n656
  competition for, 180
  Gabčíkovo–Nagymaros Project, 72
  Soviet Military Legacy, 32
  see also EBRD, EIB, GEF, IMF, USAID
Western Europe
  Hungarian trade with, 54
  Russian oil and gas transshipment, 34
  see also European Union, and
    specific West European states
western investment
  Romanian–Bulgarian Danube River
    Gauntlet, 38
Westinghouse
  Temelín dispute, 26
  Romanian–Bulgarian Nuclear
    Corridor, 41
World Charter for Nature, n983
World Health Organization (WHO),
  36
WHO Case, n361, n602

Zaire nad Hormone, Slovakia, 143
zero-variant agreement, 33